An Introduction to the Geography of Tourism

An Introduction to the Geography of Tourism

Second Edition

Velvet Nelson

ROWMAN & LITTLEFIELD
Lanham • Boulder • New York • London

Published by Rowman & Littlefield
A wholly owned subsidiary of The Rowman & Littlefield Publishing Group, Inc.
4501 Forbes Boulevard, Suite 200, Lanham, Maryland 20706
www.rowman.com

Unit A, Whitacre Mews, 26–34 Stannary Street, London SE11 4AB, United Kingdom

First edition 2013.

British Library Cataloguing in Publication Information Available

Library of Congress Cataloging-in-Publication Data
Names: Nelson, Velvet, 1979– author.
Title: An introduction to the geography of tourism / Velvet Nelson.
Description: Second Edition. | Lanham : ROWMAN & LITTLEFIELD, [2017] | "First edition 2013"—T.p. verso. | Includes bibliographical references and index.
Identifiers: LCCN 2016048955 (print) | LCCN 2016049105 (ebook) | ISBN 9781442271074 (cloth : alk. paper) | ISBN 9781442271081 (paper : alk. paper) | ISBN 9781442271098 (electronic)
Subjects: LCSH: Tourism—Environmental aspects. | Geographical perception.
Classification: LCC G156.5.E58 N45 2017 (print) | LCC G156.5.E58 (ebook) | DDC 338.4/791—dc23
LC record available at https://lccn.loc.gov/2016048955

∞™ The paper used in this publication meets the minimum requirements of American National Standard for Information Sciences—Permanence of Paper for Printed Library Materials, ANSI/NISO Z39.48-1992.

Printed in the United States of America

For Kai, my Traveler 2.0

Brief Contents

Contents

Illustrations

Maps

Tables

Figures

Boxes

Preface to the Second Edition

When I first undertook this project, I wondered if I would have enough material to fill a textbook. However, once I started working on it, I quickly realized that this concern was misplaced. Instead, I began to wonder how I was going to fit all of the material I wanted to cover into one textbook. The possibilities for subjects to be discussed in this book are endless. Making choices about what to include was made even more difficult by the fact that every day I would read about some new product, destination, issue, or trend in tourism. This has not gotten any easier with the second edition.

My goal for this textbook was to provide a broad overview of tourism from a geographic perspective. I wanted it to help students in geography use the foundation that they had been building to learn about a new topic and to help students in tourism look at their topic from a new perspective. I drew from the ever-growing literature on tourism in general, and tourism geography specifically, and covered a wide range of subjects and approaches. But there is much more that could be done. For example, in chapter 3, I introduce a number of tourism products. For the most part, these are products that are useful in framing the discussions in subsequent chapters, such as the economic, social, and environmental issues associated with tourism. But these are only a small sampling. New products, designed for various purposes, needs, and/or special interests, are emerging all the time, including things like medical/dental tourism, slow tourism, slum tourism, industrial tourism, creative tourism, tourism for peace, birth tourism, and more. Each new product has its own implications, many of which still remain to be seen.

Tourism is clearly a dynamic industry, and the geography of tourism is an exciting field of study. This book is just a place to start. I encourage students to use this introduction as an opportunity to decide what part of the topic interests them most and learn more about it. For additional secondary research, begin by checking out the sources listed at the end of each chapter. However, in my opinion, a more in-depth knowledge of tourism geography requires *primary* research. For this, I recommend the participant observation methodology—get out there and experience it yourself!

There are many people to thank for their invaluable contributions to this project. In particular, I would like to thank Susan McEachern, editorial director at Rowman & Littlefield, for her work on this second edition. Big thanks go to Carolyn Nelson for being my first reader. Thanks also to Tom Nelson for answering all of my "give me an example of…" questions and to Kim Sinkhorn for being my expert on anything related to parks and protected areas. I am indebted to Scott Jeffcote for supplying me with many of the photographs used in this edition and for editing all of the rest. I hope I have given you an incentive to travel more often, Scott. I also appreciate the work of my colleagues, Gang Gong and Samuel Adu-Prah, for producing many of the maps used in this edition. Special thanks go to the River Road research team (Derek Alderman, Candace Bright, David Butler, Perry Carter, Stephen Hanna, Arnold Modlin, and Amy Potter) for giving me permission to reproduce their map of Southern plantation and slavery museum sites. Their work has been of great inspiration to me. Thanks to Leo Zonn for sending articles and other items of interest my way. They have given me some great ideas. I am grateful to all of the family, friends, and friends-of-friends who shared their travel photos and stories with me. Your experiences—sometimes typical and sometimes extraordinary—help me, and hopefully the readers of this book, to think about things in new ways. Thanks also to the reviewers of the first edition, whose comments guided my approach to this edition. Finally, thanks to Brian Cooper for continuing to support my work and to Barret Bailey for everything else.

Abbreviations

AAG	American Association of Geographers
BTA	Barbados Tourism Authority
CTO	Caribbean Tourism Organization
eWOM	electronic word-of-mouth
FSC	full service carriers
GIS	geographic information system
IGU	International Geographical Union
IOC	International Olympic Committee
ISIL	Islamic State of Iraq and the Levant
LCCs	low cost carriers
MDGs	Millennium Development Goals
MICE	meetings, incentives, conventions, and exhibitions
NCGE	National Council for Geographic Education
NGO	nongovernmental organization
NGS	National Geographic Society
PATA	Pacific Asia Travel Association
PPT	pro-poor tourism
RGS	Royal Geographical Society
SDGs	Sustainable Development Goals
SIDS	small island developing states
STEP	Smart Traveler Enrollment Program
TALC	tourist area life cycle
TIES	The International Ecotourism Society
TRA	tourism resource audit
UGC	User-Generated Content
UNESCO	United Nations Educational, Scientific, and Cultural Organization
UNODC	United Nations Office on Drugs and Crime

UNWTO	United Nations World Tourism Organization
VFR	visiting friends and relatives
WOM	word-of-mouth
WTP	willingness to pay
WWF	World Wildlife Federation

Part I

THE GEOGRAPHY OF TOURISM

While the study of tourism has at times been dismissed as the study of fun, tourism has an undeniable social, cultural, political, economic, and environmental impact on the world today. In fact, tourism has never been more important. The value of the tourism industry continues to increase. In 2014, international tourism receipts exceeded US$1.245 trillion. At the same time, more people are participating in tourism than ever before. In the same year, international tourist arrivals reached 1.133 billion and domestic tourists were estimated at five to six billion. This perhaps surprisingly complex global phenomenon is naturally a topic of geographic inquiry, and geography has much to contribute to our understanding of tourism.

This section establishes the framework for our examination of the geography of tourism. Chapter 1 introduces the relationship between geography and tourism and outlines the thematic approach that will be used throughout the text. Chapter 2 lays the foundation for our discussion of tourism: it discusses the basic terminology of tourism and key concepts from the perspective of both the demand side of tourism and the supply side. Chapter 3 explores the concept of tourism products and introduces several products that will be referenced in the remaining sections.

CHAPTER 1
Geography and Tourism

At first glance, "the geography of tourism" appears to be a statistically improbable phrase. Individually, neither "geography" nor "tourism" is remarkable. After all, most people have some idea—albeit not always an accurate one—of each. It is the combination of the two that is unexpected. Yet, if we look closer, we see that there is really nothing improbable about it. Admittedly, the phrase is not likely to become part of our everyday vocabulary anytime soon. Nonetheless, as we seek to develop a greater appreciation for and understanding of tourism, we will find the geography of tourism provides a powerful approach to the astonishingly complex phenomenon.

We typically think of tourism in terms of our own experiences. Tourism is something that we *do*: the vacation we took, the site we visited, or the place to which we have always wanted to go. Of course, we may not want to admit that we are tourists. We are all too familiar with the highly satirized images that have appeared in everything from classic literature to popular films in which tourists are characterized as overly pale (or conversely, badly sunburned), wearing loud print shirts and black socks with sandals, wielding cameras, brandishing maps and guidebooks, talking loudly, and eating ice cream. In this respect, we may actually think of tourism as something that we do *not* want to do: visit an overcrowded place filled with . . . tourists.

Beyond our own perspective, we may also think of tourism as an enormous global industry. Nearly every country in the world is now trying to get a piece of this trillion-dollar business. It has become so economically significant that scarcely an event occurs where we do not hear about its potential impact on tourism.

There are many aspects of tourism that we generally do not think about. However, if we are to truly understand tourism, we must consider everything from the characteristics of the places that tourists are coming from to the characteristics of the places that they are visiting, how they are getting there, what they are doing, and what effects they have. Once we start to think about all of these things—and more—we begin to appreciate that this idea of tourism is far more complicated than we ever realized.

In recognition of this complexity, tourism studies have grown exponentially in recent years. Although new departments, schools, and faculties dedicated to tourism have been developed around the world, many programs—particularly in the United States—are housed in other schools. For example, it may be a part of a business school (e.g., an emphasis on tourism as an industry) or various health, leisure, and recreation departments (e.g., an emphasis on tourism as an activity). However, there is a largely underexplored and underutilized alternative: geography. Geography can provide the framework to help us understand this often-confusing mix of aspects, activities, and perspectives that constitute tourism.

Simply put, tourism is the subject of this textbook, and geography is the approach. In the chapters that follow, we will make an in-depth examination of tourism. We will first introduce the subject and then break it down with the use of different geographic themes or topics.

What Is Geography?

While the question "what is geography?" seems simple enough, the answer often proves surprisingly elusive. We have heard the term "geography" all our lives. It is typically part of the primary school social studies curriculum. It is inherently tied to the popular *National Geographic* media. It is even a category in *Trivial Pursuit.* Yet, ideas that come from these sources—and others—do not make it any easier to answer the question. In fact, the more you know about the topic, the harder it becomes to produce a neat, concise definition that encompasses everything geography is and geographers do.

From the literal translation of the original Greek word, *geography* means "writing about or describing the earth." People have always had a desire to know and understand the world they live in. Particularly during the ages of exploration and empire, there was a distinct need for the description of new places that people encountered. People wanted to know *where* these new places were, but they also wanted to know *what* these places were like. This included both the physical characteristics of that place and the human characteristics. They wanted to know how these new places were similar to and different from those places with which they were familiar. Therefore, the description of places—where they were and what was there—provided vast amounts of geographic data.

Thus, geography and travel have long been interconnected. The fundamental curiosity about other places, and the tradition of travel to explore these places, continued with scientific travelers of the eighteenth and nineteenth centuries. Although Charles Darwin is the most famous of these, the German Alexander von Humboldt (1769–1859) was one of the most notable geographers. Geographic historian Geoffrey Martin argues that von Humboldt was one of the figures who played an important role in the transition between the classical era of geography and what geography would become in the modern era.[1] Von Humboldt traveled extensively in Eurasia and the Americas and produced a tremendous body of work based on his observations. Through his descriptions of traveling in and experiencing new places, he generated significant interest in geography and inspired subsequent generations of travelers, including Darwin himself.

By the contemporary period, geography had outgrown the classical tradition of description. There were few places left in the world that were uncharted. Although the need for description was diminished, the need to understand the world persisted. In fact, this need seemed greater than ever before. The world was changing. Countless new patterns were emerging, new problems had to be faced, and new connections were forged between places. Having established the where and the what, geographers turned their attention to the *why*. Today, geographers continue to seek an understanding of the patterns of the world, everything from the physical processes that shape our environment to the various patterns of human life, and all of the ways in which the two—the physical and the human—come together, interact, and shape one another. Obviously, this is an enormous undertaking that involves an incredible diversity of work by geographers on topics like the processes of chemical weathering in the Himalayas, spatial modeling of deforestation in West Africa, natural resource management in Australia, the effect of extreme weather events on Central American coffee growers, the quantification of India's urban growth through the use of remote sensing data, American immigration patterns, the geographical dimensions of global pandemics, and more.

When we look at it this way, it shouldn't be surprising that it is no easy task to craft a definition that, in a few words, summarizes all of this. Over time, geographers and geographic associations have proposed compound definitions and frameworks to provide a mechanism for organizing the field. In one example, the Joint Committee on Geographic Education of the National Council for Geographic Education (NCGE) and the American Association of Geographers (AAG) proposed the following five themes in geography: (1) location, (2) place, (3) human–environment interactions, (4) movement, and (5) regions.[2] Frameworks such as this highlight some of the important concepts in geography and provide a starting point for discussing the ways in which geographers view the world. However, there is still no one universally accepted definition or set of concepts for geography.

Taking a different approach, we can return to geography's broad mandate of understanding the world. While this is an admirable goal, it is certainly an unrealistic task for any one geographer to take on alone. Therefore, geographers must necessarily break things down to understand particular parts of the world. On one hand, we can focus on the events and patterns of specific places or regions. Both place and region are identified as key themes in geography. **Regions** have long been a fundamental concept in geography as a means of effectively organizing and communicating spatial information. Essentially, regions help us break down the world into more manageable units. We can determine those areas of the earth's surface that have some commonality—based on a specific physical or human characteristic, like climate or religion, or a combination of characteristics—that distinguishes them from other parts of the world. Regions are still very much a part of the world that we live in. Not only do we continue to conceptualize the world in regional terms (e.g., Eastern Europe or the Middle East), but we have also seen an unprecedented rise in regional organizations in recent years (e.g., the North American Free Trade Agreement or the European Union). However, the concept of regions has, to some extent, been superseded by place in geography in recent years. **Place** generally refers to parts of the earth's surface that have meaning based on the physical and human features of that location.

On the other hand, we can focus on the issues associated with specific topics in geography. For example, figure 1.1 provides a graphic illustration of the discipline of geography. This diagram demonstrates a hierarchy of just some of the topics studied

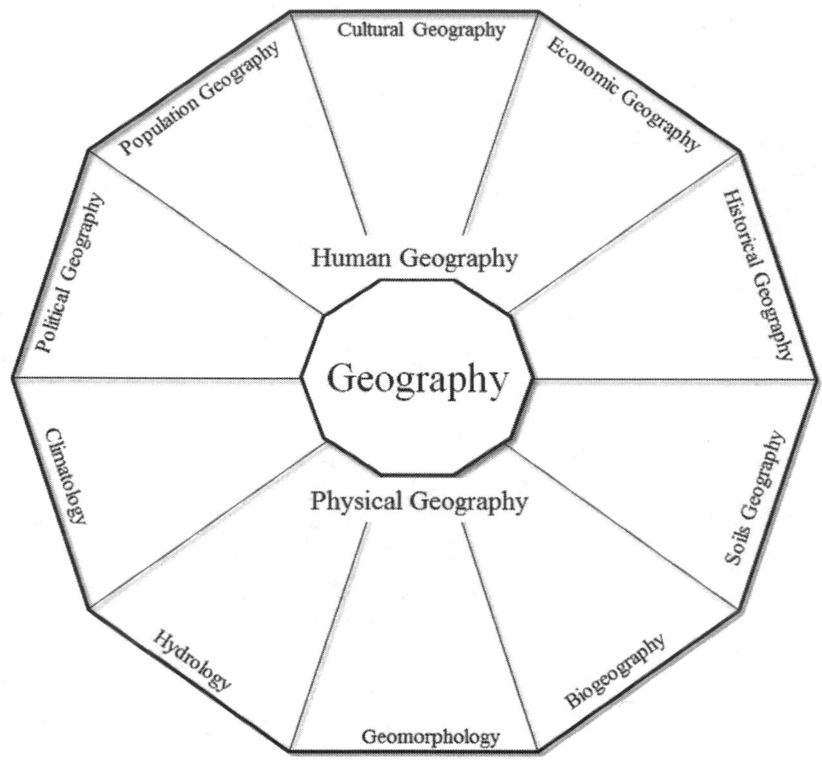

Figure 1.1. This graphic representation of geography illustrates the topical approach in which the discipline is broken down based on various topics in human and physical geography. Topical geography allows us to understand a particular aspect of the world.

in geography. Geography as a whole is broken down into two principal subdivisions. **Human geography** is the subdivision that studies the patterns of human occupation of the earth, while **physical geography** is the subdivision that studies the earth's physical systems. These subdivisions are further broken down into the topical branches of geography. The subjects addressed by each of these topical branches, including everything from climate to culture, may be examined through the key concepts in geography.

Thus, **regional geography** studies the varied geographic characteristics of a region, while **topical geography** studies a particular geographic topic in various place or regional contexts. Both provide a means of helping us work toward the goal of understanding the world. That an understanding of the world is important should need no explanation. Yet, the widely publicized abysmal results of American eighteen-to-twenty-four-year-olds on the National Geographic-Roper Survey of Geographic Literacy in both 2002 and 2006 indicate that a knowledge of the geographic context of the world is sorely lacking.[3] This context is fundamental if we are to understand the varied and complex relationships between people and place in the ways that matter most. These may be the factors that affect our day-to-day lives or that shape the big topics of the world today, whether we are considering food security, international trade relations, the spread of AIDS, patterns of migration, global environmental change, or the massive human phenomenon we know as tourism.

What Is Tourism?

The classic United Nations World Tourism Organization (UNWTO) definition (now also recognized by the United Nations Statistical Commission) considers **tourism** to be "the activities of persons traveling to and staying in places outside of their usual environment for not more than one consecutive year for leisure, business, and other purposes."[4] This is a broad definition that includes movement from one place to another, accommodation at the place of destination, and any activities undertaken in the process. Moreover, it accounts for different purposes. Leisure activities are most commonly associated with tourism, but this definition allows for business to be the primary motivation for travel, as well as "other purposes," which may include health, education, or visiting family and friends. While any of these may be the explicit reason for travel, there are also secondary reasons for both travel and any tourism activities undertaken at the destination.

While providing an encompassing description, this definition has limitations. It indicates that only movement that takes people away from their home environment for at least a day—and lasting a variable amount of time up to one year—would be considered tourism. As such, local and day trip activities would be classified as part of normal recreation activities undertaken in our **leisure time**, or the free time that we have left over after we have done what is necessary—from work to household chores to sleep—and during which we can do what we choose. Nevertheless, these people are often participating in the same activities that would be considered "tourism" for someone coming from farther away. Depending on one's starting location, day trips can even take place over international borders.

In addition, this definition focuses more on tourists and their activities—essentially the demand side of tourism. The concept of supply and demand has, of course, been borrowed from the field of economics. Adapted for the purpose of tourism, *demand* is defined as "the total number of persons who travel, or wish to travel, to use tourist facilities and services at places away from their places of work and residence."[5] When we think about experiences of tourism, we are thinking about the demand side of tourism. This is a fundamental component of tourism: essentially tourism would not exist without tourists and the demand for tourism experiences. *Supply* is defined as "the aggregate of all businesses that directly provide goods or services to facilitate business, pleasure, and leisure activities away from the home environment."[6] When we think about tourism as an industry, we are thinking about the supply side of tourism. Just as demand is a fundamental component of tourism, so is supply. Tourism necessarily involves the production of services and experiences. Although we are generally familiar with this side of tourism from participating in tourism in one form or another, few of us ever consider tourism beyond our own interests and experiences. In fact, the very nature of tourism often means that we do not *want* to have to think about all of the things that comprise the supply side of tourism and make our experiences possible.

Depending on our perspective, we may focus more on **tourism demand** than **tourism supply**, or vice versa. However, both are fundamental components of tourism contingent upon and shaped by each other, and therefore both must be considered if we are to understand the whole of tourism.

How Is Tourism Geographic?

Tourism is inherently geographic. As we put together the components of tourism, we can begin to conceptualize tourism as a geographic activity. Tourism is fundamentally based on the temporary movements of people across space and interactions with place. Thus, basic concepts in geography can contribute to our understanding of tourism.

Like place, discussed earlier, space is an essential geographic concept. **Space** may be defined as locations on the earth's surface, and the related concept of **spatial distribution** refers to the organization of various phenomena on the earth's surface. To break this down further, location is one of the geographic themes. Location can refer to one of two things. Absolute location is the exact position of a place based on some type of structure or grid, such as longitudinal and latitudinal coordinates. For our purposes, however, relative location is more important. **Relative location** is the position of a place in relation to other places and how they are connected. This is related to movement, another geographic theme. Movement is, of course, a fairly self-explanatory concept that allows geographers to explore the ways in which places are connected. Thus, tourism involves clear spatial patterns, including not only where people are coming from and where they are going, but also how they are getting there.

A destination that has a good location relative to large tourist markets has a distinct advantage over one that is much farther away. However, accessibility can help equalize this factor. **Accessibility** is the relative ease with which one location may be reached from another. For example, a direct flight or a high-speed train increases the accessibility of a place. Yet, it is often the case that remote locations are harder to get to and from because of fewer transportation connections, longer travel times, and frequently higher transportation costs. As such, remote places are less likely to develop into large destinations in terms of the quantity of tourists received. Of course, that doesn't mean these locations cannot develop tourism, just that they are likely to develop in a different way than one that is more accessible. Such destinations may receive smaller numbers of tourists who stay at the destination for longer periods of time and spend more money.

Think, for example, of the difference between the Bahamas and the Seychelles. Both are small tropical island destinations with attractive beach resources, among others. Just off the coast of Florida, the Bahamas are strategically located relative to the large North American tourist market and well connected via transportation networks. As such, they have a well-developed tourism industry. With 1.6 million stay-over tourists and 3.4 million cruise ship passengers, the majority of whom are coming from the United States and Canada, the islands receive over US$2 billion in tourist spending annually. Yet, these tourists are staying, on average, less than five nights at the destination. In contrast the Seychelles, located nearly 1,000 miles off the coast of eastern Africa in the Indian Ocean, is far from the principal tourist markets in North America, Europe, and even East Asia. Despite significant tourism resources, these islands receive only 129,000 stay-over tourists, 6,000 cruise passengers, and US$192 million in tourist spending annually. These tourists average at least seven nights.[7]

Place refers to parts of the earth's surface that have meaning based on the physical and human features of that location. Destinations are the places of tourism. The ideas and meanings attached to these places create a demand for experiences in these places.

For example, we could reduce Paris to an absolute location at 48°50'N latitude and 2°20'E longitude, but it is, of course, far more than that. As a place, it is associated with the physical characteristics (e.g., the Seine, the many architectural sites) and the human characteristics of a well-known tourism destination (e.g., culture, history, an atmosphere for romance).

Tourism occurs at different geographic scales. **Scale** generally refers to the size of the area studied. Increasingly, we think of tourism in global terms. The tourism industry has become increasingly globalized with things like global airline alliances and multinational hotel chains. As a result, tourism activities have also become more global. In 2014, international tourist arrivals exceeded 1.1 billion,[8] a number that has been steadily increasing (with some temporary fluctuations) for more than half a century. Yet, this movement of people across space creates connections between places, and tourism involves distinctly local, place-based activities. These activities depend on the unique physical and/or human characteristics of that place. In fact, tourism is often used to highlight and promote unique local resources.

There has been much debate about the effect on local places of **globalization** and the increasing interconnectedness of the world. One argument maintains that places are becoming more similar with the forces of globalization, such as the diffusion of popular culture through media and the standardization of products from large multinational corporations. Yet another argument suggests that, in light of globalization, it is more important than ever to create or reinforce a sense of distinctiveness at the local or regional scale. Tourism has been recognized as an extraordinarily important component in creating and/or promoting a sense of distinctiveness to raise awareness about that place or enhance its reputation.

Finally, tourism provides unique opportunities for interactions between tourists and the peoples and environments of the places they visit. Tourism may be considered one of the most significant ways in which people know places that are not their own. It creates connections between geographically distinct groups of people, people who otherwise might have little knowledge of or contact with one another. It also offers people the potential to explore new environments that are different from the ones with which they are familiar. At the same time, these interactions between tourists and places have specific effects for both the peoples and environments of those places. Tourism can actively play a role in shaping the world in which we live. Thus, issues of sustainability in tourism—economic, social, and environmental—have gotten greater attention.

For example, the densely populated—not to mention well-connected—urban areas of the Northern Hemisphere, such as the North American megalopolis (i.e., the large urbanized area along the Northeast coast stretching from Boston to Washington, D.C.), constitute a significant tourist market. Although there are always exceptions, cold-climate city dwellers have an interest in environments vastly different from their own, such as the tropical rain forest climate and biome in places like Hawaii, Costa Rica, or Thailand. Recognizing the attractiveness of such destinations, countries possessing these environments have a clear incentive to protect these forests as parks and preserves instead of developing them in more environmentally destructive ways. Yet, as more and more tourists come to visit the park, the overcrowding overwhelms the infrastructure, paths are degraded, natural features are vandalized, waste builds up, and so on.

Box 1.1. Case Study: Geotourism as an Approach to Tourism in the Four Corners

Geotourism has evolved over the past two decades and across two disciplines. In the mid-1990s, English geologist Thomas Hose first proposed the idea of geotourism as "the provision of interpretive and service facilities to enable tourists to acquire knowledge and understanding of the geology and geomorphology of a site (including its contribution to the development of the Earth sciences) beyond the level of mere aesthetic appreciation."* This geologic tourism is recognized internationally and facilitated by the 120 United Nations Educational, Scientific and Cultural Organization (UNESCO) Global Geoparks, encompassing geologically significant landscapes, in thirty-three countries.† From this tradition, geotourism is often viewed as one of the special interest tourism products that have emerged with the expansion of the global tourism industry (see chapter 3). It has also been criticized for appealing to a relatively small tourist market.‡

Tourism consultant Jonathon Tourtellot proposed the second idea of geotourism as an *approach* to tourism rather than as a specific product. Adopted by the National Geographic Society's Geotourism Charter, this definition identifies **geotourism** as "tourism that sustains or enhances the geographical character of a place—its environment, culture, aesthetics, heritage, and the well-being of its residents."§ From this perspective, a geotourism strategy should highlight and enhance the human and physical characteristics of a place (including its geology) that make it unique. Likewise, it should be economically profitable, to contribute to the conservation of those characteristics.

National Geographic's geotourism program combines global sustainable tourism expertise with local knowledge (via a local Geotourism Stewardship Council) to create an approach to tourism at the destination that meets the established principles of geotourism. One of the outcomes of this program is the Geotourism MapGuide distributed via print, web, and mobile media. MapGuides work to educate tourists about geotourism to set appropriate expectations for the destination experience and influence patterns of behavior at the destination. Guides allow a destination to benefit from the visibility of the National Geographic brand while highlighting local attractions and encouraging visitors to experience place.

The Four Corners Region—at the intersection of Colorado, Utah, New Mexico, and Arizona—joined the program in 2010. Geotourism is a fitting approach for this area with its distinct geographic character, from the unique landscape of the Colorado Plateau to the rich Native American cultural heritage. Many of the region's sites, protected under the auspices of the national park service, offer this combination of physical and human characteristics of place. For example, Mesa Verde National Park (also a UNESCO World Cultural Heritage Park) was created to protect approximately 5,000 archeological sites dating back to 600 BCE, but it also includes distinctive geologic formations, microclimates, and biodiversity.** In addition, tourism constitutes an important part of the region's limited economy, where both unemployment and poverty rates are some of the highest in the nation.

---

* Thomas A. Hose. "Selling the Story of Britain's Stone." *Environmental Interpretation* 10 (1995): 16–17.

† Ralf Buckley. "Environmental Inputs and Outputs in Ecotourism: Geotourism with a Positive Triple Bottom Line?" *Journal of Ecotourism* 2, no. 1 (2003): 76–82.

‡ United Nations Educational, Scientific, and Cultural Organization. "UNESCO Global Geoparks," *Earth Sciences*, accessed May 17, 2016, http://www.unesco.org/new/en/natural-sciences/environment/earth-sciences/unesco-global-geoparks/.

§ National Geographic Society, "The Geotourism Charter," accessed July 14, 2016, http://travel.nationalgeographic.com/travel/sustainable/pdf/geotourism_charter_template.pdf, 1.

** National Park Service, "Nature," *Mesa Verde*, accessed May 17, 2016, https://www.nps.gov/meve/learn/nature/index.htm.

The Four Corners Region Geotourism MapGuide includes the descriptive and logistical information about the destination that one would expect to find on a promotional site. However, it also includes a "local voices" section in which area residents talk about the region from their lived experiences to create a richer narrative of place. Finally, the site features an interactive map based on a geographic information system (GIS). This tool allows potential visitors to explore the area's natural and cultural attractions, events, and even opportunities to get involved in the community through volunteering. In addition to well-known attractions, area residents have a chance to nominate their favorite sites for inclusion in the guide. This approach allows local people to play an active role in representing what they feel makes that place unique. It also helps visitors learn about little-known attractions, get "off the beaten path," and experience place in a way they would not be able to on their own.

Discussion topic: If geotourism is intended to highlight the unique geographic character of a place, do you think all tourism might be considered geotourism? Why or why not?

Tourism on the web: National Geographic, "Four Corners Region Geotourism MapGuide," at www.fourcornersgeotourism.com

What Is the Place of "the Geography of Tourism" in Geography?

As we recognize that travel has long been a part of geography and that tourism is an inherently geographic activity, "the geography of tourism" should seem less and less improbable. In recent years, the field has seen considerable growth. Major academic geographic associations now have special groups or commissions devoted to the topic, including the Recreation, Tourism, and Sport specialty group of the American Association of Geographers (AAG), the Geography of Leisure and Tourism research group of the Royal Geographical Society (RGS), and the Commission on the Geography of Tourism, Leisure, and Global Change of the International Geographical Union (IGU). Research on topics in the field is published in journals across both geography and tourism studies, including the dedicated journal *Tourism Geographies*. Yet, the place of the geography of tourism within the field of geography is still not widely understood and could use some further discussion. In particular, if we return to our introduction of geography, we see that we can approach the subject regionally or topically.

TOURISM AND REGIONAL GEOGRAPHY

The concept of regions has long been considered an effective means of organizing and communicating spatial information, especially to nongeographers. As such, regions are applied in the context of tourism in a number of ways, not the least of which is the study of tourism generally and the geography of tourism specifically. Many tourism geography textbooks use a regional approach to examine circumstances of tourism in different parts of the world.

The concept of regions may be used to explain patterns or trends in tourism. For example, **tourist-generating regions** are source areas for tourists, or where the largest numbers of tourists are coming from. We can identify characteristics of these regions that stimulate demand for tourism, such as an unfavorable climate or a high level of economic development. Likewise, we can identify characteristics of regions that would facilitate demand, such as a good relative location and a high level of accessibility. Tourist-generating regions are important in helping us understand why certain people may be more likely to travel and where. Theoretically, this information may be used to create new opportunities for people to travel. Specifically, if we understand the barriers to travel for a particular region, we can begin to develop strategies to overcome these barriers. In practical terms, tourism marketers use this information. If a destination identifies its largest potential tourist market, then it will be able to develop a promotional campaign targeted at that audience.

Conversely, **tourist-receiving regions** are destination areas for tourists, or where the largest numbers of tourists are going. We can identify characteristics of these regions that contribute to the supply of tourism. Again, a good relative location and a high level of accessibility are important, as well as the attractions of the region and a well-developed tourism infrastructure. Tourist-receiving regions are important in helping us understand why certain places have successfully developed as destinations. This information may be used as an example for other places also seeking to develop tourism.

International agencies such as the UNWTO use regions to examine trends in the global tourism industry. The UNWTO identifies Europe as both the single largest tourist-generating region and the largest receiving region. As of 2014, the European region accounted for approximately 51 percent of each international tourists and international tourist arrivals (map 1.1).[9] This is attributed to a range of factors, including a diverse set of attractive destinations, high levels of accessibility, a well-developed tourism infrastructure, and a long tradition of travel. Yet, long-standing trends in international tourism have been changing in recent years. The importance of Europe as both a generating and a receiving region has been declining with the emergence of new tourists and new destinations.

Destinations also use regions to present information to potential tourists and to create unique tourism experiences (see box 1.2). In some cases, a national destination will use the concept to organize smaller destination regions. This allows tourists searching for a destination to match their interests or requirements to a particular place within that country. For example, the official website of India's Ministry of Tourism, Incredible India, invites potential visitors to discover destinations in six distinct geographic regions of the country based on different resources and experiences. In other cases, several nations will work together to generate interest in and awareness of themselves as a destination region. For instance, the mission of the Pacific Asia Travel Association (PATA) is to act "as a catalyst for the responsible development of the Asia Pacific travel and tourism industry" and to "enhance the sustainable growth, value and quality of travel and tourism to, from and within the region."[10]

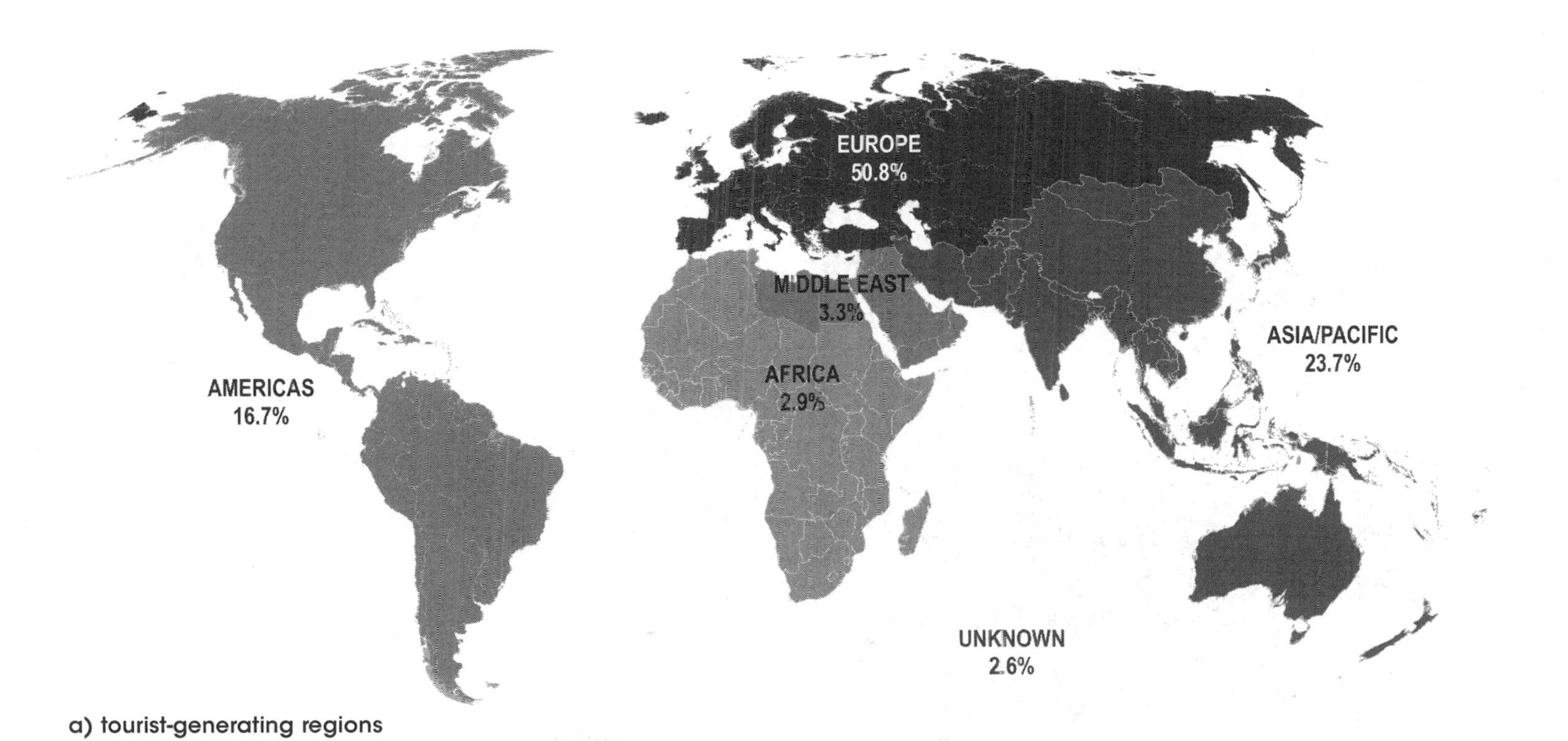

a) tourist-generating regions

Map 1.1. International tourist arrival data allows us to understand (a) where tourists are coming from (i.e. tourist-generating regions on page 13) and (b) where they are going (i.e. tourist-receiving regions on this page). Europe continues to be the world's most significant tourist region with a majority of both international tourists and international tourist arrivals. *Source*: Gang Gong

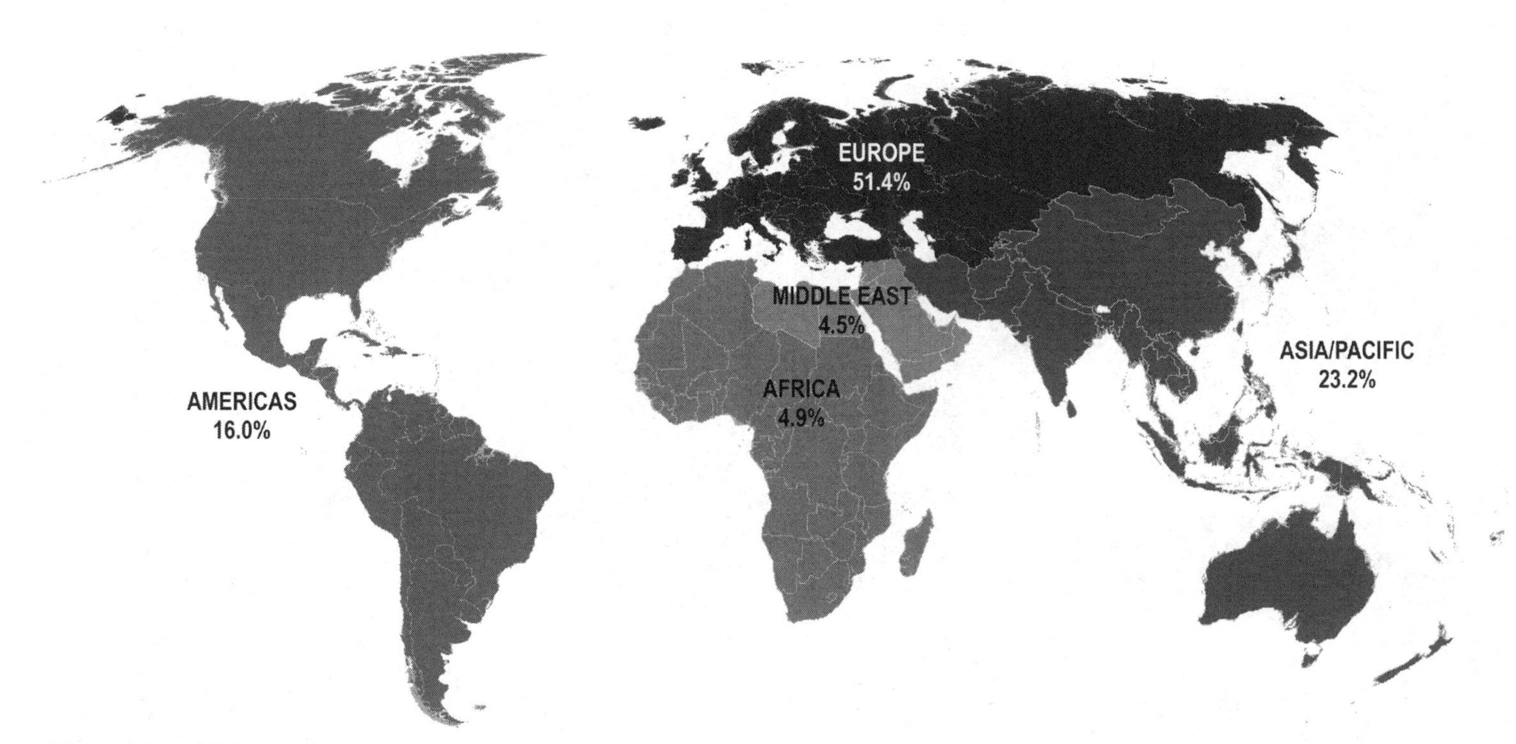

(b) tourist-receiving regions

Map 1.1. (*continued*)

Box 1.2. In-Depth: Cross-Border Destination Regions

The concept of regions has a very long tradition in geography. If we apply the definition of a region to tourism, we could consider destination regions to be those areas of the earth's surface with similar resources for tourism, whether this is a specific characteristic in the physical or human geography of that place or a combination of characteristics. For large countries, destination regions play a significant role in organizing and coordinating tourism efforts in different geographic areas. For diverse countries, regions provide an important means of communicating information about distinct areas and helping potential tourists match their interests with what the destination has to offer.

Ranking 155th in the world in terms of total area, the Central European nation of Slovenia covers an area of 20,273 square kilometers (slightly smaller than the state of New Jersey).[*] Despite this size, the landscape is so diverse that the Slovenian Tourist Board identifies no less than twelve different regions for visitors.[†] Slovenia's neighbor, Croatia, ranks 127th in the world with an area of 56,594 square kilometers (slightly smaller than the state of West Virginia).[‡] Similarly, the Croatian National Tourist Board identifies ten different regions.[§] Each of these destination regions has a distinctive set of natural and/or cultural resources that makes them attractive to potential tourists. Yet, these resources do not start and end at each country's political boundaries. For example, there is overlap between Slovenia's "coast and karst" region and Croatia's "Istrian" region. In recognition of this, the two countries initiated a cross-border tourism project.

Cross-border projects recognize that physical or cultural regions, including the unique resources on which tourism is based, may span two or more countries. In the past, this would have been a barrier to cooperation, but countries are increasingly recognizing the value of working together to improve the position of both. Cross-border tourism projects raise the profile of a region and its destinations and create a sense of differentiation from other destinations. This increases the destinations' competitive advantage by appealing to special interests (e.g., specialized routes or itineraries throughout a region) and creating distinctive experiences (e.g., the opportunity to participate in activities in more than one country). Complementary destinations that might once have been seen as competitors become partners as they seek to bring in visitors who might be interested in both places. These destinations further benefit from sharing knowledge, expertise, and resources, which reduces redundancies and allows for innovation. In addition, this cooperation improves resource management and supports a more sustainable regional tourism system.[**]

Slovenia and Croatia's Project REVITAS is financed by the European Union and involves ten local partners on either side of the border. The project seeks to halt the economic and population decline and to revitalize the Istrian region through the development and promotion of an integrated cross-border destination based on unique natural features and cultural heritage. Stakeholders have established special interest cross-border routes and

[*] Central Intelligence Agency, "The World Factbook. Slovenia," accessed June 15, 2016, https://www.cia.gov/library/publications/the-world-factbook/geos/si.html.

[†] Slovenian Tourist Board, "Regions," accessed June 15, 2016, http://www.slovenia.info/en/Regions.htm?_ctg_regije=0&lng=2.

[‡] Central Intelligence Agency, "The World Factbook: Croatia," accessed June 15, 2016,https://www.cia.gov/library/publications/the-world-factbook/geos/hr.html.

[§] Croatian National Tourist Board, "Regions," accessed June 15, 2016, http://www.croatia.hr/en-GB/Destinations/Regions.

[**] Ksenija Vodeb, "Cross-Border Regions as Potential Tourist Destinations along the Slovene Croatian Frontier," *Tourism and Hospitality Management* 16 (2010): 219–21.

itineraries centered on natural attractions, historic towns, or churches and religious art. To ensure that economic, social, and environmental objectives are met, the project prioritizes sustainable tourism development. In recognition of the demands of modern tourists, the project also works to enable ready access to information throughout the destination region by expanding information and communication technologies.[††]

Tourism is an attractive strategy for places in often-peripheral border regions, but small destinations can struggle to compete on their own. Countries have viewed projects to establish cross-border destination regions as an opportunity in recent years. Nonetheless, any project involving international collaboration has potential challenges, and the extent of cooperation can range from a simple exchange of information to complete integration of the tourism system. Geography provides us with insight in every step of the process.

Discussion topic: What do you think might be some of the challenges associated with cross-border tourism projects?

Tourism on the web: Operativni Program Slovenija-Hrvaška 2007–2013, "Project REVITAS," at http://revitas.org/en/

[††] Project REVITAS, "Project Description," accessed June 15, 2016, http://revitas.org/en/project/project-description/.

The regional approach to the geography of tourism is particularly useful for examining cases of tourism within different regional contexts. Moreover, there are distinct applications for critical regional geography in research on the geography of tourism. Critical regional geography is based on the idea that regions are "social constructions." This means that regions do not just exist in the world—people define them, create boundaries for them, and give them meanings.[11] These meanings shape how people think about potential destinations and therefore have implications for the development of tourism. However, this approach is limited in its potential to fundamentally unpack the concept of tourism. Instead, we will be using a topical approach throughout this textbook.

TOURISM AND TOPICAL GEOGRAPHY

If we return to the graphic depiction of geography in figure 1.1, we can see how topical branches fit together to comprise the subdivisions of human geography and physical geography, as well as geography as a whole. Yet, fitting the geography of tourism into this picture is no easy task. If pressed, most geographers would probably consider the geography of tourism to be a branch of human geography. Certainly tourism is a human phenomenon, and much of the focus in the geography of tourism is on human ideas and activities. Likewise, the majority of geographers who study the geography of tourism are, in fact, human geographers. This would suggest that we could insert a new "wedge" into the pie for tourism geography, and it would largely go unquestioned (figure 1.2).

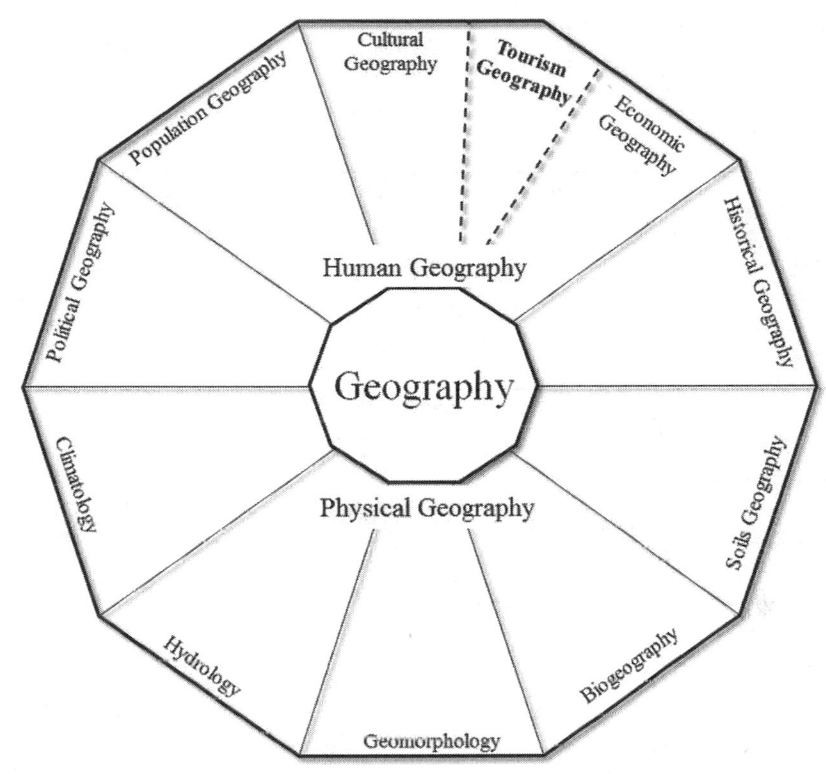

Figure 1.2. We can try to fit the topical branch of tourism geography into our graphic representation of the discipline. Based on what we know about tourism so far—that tourism is often seen as an activity or an industry—the argument could be made that the geography of tourism has a place in between the major topical branches of cultural geography and economic geography.

This kind of conceptualization may be useful in showing that the geography of tourism is a topical branch that coexists with the others at the center of geography. However, it is less useful in helping us understand how to approach its study. As a new space for the geography of tourism is created, it may be tempting to come to the conclusion that the topic can stand alone. To some extent, overlap exists between the topical branches. Yet, this goes beyond mere overlap in the case of the geography of tourism. All of these other areas—cultural geography, economic geography, population geography, political geography, etc.—have much to contribute to the study of tourism through the lens of geography. By tracing the geography of tourism through the human side, we lose some of the components in physical geography—geomorphology, climatology, hazards, etc.—that also play extraordinarily crucial roles in shaping tourism destinations and activities. Furthermore, we cannot truly separate the human and physical divisions, as much of tourism involves interactions between people and the environments of the places they visit.

Rather than thinking of the geography of tourism as part of this hierarchy of topics, it may be more productive for our purposes to think of the geography of tourism

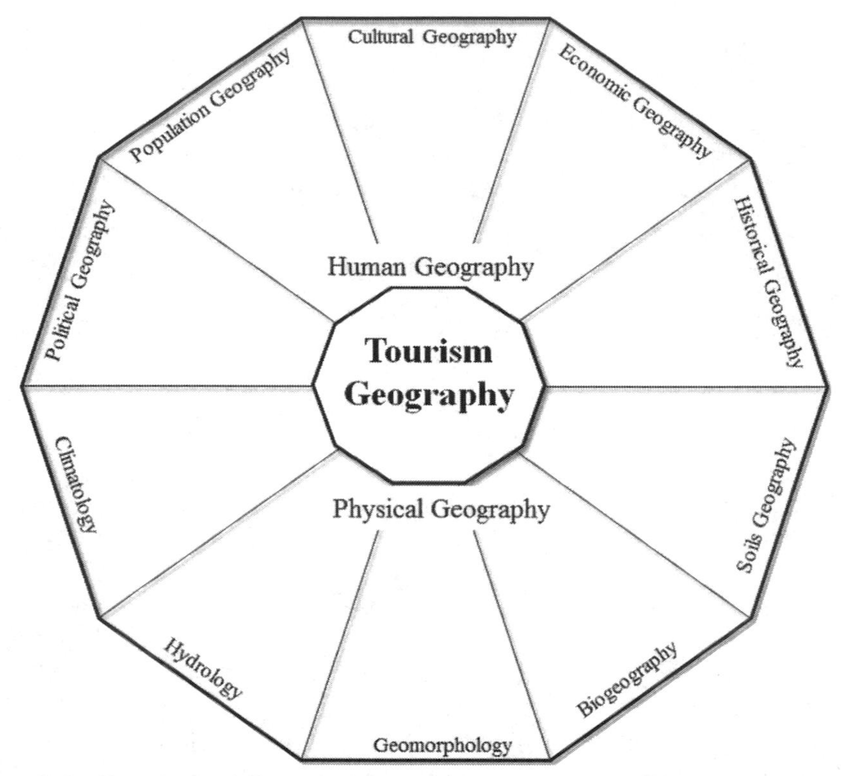

Figure 1.3. If we adapt the graphic representation of geography for the geography of tourism, we can begin to appreciate all the components of tourism. Likewise, we can see how the topical structure of geography will allow us to break down and investigate this complex phenomenon.

in the same way geography as a whole is conceptualized. With the geography of tourism in the center of the schematic, we can recognize that there are both human and physical components at work in tourism, and each of the topical branches can help us understand a different part of the complex phenomenon that is tourism (figure 1.3).

For example, we can use the tools and concepts of climatology to help us understand tourism. Patterns of climate provide insight into things like tourism demand and supply and, by extension, tourist-generating and tourist-receiving regions. Winter vacations are often popular among people who live in the higher latitudes because long, cold winters generate a demand for the experience of a warm, sunny place. As such, cold climates are significant source areas for tourists, while tropical climates have long been significant destinations. Likewise, we can use the framework of political geography to provide insights into patterns of tourism. On a routine basis, politics and international relations create barriers to tourism between two places through visa requirements, complicated permits, and so on. Conversely, the removal of these barriers can create new opportunities. While geopolitical events like terrorism and armed conflicts will obviously have a direct impact on tourism for that destination, they can also have ripple effects on tourism throughout the world. After the events of September 11, 2001, there was an immediate decline in tourism to New York City and Washington, D.C., as well as a general decline in travel globally.

Box 1.3. Terminology: *Affect* and *Effect*

Affect. Effect. The two words may sound the same when they are pronounced. There is only a difference of one letter in the spelling. It probably doesn't help that one is often used in the definition of the other. But that doesn't mean they can be used interchangeably. They do, in fact, have different meanings, and they have distinct implications for our purposes in the geography of tourism.

We will be using *affect* as a verb. To **affect** is to act on or produce a change in something. We can use the topical branches of geography—on both the human and physical sides—to understand the factors that affect the tourism industry. For example, our understanding of climatology or political geography can help us understand how climatic hazards (e.g., hurricanes and cyclones) or geopolitical events (e.g., war and terrorism) have the potential to affect the tourism industry—that is, to act on or produce a change in tourism. Any of these events have the potential to destroy the tourism infrastructure and prevent people from visiting a destination, at least for a while.

We will be using *effect* as a noun. An **effect** is something that is produced by an agency or cause: it is a result or a consequence. Again, we can use the topical branches of geography to understand what kinds of effects the tourism industry has. For example, our understanding of economic geography or environmental geography can help us understand the effects of tourism—that is, the results of tourism. The flow of income and investment into a place from the tourism industry may act as a catalyst for other types of development, or increased tourism development in fragile natural environments may cause environmental degradation.

Discussion topic: Pick a tourism destination and identify three factors that you think might *affect* tourism at that destination and three *effects* that you think tourism might have on that destination. What topical branches of geography would you use to examine each of these factors and effects?

Who Are Tourism Geographers?

Tourism geographers are geographers. Geography provides us with the flexibility to study an incredible diversity of topics from a variety of perspectives. Although geographers specialize both regionally and topically in order to make the task of understanding the world more manageable, many geographers shun labels. Therefore, regardless of what we study, where, or how, we are geographers above all else. Geography provides the framework, or lens, through which we can view, explore, and understand various phenomena of the world in which we live. Second, labels are often unnecessarily restrictive. As discussed above, topical areas in geography do not stand alone: there is considerable overlap between them.

For some geographers, tourism is the primary theme in their research. Yet, these researchers will draw on various perspectives from different topical branches. These geographers are just as likely to be called cultural geographers, economic geographers, environmental geographers, or any other type of geographer for that matter, as they are to be called tourism geographers. In his report on the geography of tourism, Chris Gibson compiled a bibliography of academic articles on tourism written by geographers.[12] His study found that the most common themes in these articles came from the areas of environmental geography, historical geography, and cultural geography.

For others, tourism may not be the primary theme or object of their work, but it is still a topic that has a distinct part to play. This is indicative of the fact that tourism is such a far-reaching phenomenon in today's world. These geographers may never be called tourism geographers: however, their contributions to the geography of tourism should be considered important nonetheless. Gibson's findings confirm that some of the most widely cited authors of papers on issues related to the geography of tourism do not list this as one of their topical specialties. He argues that geography is a discipline that allows researchers to work on some aspect of tourism, as it is situated within wider issues such as sustainability, poverty, changing patterns of land use, the rights of indigenous peoples, and others.

Finally, Gibson's study explores where this research is coming from. The so-called Anglo-American regions of the world, particularly the United States and the United Kingdom, have dominated published geographic research. Although much research in the geography of tourism does, in fact, come from these areas, the proportion is considerably smaller than for geography as a whole. In contrast, the Australasia region, including Australia, New Zealand, and Singapore, has made some of the greatest contributions to the geography of tourism. Similarly, parts of Europe have also recognized the value of this research and have made key contributions to the field.

Box 1.4. Experience: The Life of a Tourism Geographer

Geographers are interested in so many different topics that it can be hard to believe we are all working in the same field. However, one thing we have in common, whether we are physical geographers or human geographers, is a fundamental curiosity about the world. We want to learn about other places. Tourism geographers know better than anyone that the best way to do this is to experience those places for ourselves, and many of us take advantage of every opportunity we get. American tourism geographer Dr. Nick Wise exemplifies this mindset. While his extensive travel experiences across 45 states and 91 countries are far from the norm, his account below shows us how travel is an integral part of the work—and life—of a tourism geographer.

Travel is work and work is travel. As a geographer who teaches and conducts research in the area of tourism, I spend a lot of time visiting different places around the world so that I can bring my experiences into the classroom. I travel for many reasons: conferences, guest teaching, leading field studies, research, volunteering, visiting family, and of course, fun. I grew up in Pennsylvania and always wanted to learn about how people and places differ from my home and familiar surroundings. My early travels took me around the United States and to destinations in the Caribbean on family vacations. I was not one for hotels or resorts so I often found myself, even at a young age, just wandering through local areas because I wanted to see how people lived. Years later, I was fortunate enough to spend my last semester of high school in Springwood, New South Wales, Australia. It was from this experience that I decided I wanted to be a geographer. Since then, I have travelled extensively through more than ninety countries (map 1.2).

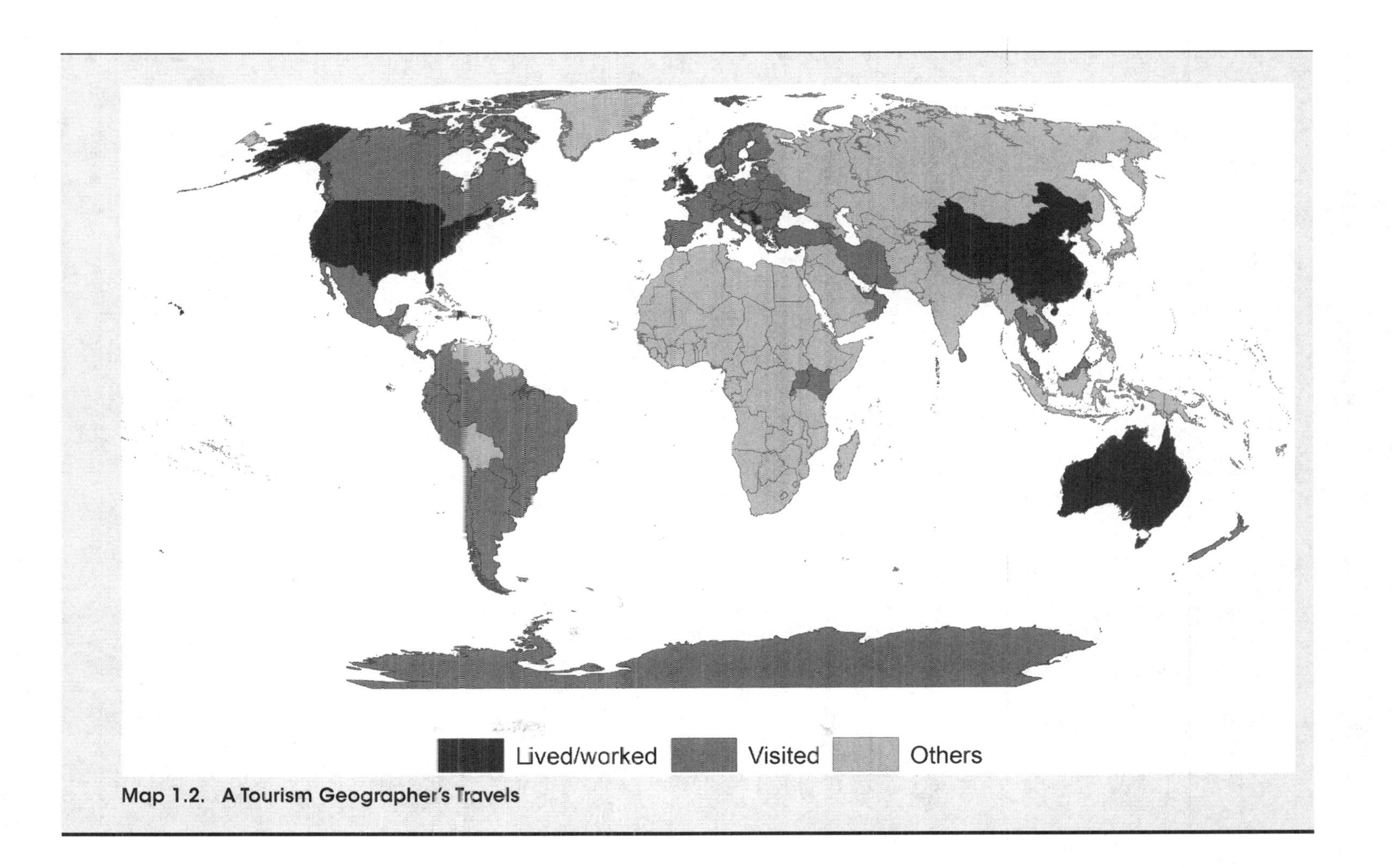

Map 1.2. A Tourism Geographer's Travels

For me, travel is a learning experience. As an academic, the beauty of the job is I teach, I research, and—most importantly—I never stop learning. Travel fulfils this desire to learn about new places and seek understandings from local perspectives. During my undergraduate studies, I planned ahead early in my degree to be able to participate in semester-long study abroad programs in three different places—Beijing, China; Plymouth, England; and Melbourne, Australia. As I worked toward my master's degree, I participated in shorter international trips in the Dominican Republic, Costa Rica, and Cyprus. Later, my PhD fieldwork took me back to the Dominican Republic, and I found opportunities to present my research in the United States as well as in countries like Mexico, Colombia, Serbia, and Bosnia and Herzegovina. A personal goal of mine was to develop opportunities for other students to gain experiences abroad, so I also co-organized field study tours to Costa Rica and Peru.

Since completing my PhD in geography with a focus on tourism, I have worked in an Events, Sport, and Tourism Management program in Scotland and a Department of Leisure and Recreation Management in Taiwan. Teaching in Europe in particular gave me additional opportunities to visit and teach at partner universities in places like Germany and Switzerland through the Erasmus Teaching Mobility and Lifelong Learning Program. I am also involved in research projects in Croatia and Serbia. I am primarily interested in the social impacts of tourism in communities and helping organizations develop business models that focus on experience and value delivery for a wide range of participants and visitors.

I believe that travel experiences complement course content. Geography educators need to motivate and encourage students to learn more about the world. When I teach, I expect my students to read assigned journal articles or textbook chapters. But when I outline the content in the class, I link course materials to my experiences around the world to give students other perspectives to take in. I want them to consider differences and to engage with the content opposed to just reviewing it. Thus, even the places around the world I have visited for leisure, or simply to learn about a place I previously knew little about, is part of my job. Although such travel does not necessarily seem like work, it fulfils that desire to learn and to acquire new content for my tourism and geography classes.

In my lectures, I talk about the experience of hiking through remote forests in Uganda to see gorillas. In the process, I learned about nature preservation techniques and how tourism is used to educate locals and deter poaching. I talk about post-conflict tourism based on what I learned about war and tragedy in Vietnam and Cambodia. I talk about environmental sustainability in the context of remote destinations based on what I observed on a trip to Antarctica, which was truly a once-in-a-lifetime opportunity (figure 1.4). I talk about the rise of destinations in the Middle East such as Dubai, United Arab Emirates and Doha, Qatar, and how neighboring countries such as Oman are trying to compete by comparing my observations of tourism developments in each country. I talk about destination images and place branding based on my conversations with destination managers in Serbia to better understand how they are trying to attract tourists by hosting popular music events and festivals.

I tend to avoid all-inclusive travel. I want to feel that my travels are benefitting a community by staying in family-run hotels and eating where locals eat. By traveling away from all the known tourist paths, I can better experience real, everyday life in the very places people call "home." By doing this I have experienced a Ramadan Breakfast in Hulhumale, Maldives, with the family that ran a small hotel. A family in San Pedro de Atacama, Chile, told me about the challenges of living in the world's driest desert. When I visited Iran with my wife, we travelled to her hometown of Birjand in the east of the country where people generously welcomed me in the markets and at restaurants. I have enjoyed local foods from around the world, most notably Pupusas in a small restaurant in San Salvador, El Salvador; Lamprais in Bandarawela, Sri Lanka; Stinky Tofu in the Night Markets of Taipei, Taiwan; and

Figure 1.4. With a trip to Antarctica in 2013, Nick achieved his goal of visiting all seven continents by his thirtieth birthday. *Source:* Nick Wise

I always make sure I have a Philadelphia Cheese Steak when I am back home visiting family in Pennsylvania.

Travel opportunities are vast, and we all travel for different reasons. My motivation to explore the world stems from my desire to expand my knowledge about places I know little about and to bring new experiences and perspectives on geography and tourism into the classroom.

—*Nick*

Conclusion

Geography has a long tradition based on the fundamental human desire to understand the world, and the modern discipline provides us with the tools and concepts to explain the patterns and phenomena that comprise the world. Although geography and tourism may not automatically be associated with one another, the relationship is undeniable. As such, geography is particularly well suited to provide the framework for exploring the massive worldwide phenomenon of tourism. In particular, we will use a topical approach in geography to break this complicated concept down into more manageable pieces.

This textbook is intended to be precisely what it says it is: an introduction. It is not, and cannot be, comprehensive. Any one of the topics discussed in the chapters of this text could very well merit an entire text of its own. In fact, there are many excellent examples available that discuss such specific topics in much greater depth than what has been done here. At the same time, there are many other topics that could have just as easily been included. The fact that they were not is more a function of a lack of space than a lack of importance. This text is but a beginning, a starting point.

This first chapter briefly discussed geography and tourism for the purpose of introducing this idea of a "geography of tourism." The remaining chapters in Part I continue to develop a basis in tourism that will allow us to subsequently examine key issues through the framework of geography. Specifically, chapter 2 ("Basic Concepts in Tourism") introduces some of the terminology and ideas in tourism that will provide the foundation for discussions in the remaining chapters, while chapter 3 ("Overview of Tourism Products") provides a brief overview of the types of tourism experiences (i.e., the "products" of the tourism industry) that are offered by destinations around the world.

Key Terms

- accessibility
- affect
- effect
- geotourism
- globalization
- human geography
- leisure time
- physical geography
- place
- region
- regional geography
- relative location
- scale
- space
- spatial distribution
- topical geography
- tourism
- tourism demand
- tourism supply
- tourist-generating regions
- tourist-receiving regions

Notes

1. Geoffrey J. Martin, *All Possible Worlds: A History of Geographical Ideas* (New York: Oxford University Press, 2005).

2. David A. Lanegran and Salvatore J. Natoli, *Guidelines for Geographic Education in the Elementary and Secondary Schools* (Washington, DC: Association of American Geographers, 1984).

3. National Geographic Education Foundation, "Survey Results: U.S. Young Adults Are Lagging," accessed July 14, 2016, http://www.nationalgeographic.com/geosurvey/highlights.html: John Roach, "Young Americans Geographically Illiterate, Survey Suggests," *National Geographic News,* May 2, 2006, accessed July 14, 2016, http://news.nationalgeographic.com/news/2006/05/0502_060502_geography.html.

4. World Tourism Organization, *Collection of Tourism Expenditure Statistics, Technical Manual No. 2* (Madrid: World Tourism Organization, 1995), 9.

5. Alister Mathieson and Geoffrey Wall, *Tourism: Economic, Physical, and Social Impacts* (London: Longman, 1982), 1.

6. Stephen L.J. Smith, "Defining Tourism: A Supply Side View," *Annals of Tourism Research* 15, no. 2 (1988): 183.

7. Jerome L. McElroy and Courtney E. Parry, "The Characteristics of Small Island Tourist Economies," *Tourism and Hospitality Research* 10, no. 4 (2010): 319–20.

8. United Nations World Tourism Organization, "International Tourist Arrivals Up 4% Reach a Record 1.2 billion in 2015," January 18, 2016, accessed June 8, 2016, http://media.unwto.org/press-release/2016-01-18/international-tourist-arrivals-4-reach-record-12-billion-2015.

9. United Nations World Tourism Organization, *Tourism Highlights 2015 Edition* 8, no. 1 (2015), accessed June 8, 2016, http://www.e-unwto.org/doi/pdf/10.18111/9789284416899.

10. Pacific Asia Travel Association, "About PATA," accessed July 14, 2016, https://www.pata.org/about-pata/.

11. Paul Claval, "Regional Geography: Past and Present (A Review of Ideas, Approaches and Goals)," *Geographia Polonica* 80, no. 1 (2007): 25–42.

12. Chris Gibson, "Locating Geographics of Tourism," *Progress in Human Geography* 32, no. 3 (2008): 407–422.

CHAPTER 2

Basic Concepts in Tourism

The concept of tourism means something different to all of us because we have different perspectives and experiences. For example, people in significant tourist-generating regions may think of tourism as something that they have done in the past and that they would probably like to do again sometime in the future. This is a demand-side perspective. In contrast, people in significant tourist-receiving regions may associate tourism with all of the tourists who come and go during the course of a season. This is a supply-side perspective. Both are fundamental in understanding tourism.

In this chapter, we will discuss some of the key terms and concepts from the perspective of both the demand side of tourism and the supply side. In particular, we will consider what tourism means from the demand side, who tourists are, and what geographic factors motivate them and affect their demand for travel and tourism. We will also examine what types of tourism are provided on the supply side, what characteristics of places create tourism attractions, and what constitutes the tourism industry.

Box 2.1. Terminology: Tourism

In chapter 1, we discussed the classic UNWTO definition of tourism. But because tourism can be approached from different perspectives, some additional terminology is useful. **Inbound tourism** is where tourists from somewhere else, typically another country, are traveling to that destination. **Outbound tourism** is where tourists are traveling from a place to a destination, again typically in another country. This marks a distinction between **domestic tourism**, which includes those tourists traveling within their own country, and **international tourism**, which includes those tourists traveling to another country. Additional distinctions may be made between short-haul tourism and long-haul tourism. This is based on either distance or travel time by a particular mode, or type, of transport. For example, a short-haul flight is generally considered to be less than three hours, while a long-haul flight takes longer than six hours. However, there is no standardized measure for how these categories are actually defined.

Discussion topic: Find an example of a short-haul international trip and a long-haul domestic trip.

The Demand Side

One approach to tourism is from the demand side, with a focus on tourists. This is, of course, a fundamental component of tourism: tourism would not exist without tourists and the demand for tourism experiences. Interestingly, however, the demand side has often been a less studied component in the geography of tourism. Instead, this approach has been seen as a topic more in the realm of psychology, sociology, or anthropology. Yet, the demand side has distinct implications for our understanding of geographic patterns in tourism. The first half of this chapter introduces some of the theories and concepts that have been put forth to help us understand tourism from the demand side.

TOURISM

In the previous chapter, we began to consider the different ways we think of tourism. When we think about our experiences, we are thinking about the demand side of tourism. Therefore, one of the easiest ways for us to conceptualize tourism is as a process with a series of stages (figure 2.1).

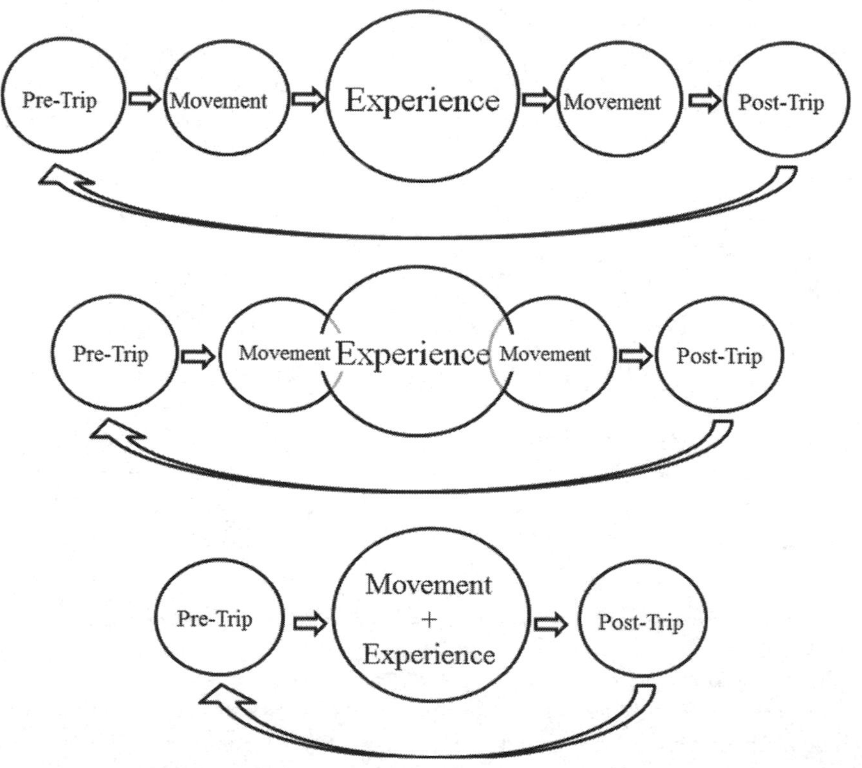

Figure 2.1. This conceptualization approaches tourism from the demand side and takes into consideration the stages that contribute to the overall process of tourism. These stages do not necessarily occur in a linear fashion but may overlap and influence the others.

This process begins in the **pre-trip stage**, when we think about traveling and consider our options. We might evaluate different destinations in terms of the resources and attractions of each place, the tourism products (i.e., the type of experiences) offered there, the level of infrastructure (e.g., accommodations or transport accessibility), and how they match up with our interests and expectations. Likewise, we will consider the overall cost of a trip to these places in relation to our budget. We have access to a tremendous amount of information to aid us in this evaluation and decision-making process. We rely on our previous experiences, input from family and friends, the ideas and images of destinations that come from a range of popular media sources (e.g., news, movies, popular television shows, travel-related television shows, and tourism guidebooks), and, increasingly, online media. We use destination-specific websites as well as social media to get inspiration and information, and we use online booking sites like Travelocity or Hotels.com to make our travel plans. In general, the Internet has made this stage easier—at least in the parts of the world with widespread Internet access—in that almost all aspects of pre-trip planning may be completed online.

The next three stages comprise the trip itself. For many trips, these stages will occur one after the other. In the **movement stage**, we use some form of transportation to reach the place of destination. At the destination, the **experience stage** is the main component in the process, in which we participate in a variety of activities. Then we repeat the movement stage as we return home again. For example, the Midwestern family taking a trip to Disney World may fly from their nearest airport to Orlando, Florida (movement), spend a few days at the resort and/or theme parks (experience), and then fly home (movement). In this case, movement is simply a means to an end to get to the destination and the experience stage. However, these stages are not always distinct, and the act of traveling can be an integral part of the experience stage. For example, the Midwestern family on a road trip may take a scenic route and visit any number of tourist attractions over the course of the trip. In this case, the movement stage lasts the duration of the trip, from the time they leave home to the time they return. The experience stage takes place concurrently with the movement stage.

The final stage is the **post-trip stage**, which occurs after we return from our trip. We relive our trip through memories and conversations about the trip, as well as through tangible products of the trip, like pictures and souvenirs. These memories can be positive or negative, depending on what happened during the three principal stages of the trip. This stage is typically most intense in the period immediately following the trip, and although it diminishes over time, many things can trigger memories for a long time afterward. The tourism process then becomes circular, when we tap into these memories and past experiences to help us make decisions as we start planning our next trip (i.e., the pre-trip stage).

One of the principal advantages of this demand-side conceptualization of tourism is that it is readily understood and does not complicate something that should be relatively straightforward. In addition, it takes into consideration the role of pre-trip planning and the decision-making process, which are neglected in typical definitions of tourism that focus only on travel to a destination and activities undertaken there.

TOURISTS

Based on the United Nations World Tourism Organization's (UNWTO) definition of tourism as quoted in chapter 1, a tourist could be defined as a person who travels to and stays in a place outside of his or her usual environment for not more than one consecutive year for leisure, business, or other purposes. While this official-sounding definition is often used to identify tourists for the purpose of record keeping and statistics, it holds relatively little practical meaning to help us conceptualize who tourists are, as it is broad enough to encompass anything from children on vacation with their parents to adults traveling for work, and from week-long spring break partiers to students spending a semester studying abroad. Instead, it is popular ideas and stereotypes that have long been more influential in shaping our ideas about tourists.

The term *tourist* came into widespread use in the nineteenth century, and even then there were clear—and not always flattering—connotations. Up to this time, explorers were recognized to be individuals who traveled to places that had not previously been extensively visited or documented by others from their society. Likewise, travelers were considered to be those who traveled for a specific purpose, such as business enterprises or official government functions. The new category of "tourists," however, was different from either of these. Unlike travelers, tourists were regarded as individuals who did not travel for any purpose other than the experience of travel itself and the pleasure they derived from that experience.[1] Unlike explorers, tourists were often criticized for traveling to the same places and having the same experiences as all of the explorers, travelers, and even other tourists who came before them.[2]

From this time, highly satirized representations of tourists began to appear in various media, from newspapers to novels. Today, these ideas are more widespread—and more exaggerated—than ever. Take, for example, the opening scene of the 2010 family-friendly animated comedy *Despicable Me*, in which the unruly child of the loud, overweight, brightly dressed, camera-wielding American tourists accidentally exposes the villainous theft of the Great Pyramid of Giza. It is this well-known concept of tourists as a category of people that provokes the sort of reaction expressed in the hilarious *Uncyclopedia* entry titled "Tourist—The Stereotype": "Tourists. We all have 'em. They infest every corner of the globe. Korea, with admirable common sense, arrests all tourists at the border and nukes 'em. . . . Most other nations on Earth, sadly, tolerate them."[3]

Although it has long been easy to make fun of tourists based on stereotypes, it is far more difficult to make generalizations about tourists in reality. Of course, there are always tourists who continue to fuel the stereotypes. Yet, there are also those who may be closer in spirit to the early explorers or have motivations in common with travelers (as even the official UNWTO definition explicitly includes business as well as other purposes). To accommodate the differences that exist between tourists, scholars have developed **tourist typologies** to identify categories (or types) of tourists. Typologies have used many different variables to categorize tourists, such as motivations and behavior as well as demographic characteristics, lifestyle, personality, and more. This type of framework is typically conceptualized as a spectrum or continuum of tourists, in which several important categories are identified and defined. These categories merely identify some of the characteristics of tourists at certain points on the continuum. Not all tourists can be grouped into these defined categories but instead will fall at various points along the continuum between categories.

While many different typologies have been proposed, the following simplified framework is often used as a summary of key categories from the most influential typologies. To some extent, this framework is similar to the earlier distinctions made between explorers, travelers, and tourists; it divides tourists into four broad types based on factors such as the purpose of travel and the type of experience sought.[4]

The **drifter** occupies one end of the spectrum. Drifters are tourists who likely do not consider themselves tourists. Like explorers from an earlier era, this category of tourists may be characterized as a pioneer who is the first to "discover" new and developing destinations. They seek out these destinations in an effort to avoid other tourists. Such places may have little in the way of a dedicated tourism infrastructure or tourism services. As a result, these tourists may stay in local guesthouses or private homes, use local transportation, shop at local markets, and eat at local restaurants and kitchens. Whether it is out of interest—or necessity, given the nature of these destinations—drifters immerse themselves in the local culture. For some, this is a process of education and self-exploration. For others, it is about doing something different, something not usually done.

The **explorer** bears resemblance to the earlier definition of a traveler. This category of tourists may have motivations for travel other than simply diversion, whether education, religious enlightenment, mental or physical well-being, or other specific types of experiences at the destination. These tourists look for unusual types of experiences and greater contact with the local population than just interacting with the people who hold service positions in the tourism industry, such as front desk clerks, restaurant servers, or housekeeping and maintenance staff. Explorers typically make their own travel arrangements and rely on a combination of both the tourism infrastructure and the local infrastructure. For example, these tourists may arrive at the destination by the same means as other categories of tourists, but instead of taking a tour bus or hiring a private taxi to explore the destination, they use the local transportation infrastructure.

In this typology, the traditional "tourist" category is divided into two different types. The next type along the continuum is the **individual mass tourist**. For these tourists, the primary motivation is typically some form of relaxation, recreation, or diversion, and they have some desire for things that are familiar and comfortable. They are generally dependent on the tourism infrastructure for getting to and staying at the destination, and they may use tourism industry services for at least part of their trip, such as taking a guided tour at the destination. However, these tourists are also interested in having experiences at the destination that would not be available to them in their home environment, and they will seek the opportunity to explore the destination, albeit in a relatively safe manner.

Finally, the **organized mass tourist** occupies the position at the opposite end of the continuum. These tourists are primarily interested in diversion and escaping the boredom or repetition of daily life. They place a high emphasis on rest and relaxation and enjoying themselves with good food and/or entertainment. These tourists are less interested in unique experiences of place and are more likely to travel to destinations that are familiar or have characteristics that are familiar. Therefore, even if they travel to a foreign destination, they will stay in recognized brand name (i.e., multinational) resorts. These facilities are designed to provide the standard of accommodation, services, or types of food that such tourists are accustomed to at home. Organized mass tourists are highly dependent on the tourism infrastructure and services to structure their vacation. This may be a package that bundles services together at competitive prices, whether it is a comprehensive guided tour (with all transportation, accommodation, most meals, and tour services included) or a resort package (accommodation, some or all meals, and airport transfers included). As a result, little additional planning for the trip is necessary, there is little uncertainty about what will happen on the trip, and there may be little incentive to stray from the confines of the tour bus or resort complex. Thus, there is little to no interaction with the people or the place of the destination.

TOURIST MOTIVATIONS

Clearly these types of tourists have different motivations for and interests in tourism experiences. In the geography of tourism, we need to know what factors cause people to temporarily leave one place for another. If we understand these factors, we can begin to explain why certain places developed as significant tourist-generating regions and why others became significant receiving regions. Likewise, it helps destinations to better know where their potential tourist markets are by matching up what they have to offer with the places where the demand for that product is greatest. However, motivations may be complicated, and it is rarely just one thing that causes people to seek tourism experiences.

The motivation that has long been most commonly associated with tourism is the pursuit of pleasure. However, implicit in this motivation is the real or perceived need for a temporary change of setting. This may be considered a geographic **push factor**, or something that impels people to temporarily leave home to travel somewhere else. We may think of this as an escape from the routine of daily life with the associated home and work issues, or boredom with familiar physical and social environments. Correspondingly, it is assumed that there is something that can be obtained at the destination that cannot be obtained at home. This may be considered a geographic **pull factor**, or something that attracts people to a particular destination. The pull may be something tangible that may be obtained at the destination, like being able to buy certain types of local products or eat authentic local cuisine. In most cases, however, it is an intangible, like having the opportunity to interact with new people, getting a week's worth of sunny 80°F weather in the middle of winter, or having access to fresh snow at a prime ski resort. For both the push and the pull, this "something" will be different for everyone.

Borrowing from one of geography's related disciplines—anthropology—we can see how these motivations have been laid out in Nelson Graburn's concept of **tourist inversions**.[5] In this theory, the experience we seek in our temporary escape is one of contrasts. Much of this involves a shift in attitudes or patterns of behavior away from the norm to a temporary opposite. A common example is the inversion from work and stress to peace and relaxation. For example, when we spend a long period of time working hard at school or at a job (or, in some cases, both simultaneously), tourism becomes our means of seeking the opposite: going on vacation for a period of rest and relaxation away from the stresses of what occupies us in our daily lives. Likewise, the shift from economy to extravagance is another common inversion that applies to many of us. We often have to budget our money in the course of our daily lives, but we will save up and splurge on a vacation. During these few days, we may spend more on food, drinks, entertainment, and other activities than we normally would.

In some cases, these inversions in behavior contribute to the generally poor reputation of tourists in many parts of the world. In particular, many inversions go from moderation to excess. Graburn suggests that overindulgence in food is the product of one tourist inversion. The same idea applies to overindulgence in alcohol and drugs. This inversion, as highlighted by popular media, is the one that gives spring break tourists—and, by extension, spring break destinations—a bad name. In the case of this inversion, students who usually go to class, study, work, party occasionally, and generally live within the norms of society travel to a spring break hotspot during the designated semester break and party to excess, with all that it entails.

There is also a geographic dimension to tourist inversions, in terms of a shift away from the tourist's home and community toward a temporary opposite. This shift is much more locally contingent, and the inversions may, in fact, work both ways. One of the most common inversions of this type involves the movement from cold climates to warm ones. People in middle and upper latitudes who experience long, cold winters may seek to escape that weather—and the associated symptoms of seasonal affective disorder—for a short time by traveling to a warm, sunny place in the lower latitudes. At the same time, people in warm climates may travel to colder ones to be able to participate in winter sports, such as skiing. People in densely populated urban areas may seek to escape the congestion, noise, and pollution of the city for expansive natural areas such as the national parks, although people living in rural areas or small towns may seek to get away from the insularity of that life by getting lost in a big city (figure 2.2).

These inversions will continue to provide a significant motivation for travel. However, scholars suggest that, in the modern era, tourism is increasingly becoming a part of a lifestyle rather than a contrast to daily life.[6] For example, Richard Florida describes the "creative class" as a new social class of workers in science, technology, and the arts who seek out cities with a high quality of place. This includes distinctive urban amenities, diverse peoples, and a vibrant atmosphere—in essence, the same characteristics that make an attractive destination for tourism. For this creative class, there may be little difference between expectations for the places they live and the places they visit.[7] In addition, there is a growing class of tourists who prioritize travel as a part of their lives and are interested in their next opportunity, wherever that may be.

Box 2.2. Case Study: Barbados's "Perfect Weather"

It hardly seems like one would need a reason to visit the Caribbean island nation of Barbados. That it has a reputation as a tropical island paradise is usually enough to create a distinct demand for the experience of such a destination. Yet, the island does not attract visitors from all over the world. Rather, there is a distinct geographic distribution of tourists to Barbados, as 79 percent of stay-over tourists come from only two geographic regions: North America and Europe. In particular, 34 percent come from just one country: the United Kingdom (map 2.1). Of course, there are many variables that we would need to consider if we were to fully understand these geographic patterns, such as historical relations, modern transportation connections between tourist-generating regions and the destination, levels of development in the generating regions, the type of attractions offered by the destination, and many more. However, one of the simplest explanations is weather.

Although there is relatively little seasonal temperature variation for Barbados, there is nonetheless a distinct tourism season from November through April. On average, two-thirds of total tourist arrivals, including over three-fourths of cruise passenger arrivals, are concentrated in this six-month time period, consisting of the most difficult winter and early spring conditions for the majority of Barbados's North American and Northern European tourists. In particular, December accounts for the largest number of arrivals, with approximately 15 percent of the total. Conversely, the months that receive the lowest visitor arrivals are the summer months, when these generating regions experience the most favorable weather conditions and when Barbados experiences its least favorable conditions. At the peak of Atlantic hurricane season, October accounts for less than 5 percent of tourist arrivals.*

Among Barbados's many unique cultural and heritage attractions—and highly developed tourism amenities—one aspect of the destination that the Barbados Tourism Authority (BTA) emphasizes most is the island's weather. In recognition of weather as a geographic push factor for their primary tourist market in Northern Hemisphere countries such as the United Kingdom, the United States, and Canada, the BTA has worked to ensure that these tourists are well aware that Barbados has the corresponding pull factor. Visit Barbados, the official website of the BTA, explicitly highlights the island's "perfect weather." On a designated page, the BTA claims, "With an average daily high temperature of 78°F/26°C, an average daily rainfall of less than ¼ inch and 3000 hours of annual sunshine, it's hard to imagine a place that enjoys weather any more perfect than in Barbados."† At the time of this writing—in February with overcast skies, a chance of wintery mixed precipitation, and a current local temperature of 28°F (in Texas, no less)—the page's "perfect weather" weather report showed sunny skies and temperatures ranging from 82°F to 84°F for the following four days.

Discussion topic: What do you consider to be the greatest geographic push factor(s) for tourism from your home environment and why? What destination(s) do you think have the corresponding pull factor?

Tourism on the web: Barbados Tourism Authority, "Visit Barbados," at http://www.visitbarbados.org/

---

* Caribbean Tourism Organization, "Country Statistics and Analysis: Anguilla, Antigua & Barbuda, Aruba, The Bahamas, Barbados, Belize, Bermuda, Bonaire," accessed June 9, 2016, http://www.onecaribbean.org/content/files/Strep1AnguillaToBonaire2010.pdf.

† Barbados Tourism Authority, "Perfect Weather," accessed July 14, 2016, http://www.visitbarbados.org/perfect-weather.

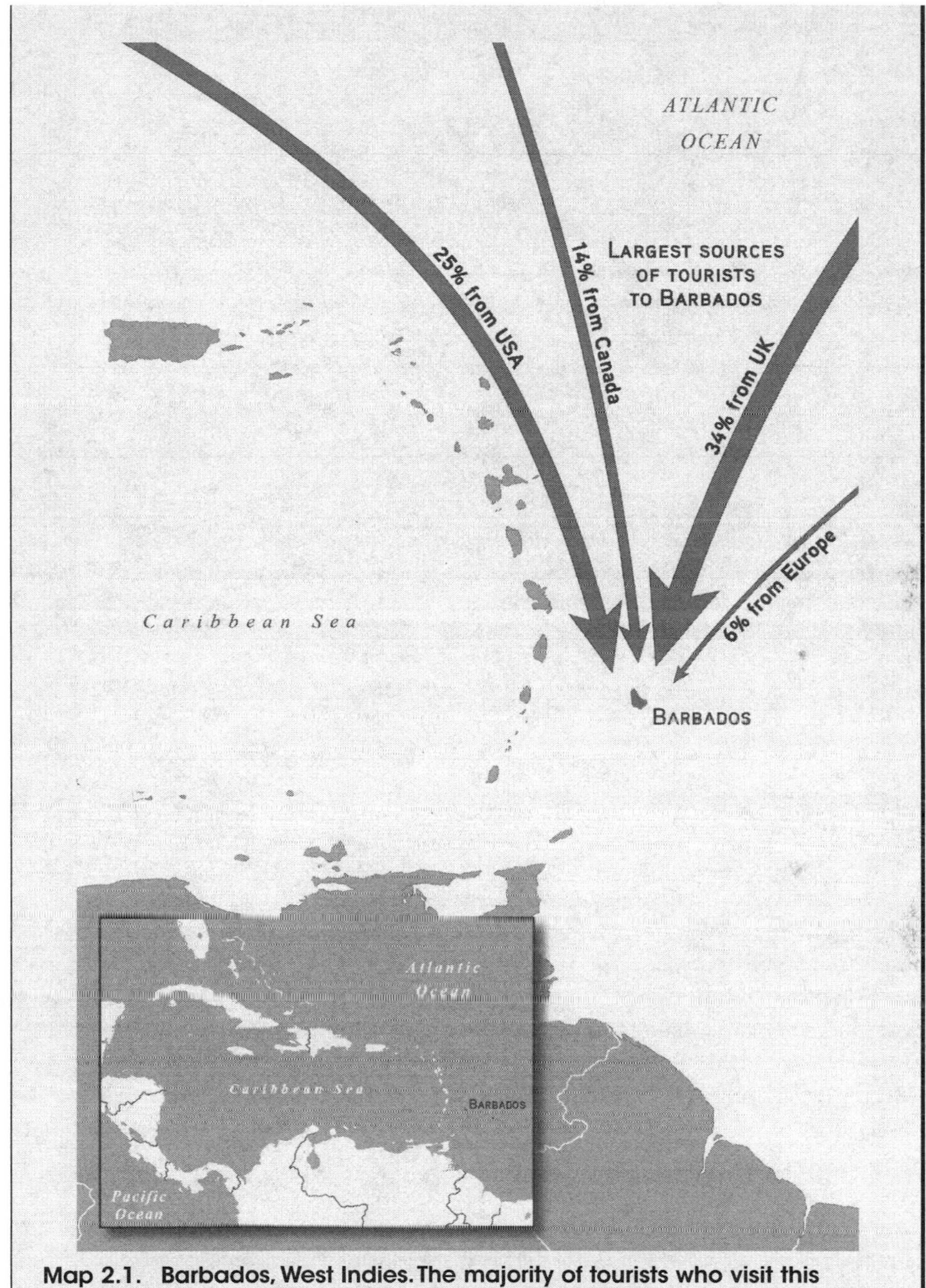

Map 2.1. Barbados, West Indies. The majority of tourists who visit this Caribbean island destination come from the significant generating regions in Europe and North America. *Source*: XNR Productions

Figure 2.2. Times Square in New York City is known for its bright lights and crowded streets. While this type of experience may not appeal to everyone, it may exert a strong pull force on someone from a rural or small town area looking to "get lost" in the big city. *Source:* Scott Jeffcote

TYPES OF DEMAND

In the previous chapter, our definition of demand included those who travel and those who wish to travel. Consequently, we need to distinguish between different types of demand, including effective demand, suppressed demand, or no demand. **Effective demand** is the type of demand we typically think of, as it refers to those people who wish to and have the opportunity to travel. We can measure effective demand relatively easily with tourism statistics like visitation rates and participation in certain tourism activities.

However, this does not give us a complete picture, as participation is not always reflective of desire. How many of us have wanted to travel (i.e., have had a demand for travel) at some point in our lives but have not been able to, for one reason or another? **Suppressed demand** refers to the people who wish to travel but do not. It is much more difficult to measure the number of people who simply *want* to travel. Moreover, there are many reasons why people who wish to travel do not, so we can break this category down even further. **Potential demand** is a type of suppressed demand that refers to those people who want to travel and will do so when their circumstances change. For example, students often have a potential demand for tourism. This means that they may have an interest in (or a perceived need for!) tourism experiences, but they may not have the **discretionary income** (i.e., the money that is left over after taxes and all other necessary expenses of life like rent, food, transportation, clothing, tuition, and books have been taken care of) to travel.

Deferred demand is a type of suppressed demand that refers to those people who want to travel but have to put off their trip, not because of their own circumstances but because of some problem or barrier in the supply environment. This could be a problem—or even a perceived problem—at the desired destination. For example, after the April 2010 explosion on the Deepwater Horizon drilling rig and the subsequent oil spill in the Gulf of Mexico, there was much speculation about the number of tourists who would cancel their summer vacations to the many Gulf Coast destinations due to either actual site contamination or fear of contamination. Similarly, this could include some type of a problem in the tourism infrastructure that would prevent tourists from reaching or being able to stay at their intended destination. Also in April 2010, the eruption of Iceland's Eyjafjallajökull volcano and subsequent ash cloud shut down airports across Europe and created a massive backlog of travelers. Many people who had plans to travel to a number of different destinations during this time were forced to cancel their trips. It is important to understand what factors are going to allow demand to be fulfilled as well as what factors will prevent it. If tourism stakeholders understand those factors, then they can begin to see what strategies might help people with suppressed demand get past any barriers and have the experiences they are looking for. At least theoretically, suppressed demand can be converted into effective demand if the right opportunities are presented.[8] This might involve offering discounts to students, such as Eurail discounted Youth Passes for people aged sixteen to twenty-four. Or it might involve targeted promotional campaigns, such as the US$87 million that BP paid to the Gulf Coast states for the purpose of tourism and promotion to get the word out about destinations that were not affected by the disaster.[9]

It may seem like it should be easy to assess demand because we often assume that if people are not already traveling, they probably want to. However, there is actually one additional category of demand: **no demand**. This refers to people who for various reasons really do not want to travel.

FACTORS IN DEMAND

A person's demand for tourism may be shaped by the nature of the society in which he or she lives. For example, a country's government can help generate effective demand by creating opportunities for people to travel. Government-mandated holidays will give people more time to travel. The level of development in a society is an important factor shaping demand. Generally, higher levels of development will lay the foundation for more people to translate their desire for travel and tourism experiences into effective demand. Higher levels of economic development bring an increase in both discretionary income and leisure time. Higher levels of social development bring improvements in the health, well-being, and education of the population. These things give more people within that society greater means, interest, and opportunity to travel. The more developed countries of the world continue to account for the largest proportion of international tourists. However, some of the newly developing countries like Brazil and China have been experiencing conditions that allow more people to travel, and they are quickly gaining prominence among the significant tourist-generating nations.

At the same time, individual factors—such as a person's view of the world and his or her childhood influences and experiences—as well as personality type play a distinct role in determining whether he or she has a strong desire to travel or prefers to spend his or her leisure time at home. Personal biases and even phobias—such as a fear of flying—will also shape an individual's demand. More generally, however, we can consider how a person's stage in the life cycle affects demand.

In the youth stage (i.e., children who have not yet reached the age of legal adulthood), interest in travel and tourism experiences varies. Younger children are most likely to have a demand for experiences that are specifically promoted to this demographic, like a trip to Disneyland or Disney World. Beyond this, demand is shaped by the travel patterns of family and friends. If travel is a part of their family life, they will come to expect and anticipate these experiences. Likewise, hearing about their friends' travel experiences can also stimulate demand. However, many children may have no demand for travel, not because they lack interest in other places but because they lack opportunity and therefore such experiences may not even be a part of their consciousness. Demand for travel typically increases during the teenage years with greater opportunities and a greater desire for independence. Still, decisions about whether the child's demand is effective or potential continue to be made by parents or guardians.

In the young adult stage (i.e., individuals who are legally of age but do not yet have the responsibilities associated with adulthood), there is typically a high demand for travel because of the pent-up desire for freedom and independence from the youth stage. These individuals may have fewer time constraints than adults with careers and families. Students, in particular, have long designated holidays between terms that provide the opportunity for travel. However, one of the greatest barriers to travel during this stage is financial; essentially, young adults may have less discretionary income available for travel. This may result in potential demand where they will travel in the future if their circumstances change, but many are nonetheless able to translate their desire into effective demand by using their limited disposable income to take short vacations and travel cheaply by using public transportation and staying in hostels.

The married/partnered (without children) stage can be associated with a complex set of variables that contribute to both effective demand and potential demand. With two incomes and accumulated vacation time, these couples may have both the time and money to travel. However, as they develop careers and set down roots in their home environment (e.g., buy a house, acquire pets, get involved in community activities, etc.), it may become increasingly difficult to get away for extended periods of time.

The family stage arguably has the greatest influence on whether demand is effective or potential. Couples and single parents with dependent children have increased household, childcare, educational, and other expenses and therefore less disposable income available for travel. At the same time, the cost of a trip increases, as a family must purchase more transport tickets, book a larger hotel room or suite, pay for more activities, and others. Between the parents' work schedules and the children's school and activity schedules, it may be difficult to find an appropriate time when everyone will be free to travel. Moreover, family trips require more coordination and preparation to ensure that all members of the family are ready to go on the trip and various contingencies are accounted for (e.g., packing drinks and snacks, entertainment, favorite toys or blankets, various first aid supplies in case of illness or accidents, etc.). Because of these constraints, families may have the desire for travel and tourism experiences but decide to put it off or revise their expectations by traveling to destinations closer to home, taking trips of a shorter duration, or undertaking travel to visit family, such as grandparents.

Initially, effective demand increases in the empty nest stage. Once children become independent and no longer require financial support, these individuals may experience an increase in their disposable income and an increase in leisure time. This will continue to increase after they retire from their full-time jobs. As a result, empty nesters may translate their potential demand into effective demand, not only in terms of having the opportunity to travel but also to have the type of tourism experiences they desired. As this stage transitions into the elderly stage over time, effective demand decreases again. Retirees living on a fixed income may have to make choices about the experiences they can afford. Travel may become physically more difficult, and health concerns can present a distinct challenge. Some individuals may become more easily tired by journeys. This not only affects their experience of the destination but also requires a post-trip recovery period, which will affect the way they remember the trip. Others may not be physically able to undertake long journeys. For example, individuals experiencing back pain may be unable to sit in the confines of an uncomfortable seat for a lengthy period of time. The loss of one's spouse or partner during the course of this stage has the potential to affect demand. Ultimately, suppressed demand transitions into no demand as the individual feels that the experience of tourism is no longer worth the hassles of traveling.

While there are, of course, always exceptions to these general patterns, the life cycle variable provides some insight into why demand might be effective for some groups of people within a society and potential for others. This helps tourism stakeholders develop strategies to translate potential demand into effective demand. For example, many destinations have recognized that families are a significant potential tourist market with a demand for travel, if the right opportunities are presented. As a result, the tourism industry encourages family travel with the development of family-friendly resorts that offer activities for children and/or babysitting services to allow parents some alone time. These resorts may have specially priced family packages that allow children to stay or eat at on-site restaurants for free to make such a vacation seem more affordable—and therefore more accessible—to families.

The Supply Side

Tourism may also be approached from the supply side with a focus on the mechanisms that support tourism. This, too, is a fundamental component of tourism: tourism necessarily involves the provision of services and experiences. Geography has generally had more to contribute to this side of tourism because of the discipline's focus on the places and place-based resources that play an important role in the supply of tourism. While issues of tourism resources will be the focus of the chapters in part II, the remainder of this chapter introduces some of the theories and concepts that have been put forth to help us understand tourism from the supply side.

TOURISM

From the supply-side perspective, one of the most important distinctions that we can make to help us understand many patterns in tourism is that of mass tourism and niche tourism. The concept of mass tourism is explained through Fordism, or the system of mass production and consumption, typically linked back to Henry Ford and the changes made in automobile manufacturing. Fordism refers to the manufacture of standardized goods in large volumes at a low cost. Thus, **mass tourism** is the production of standardized experiences made available to large numbers of tourists at a low cost.

At mass tourism destinations, the infrastructure is well developed to handle large quantities of tourists. There are typically good transportation links that allow people to easily reach the destination, whether it is interstate highway access, a major international airport, or a cruise terminal. There may be a spatial concentration of hotels and resorts to accommodate these tourists, as well as restaurants and entertainment facilities to meet their needs. Large multinational corporations often dominate these service providers. Whether tourists visit the Bahamas, Italy, or Thailand, they can stay at a Best Western. When they are in Orlando, Beijing, or Casablanca, they can eat at a T.G.I. Friday's. To some extent, tourists can expect similar experiences at these places regardless of where they are actually located. Because the emphasis of mass tourism is on quantity, low-cost packages may be offered to make these destinations accessible to medium- and lower-income groups. In addition, the standardization of experiences means that destinations may be considered interchangeable. This leads to competition between destinations, which contributes to a further reduction in prices.

The most prominent mass tourism destinations have been in warm climates and coastal areas. The idea of mass tourism is also associated with key inversions discussed above, like relaxation and partying. As a result, mass tourism is often seen as the worst of tourism, characterized by stereotyped tourists. Yet, mass tourism has existed since the early eras of tourism and will continue to exist because it meets certain needs. The well-developed infrastructure facilitates tourism for large numbers of people, while the competition and economies of scale allow more people to participate in tourism than would otherwise be possible. Moreover, it provides the type of experiences that many tourists continue to demand.

Mass tourism is often contrasted with **niche tourism** (also sometimes called "alternative" or "special interest" tourism), which is based on the concept of post-Fordism. This concept reflects changes in the ways in which production and consumption are understood. Post-Fordism recognizes that there is not always a single mass market in which all demands may be met through mass production. As a result, there is a need for more differentiated or specialized products targeted at specific markets. Particularly as the tourism industry has developed and more people have had the opportunity to travel to different places, there has been a growing demand for new types of experiences outside of the mainstream. Niche tourism allows destinations to exploit a particular resource that they possess and create a sense of distinction so that tourists feel they must visit that destination to have that experience. It also allows tourists to choose a vacation experience that is more tailored to their specific interests rather than a one-size-fits-all package.

Many destinations become characterized by either mass tourism or niche tourism, although a destination has the potential to tailor its offerings to meet the demands of different types of tourists. Some tourism products that will be discussed in the next chapter lend themselves more toward one type of tourism over the other, and each type will affect tourists and tourism destinations in different ways.

TOURISM ATTRACTIONS

Tourism attractions are aspects of places that are of interest to tourists and provide a pull factor for the destination. Attractions can include things to be seen, activities to be done, or experiences to be had. Some tourism attractions seem "given." For example, the most spectacular scenes of natural beauty, impressive architectural constructions, and places where significant historic events occurred are those that are natural for people to want to experience. However, these sites are attractions because they have been given meaning. This meaning may be given by the tourists themselves and the types of things they demand, but it may also be given by the tourism industry. Each potential destination has to find the attraction (or attractions) that makes it unique and will cause people to want to visit that place instead of another.

There are four broad categories of tourism attractions: natural, human (not originally intended for tourism), human (intended for tourism), and special events.[10] Natural attractions are obviously based on the physical geography of a place, such as the coast, mountains, forests, caves, inland water sources, flora, fauna, and so on. The first category of human attractions includes those places or characteristics of places that had some other purpose or function but have since become an attraction for tourism, such as historic structures, religious institutions, and aspects of local culture. The second category of human attractions include those places or aspects of places that were specifically designed to attract visitors, such as modern entertainment facilities like amusement parks, casinos, shopping centers, resorts, and museums. Finally, special events is a diverse category that can include religious and secular festivals, sporting events, conferences and conventions, and, in some cases, social events such as weddings and reunions.

Almost anything can be made into a tourist attraction, including a whole array of oddities and curiosities, such as the big (i.e., large-sized) roadside attractions described in box 2.3 and shown in figure 2.3. Even objects of dubious origins can be turned into an attraction. The Blarney Stone of Blarney Castle in Ireland has become a well-known tourism attraction visited by an estimated 400,000 people annually. Legend has it that those who kiss the Blarney Stone will gain the gift of eloquence, and visitors have reportedly gone through this ritual for more than two hundred years. Early visitors were held by their ankles and lowered head first over the battlements to perform

Figure 2.3. *Shuttlecocks* is an outdoor art installation by Claes Oldenburg and Coosje van Bruggen at the Nelson-Atkins Museum of Art in Kansas City, Missouri, and identified as one of America's unusual attractions. *Source:* Scott Jeffcote

this act, but safety measures have since been put into place so that visitors only have to lean backwards while holding onto an iron railing, often with the help of a guide. The origins of the stone—and this ritual—are much debated, and some reports suggest that the stone was, in fact, once part of the castle's latrine system. Regardless of its original purpose, the Blarney Stone has recently been called the most unhygienic tourism attraction in the world.[11]

Not all attractions are created equal; some have greater pull forces than others. There are a few prominent international sites that people all over the world would like to have the opportunity to see or experience at least once in their lifetime, whether it is the Eifel Tower or the Great Wall of China. These are the attractions that have the greatest pull force. Essentially, they are one of the most important reasons people choose to visit that destination. These sites are often featured on lists like the "new wonders of the world," compiled in 2007. This type of designation only increases the desirability of such sites as tourism attractions.

A secondary tier of tourism attractions also exerts some pull. These attractions may factor into tourists' decisions to visit a particular destination and will certainly be experienced when tourists visit that place, but they are not the primary reason. In the example given above, few people are likely to visit Ireland solely because of Blarney Castle, but clearly many tourists make a point to have this experience when they are there.

There are also other attractions that may exert little pull or have little influence on tourists' decision to visit that destination. These may be attractions that people only learn about once they arrive at the destination and may be experienced only by tourists who spend more time at the destination. These tourists have the opportunity to explore the destination in greater depth and visit sites beyond those that are well known. For example, Mount Rushmore is a high-profile stop on road trips like the ones described in box 2.3. Visitors who stay in the area for a longer period of time, however, will explore other scenic and historic places in the Black Hills National Forest and Wind Cave National Park, as well as the high-quality offer of Custer State Park, South Dakota's first state park.

Box 2.3. Experience: Two American Road Trips

The American road trip is a classic family vacation. While there may be a "destination" at the end of the trip, the experience is very much tied to the journey. Tourists often plan their routes to include visits to well-known sites, while stops at roadside attractions provide an entertaining way to break up the long distances between places. Roadside attractions developed as early as the 1930s as a way to get travelers on the expanding highway system to stop and hopefully spend some money. These attractions were intended to draw attention and inspire curiosity. Some were so successful that they have become destinations in their own right. The stories below recount two road trips fifty years apart. While much has changed in terms of how we travel, many of the attractions along the way remained the same.

Georgia: In the summer of 1963, I took a summer-long cross-country road trip with my family. My parents, older sister, younger brother, a Chesapeake Bay Retriever puppy, and I all piled into a Volkswagen micro-bus home-made camper and traveled from Maryland to California. We took our time getting to California, traveled down the Pacific Coast Highway, and stayed in southern California before the return journey. We saw many major American attractions, including Mount Rushmore, Old Faithful, the Grand Tetons, the "General Sherman" tree in Sequoia National Park, Yosemite Falls, Disneyland, the Grand Canyon, Mesa Verde, and more. Of course, we did other things along the way as well. For example, after seeing the signs for Wall Drug Store (Wall, South Dakota) for hundreds of miles, we had to stop.

We stayed with relatives, at state/national parks, and at campgrounds. My parents slept in the camper. We kids slept in army-surplus jungle hammocks when we could find a place to hang them, or we just stretched out on the ground by the campfire. If there was a lake or pond at the campsite where people were fishing, my father would invite them to bring their fish to our campfire. He told them we'd do the cooking, and some took him up on the offer. For a few days, we traveled along with a fellow who was taking a family friend and her German boyfriend on an American tour. During this time, my sister rode in their car and actually picked up some German in the process. About three weeks later, we ran into them at Disneyland.

My siblings and I still have vivid memories from that trip: the clear water at Yellowstone Lake, the cold air at the Great Divide, floating in the Great Salt Lake, and doing handstands on the spot where Arizona, Utah, Colorado, and New Mexico meet at Four Corners Monument. Only years later did it occur to me how extraordinary it was for my father to take that long a time away from his business. Although it wasn't feasible to re-create this road trip with my daughter, Elizabeth, I wanted her to have some of the experiences I had.

Elizabeth: I was moving from the East Coast to the West Coast. My mom was going to make the cross-country drive with me, so we thought we should make a trip out of it. I wanted to see big American attractions—as in LARGE. I used an app called Roadtrippers to set our route and how far we were willing to venture off that route to see things. With those parameters, I could search for anything I needed on the trip. I read about some things before we left, but I could also just pull up the app on my phone from the road. We could add or remove a stop as necessary, and the app's navigation would update the route.

I used the "offbeat" attractions category to find many of the big things we saw, including the L.L. Bean Boot (Freeport, Maine), Sissie the Cow, a pink elephant wearing glasses (both in DeForest, Wisconsin), the Jolly Green Giant (Blue Earth, Minnesota), and all kinds of big things at Wall Drug. These attractions made great road trip stops. They were fun, free places to stop for a quick break from the monotony of driving. We could get out of the car and stretch our legs, take a picture, refuel, and grab a snack. The SPAM Museum (Austin, Minnesota) was one of the biggest surprises from the offbeat attractions list. It was actually really well done, and the displays were a lot of fun, especially the Monty Python SPAM sketch playing on loop. I had originally planned a route through North Dakota, but we decided to follow part of my mom's old itinerary to see Mount Rushmore (also big!) and Old Faithful at Yellowstone National Park. We capped off the trip in Washington State by driving up to "Sunrise" (the highest point at Mount Rainier National Park that can be reached by vehicle).

I could have used Roadtrippers to find lodging too, but I decided to search Airbnb. There were a few places where Airbnb didn't have any good options for us (e.g., not available at that time, too expensive, no previous reviews, etc.), but we were able to rent a room in a house in Buffalo, New York, a private apartment attached to a home in Chicago, Illinois, and an antique sheepherder's wagon on a farm in Belgrade, Montana. I actually proposed that last one as a joke. I didn't think my mom would go for it, but she said okay. Overall, Airbnb was a great experience for us. I was able to communicate with the owners via e-mail or text from the road, and since we paid online, we didn't have to worry about exchanging money. We also saved a lot of money by not staying in hotels—especially in places like Chicago. It was nice to get the owners' local insight. For example, our hostess in Buffalo gave us advice on visiting Niagara Falls and where to get the best chicken wings. And, really, where else are you going to find a unique experience like sleeping in a wagon?

—Georgia and Elizabeth

THE TOURISM SYSTEM

Attractions play an important role in creating the demand for travel, but they cannot exist alone. The services provided by the tourism industry facilitate travel to and experience of these attractions. For example, Stonehenge is a well-known United Nations Educational, Scientific, and Cultural Organization (UNESCO) World Heritage Site attraction that draws tourists from all over the world to the English county of Wiltshire. Yet, English Heritage, the organization that owns and manages the site, is not in the business of organizing trips to Stonehenge. The site itself offers only minimal options for food and drink and does not offer visitors a place to stay. Thus, other service providers in the surrounding area must meet the needs of tourists visiting Stonehenge.

Correspondingly, attractions may account for only a small proportion of income at a destination. Some attractions operate on a pay-for-participation basis, but there are just as many attractions that are free or have only a minimal admission fee. As such, it is the tourism industry service providers that generate revenues. This can be so significant that tourism is frequently described as the world's largest industry. In fact, the UNWTO estimates that the international tourism industry accounts for approximately 9 percent of global GDP (including direct, indirect, and induced contributions) and 30 percent of services exports.[12] Ultimately, however, this claim is difficult to substantiate. There is a lack of hard data regarding all aspects of the various travel, tourism, and hospitality-related economic activities ranging from transportation to accommodation, food and beverage, tours, entertainment, retail, and more. In addition, there is considerable overlap between the services provided to tourists and those provided to nontourists, and only part of tourism services takes place in the formal sector of the economy. The remainder is provided in the unregulated informal sector of the economy (e.g., tourists purchasing goods from vendors they encounter on the street or at the beach) that may not be accounted for in official statistics. To further complicate matters, we have seen the rise of peer-to-peer exchanges, in which ordinary people provide services for tourists (see box 2.4). These exchanges presently occupy a grey area in between the formal and informal economic sectors and will be subject to legal decisions in the near future.

To help us appreciate the complexity of tourism, scholars suggest conceptualizing tourism as a multifaceted and dynamic system. This recognizes that there are many interrelated elements in tourism, from attractions and infrastructure to tourism businesses, public sector organizations, members of the community, and so on. Because of the interrelationships between elements, changes in one part of the system will have implications on others—and tourism is always changing. Thus, we need a holistic view of the system to understand what is involved in tourism and how new destinations and products, technological innovations, and changing consumer preferences are continuously reshaping tourism.

Box 2.4. In-Depth: Travel 2.0

Tourism is increasing, expanding, and evolving. Technology is playing an important role in this process. The Internet changed the distribution of tourism information as well as the provision of tourism services. With Web 1.0, tourism corporations and organizations were primarily responsible for generating and publishing content that was viewed by potential visitors. Now it is changing again.

Web 2.0 provides the platform for social media applications in which users themselves are involved in creating and exchanging content.* This User-Generated Content (UGC) takes the form of blogs (e.g., TravelBlog), microblogs (e.g., Twitter), posts (e.g., Facebook), reviews (e.g., TripAdvisor), wikis (e.g., Wikitravel), photos (e.g., Flickr), and videos (e.g., YouTube). **Travel 2.0** is this new interactive approach in which tourists are both consumers and producers of travel information via the Internet.† It is influencing the ways in which tourists make decisions about where to go and what to do, interact with places, and reflect on their

* Kyung-Hyan Yoo and Ulrike Gretzel. "Use and Creation of Social Media by Travellers," in *Social Media in Travel, Tourism and Hospitality: Theory, Practice and Cases*, eds. Marianna Sigala, Evangelos Christou, and Ulrike Gretzel (Surrey: Ashgate, 2012), 189–205.

† Daniel Leung, Rob Law, Hubert van Hoof, and Dimitrios Buhalis. "Social Media in Tourism and Hospitality: A Literature Review," *Journal of Travel & Tourism Marketing* 30 (2013): 3–22.

experiences. It is also creating new opportunities—and new challenges—for tourism businesses and destination organizations. These stakeholders can gain unprecedented insight into what tourists are looking for from their experiences and what factors affect their satisfaction. This information can then be used to capitalize on the perceived strengths of the destination and to address weaknesses. However, as potential visitors increasingly look to UGC for information, destinations are losing control over their own image.

Furthermore, new websites and mobile applications have the potential to change the nature of the tourism system by promoting peer-to-peer exchanges as an alternative to traditional hospitality service providers. For example, ride sharing services like Uber and Lyft have become ubiquitous. Despite ongoing questions about legality, Uber has expanded to more than seventy countries around the world. While not necessarily targeted at travelers, they will be among the users of such services due to familiarity and comfort with the mobile, cash-free system, even in an unknown place. Accommodation sites like Airbnb.com and Flipkey.com facilitate the process of connecting those with available living spaces—whether it be a spare room, apartment, or house—with potential guests in a way that was previously unfeasible. Also facing legal issues, Airbnb now operates in more than 190 countries. Similarly, services like ToursByLocals.com promises to connect tourists with area residents to provide local insight in nearly 140 countries.

A host of increasingly specialized mobile apps also support tourists on self-guided trips. Visitors can read reviews for restaurants in the vicinity, navigate a destination, follow suggested itineraries, find off-the-beaten-path places, and even find out when is the best time to get in line for a particular ride at Disneyland. Cell phone tours are supplementing, or replacing, textual information in brochures and on displays at attractions. For the most part, these tools require cellular data packages to use while traveling. Although this can present a challenge for tourists in foreign destinations, most cellular companies offer short-term international travel plans. The rise of "smart" cities with public wi-fi networks is also helping visitors get online and take advantage of these resources.

Travel 2.0 has implications for every stage of the tourism process (figure 2.4). Travelers 2.0 are increasingly using social media to access information and make plans during the pre-trip stage. Even during the movement stage, these travelers may use crowdsourcing apps like Waze to find the most efficient route to the destination, in addition to posting social media updates regarding their progress. Travelers are able to continue to access information at the destination, make plans on the go, and yes, continue to post social media updates. For those travelers who haven't already shared their photos, reactions, and reviews during the experience stage, they will do so in the post-trip stage.

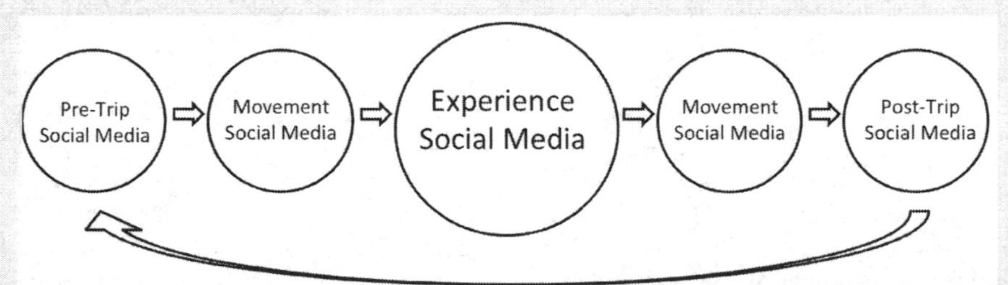

Figure 2.4. Travel 2.0 is changing the tourism process, as social media is becoming a part of every stage. We will consider the implications of Travel 2.0 in later chapters.

With these innovations and implications, the tourism system approach is vital. It helps us appreciate that tourism is complex and dynamic and that the changes we are seeing will have effects throughout the system. For example, in his article in the journal *Current Issues in Tourism*, Daniel Guttentag describes Airbnb as a "disruptive innovation." As such, the company's new business model has the potential to challenge the traditional accommodation sector and send many potential impacts—positive and negative—throughout destinations.[‡] Through both peer-to-peer exchanges and social media, ordinary people, not affiliated with the tourism industry, are taking a more active role in shaping tourism destinations and experiences. This is generating some interesting, and not altogether predictable, results. This era of Travel 2.0 is an exciting time to study tourism, and we will continue to look at these issues throughout the rest of the chapters.

Discussion topic: What are the positive outcomes of Travel 2.0? What are the negative outcomes?

‡ Daniel Guttentag, "Airbnb: Disruptive Innovation and the Rise of an Informal Tourism Accommodation Sector," *Current Issues in Tourism* 18 (2015): 1192–1217.

Conclusion

Depending on our perspective and priorities, we may focus more on tourism demand than supply, or vice versa. However, the success of tourism depends on the demand–supply match, or the ability of the tourism industry at a particular destination to provide the services and experiences that tourists demand. The demand–supply match is not going to be the same for all places, and it will change over time as conditions change on both sides. This chapter discussed some of the key concepts that will help us understand both demand and supply and how they match up, which will provide the foundation for our examination of tourism throughout the rest of the chapters in this book.

Key Terms

- deferred demand
- discretionary income
- domestic tourism
- drifter
- effective demand
- experience stage
- explorer
- inbound tourism
- individual mass tourist
- international tourism
- mass tourism
- movement stage
- niche tourism
- no demand
- organized mass tourist
- outbound tourism
- post-trip stage
- potential demand
- pre-trip stage
- pull factor
- push factor
- suppressed demand
- tourism attractions
- tourism stakeholders
- tourist inversions
- tourist typology
- travel 2.0

Notes

1. James Duncan and Derek Gregory, "Introduction," in *Writes of Passage: Reading Travel Writing*, ed. James Duncan and Derek Gregory (London: Routledge, 1999), 6.

2. Derek Gregory. "Scripting Egypt: Orientalism and the Cultures of Travel,." In *Writes of Passage: Reading Travel Writing*, ed. James Duncan and Derek Gregory (London: Routledge, 1999).

3. Uncyclopedia, "Tourist—The Stereotype," accessed July 14, 2016, http://uncyclopedia.wikia.com/wiki/Tourist_-_the_stereotype.

4. Erik Cohen, "Toward a Sociology of International Tourism," *Social Research* 39, no. 1 (1972): 167–8.

5. Nelson Graburn, "The Anthropology of Tourism," *Annals of Tourism Research* 10 (1983): 21-2

6. Greg Richards, "Creativity and Tourism: The State of the Art," *Annals of Tourism Research* 38 (2011): 1233.

7. Richard Florida, *The Rise of the Creative Class, Revisited* (New York: Basic Books, 2012).

8. Brian Boniface and Chris Cooper. *Worldwide Destinations: The Geography of Travel and Tourism*, 4th ed. (Amsterdam: Elsevier Butterworth Heinemann, 2005).

9. Kevin McGill, "Jindal: BP Funding Millions for Oil Spill Recovery," Associated Press, November 1, 2010, accessed July 14, 2016, https://www.yahoo.com/news/jindal-bp-funding-millions-oil-spill-recovery.html?ref=gs.

10. John Swarbrooke, *The Development and Management of Visitor Attractions*, 2nd ed. (Burlington, MA: Butterworth-Heinemann, 2002).

11. Paul Thompson, "Blarney Stone 'Most Unhygienic Tourist Attraction in the World.'" *Daily Mail*, June 16, 2009, accessed July 14, 2016, http://www.dailymail.co.uk/news/article-1193477/Blarney-Stone-unhygienic-tourist-attraction-world.html.

12. United Nations World Tourism Organization, *Tourism Highlights 2015 Edition* 8, no. 1 (2015), accessed June 8, 2016, http://www.e-unwto.org/doi/pdf/10.18111/9789284416899.

CHAPTER 3

Overview of Tourism Products

Eating, partying, praying, shopping, swimming, sightseeing, gambling, getting cosmetic surgery, hiking, helping, and having sex—although it may seem like these things have nothing in common, they are all activities people participate in through tourism. Tourism is not a one-size-fits-all experience. People have different reasons for traveling, and they want different things from their experiences. Consequently, there is a distinct need for different types of **tourism products**. As a service industry, the primary "products" of tourism are not tangible goods but experiences. With more people traveling than ever before, the tourism industry has developed to provide an array of increasingly diversified and specialized experiences to meet the demands of tourists across the spectrum, from organized mass tourists to drifters.

This chapter provides a brief introduction to some of the types of products that comprise the modern tourism industry. This discussion is by no means comprehensive; it is only a selection of tourism products that crosses different types of tourism and tourists. Many of these products overlap and share the characteristics of other products but have a unique emphasis or appeal to a specific market. Each product involves different resources and affects destinations in distinct ways. We will explore these issues further in the context of the thematic chapters throughout the rest of this text.

Beach Tourism or Sun, Sea, and Sand (3S) Tourism

Perhaps the most widespread and recognizable tourism product around the world is beach tourism. This is often referred to as "3S tourism" in reference to the three key resources for the product: sun, sea, and sand (figure 3.1). Sometimes additional S's are added to the mix—including sex and spirits—but for our purposes, we will consider sex tourism as a separate (albeit related) tourism product. Obviously the focal point of 3S tourism is the beach, which has served as an attraction since an early era in the modern tourism industry (see chapter 4). Yet, 3S tourism is more than just the beach. Beyond any other, this product has been used to characterize the tourism industry. Every major world region has 3S tourism destinations. Some of the largest, best-known, and most popular destinations are based on this product. Moreover, 3S tourism appeals to some of the most basic tourist motivations, including the pursuit of pleasure and self-indulgence.

Typical 3S tourism is mass tourism, which accounts for the temporary movements of large numbers of tourists from the more developed countries in the northern climates to well-established coastal destinations, often developing countries with warmer, tropical climates. This product is highly dependent on a well-developed tourism infrastructure to facilitate the mass movement of people and create the desired experience at the destination. Resorts are often a fundamental component of these destinations. They offer the comforts of home and the facilities to enjoy the three S's, including beachfront access, swimming pools, lounge chairs, water-sport equipment, and so on. Because a key goal of this product is relaxation and leisure, related facilities include restaurants, nightclubs, and other venues offering entertainment. Given these amenities, there may be little incentive to leave the resort to experience other aspects of the destination.

Characteristic of mass tourism—these destinations are relatively standardized, so there is a certain degree of interchangeability among similar destinations in different parts of the world. However, not all tourism oriented around the beach is synonymous with organized mass tourism. Destinations with beach resources may not have the capacity to develop this type of large-scale industry, and they may not want to. Correspondingly, individual mass tourists and explorers interested in vacationing at the beach may not want this type of experience. The demand–supply match allows

Figure 3.1. Sun, sea, and sand is an extraordinarily popular tourism product that can be found all over the world. *Source:* Scott Jeffcote

some destinations to maintain the natural quality of their beaches with limited infrastructure to accommodate a smaller number of tourists who appreciate the quieter, more intimate experience. In contrast with major destinations in the Caribbean basin characterized by mass 3S tourism, some of the islands with less developed tourism industries, such as Grand Turk—an island in the Turks and Caicos—offer this brand of beach tourism.

Sex and Romance Tourism

Sex tourism is a product that takes place in destinations all over the world in a variety of forms. The most commonly recognized sex tourism product involves travel to a place to engage in commercial sex (i.e., prostitution). This brand of sex tourism tends to be associated more with male tourists than female. For many destinations, sex is considered to be a byproduct of travel rather than the primary motivation (e.g., business travelers hiring prostitutes at the destination during the course of their trip). However, other destinations have become known for the availability of commercial sex or a particular type of commercial sex (e.g., homosexual or child sex) and therefore attract tourists specifically for this purpose. While this may be a one-time encounter with the prostitute, he or she may also be hired for extended periods, perhaps as a travel companion for the duration of the tourist's vacation. Additionally, these destinations may cater to those looking for the opportunity to experience things that might not be available to them in their home environment. Well-known destinations for this brand of sex tourism are in Southeast Asia, predominantly Thailand and the Philippines.

The demand generated by sex tourists has become a driving force in the commercial sex trade and consequently trade in women and children. Recently, international agencies, including the UNWTO and the United Nations Office on Drugs and Crime (UNODC), have recognized that there is a connection between the tourism industry and human trafficking.[1] In addition to sexual exploitation, trafficked persons often suffer from extreme violations of their human rights, such as the right to not be held in slavery or involuntary servitude, the right to be free from violence and cruel or inhumane treatment, and the right to health. Governmental and nongovernmental organizations (NGOs) as well as private companies have worked to develop codes of conduct for both tourism stakeholders and tourists. These codes are intended to promote responsible patterns of behavior to ensure that no one involved in the tourism industry is sexually exploited (for more on codes of conduct, see chapter 11). In particular, efforts have largely concentrated on preventing the trafficking of children and child sexual exploitation associated with tourism. Many countries like the United States have passed child sex tourism laws under which tourists who engage in sex with minors, even outside of the country, can face up to thirty years in a US prison.

Human rights organizations are now trying to harness the power of tourism to try to fight sex trafficking. For example, TraffickCam encourages tourists to take photos of their hotel room and upload them to the website or mobile application. When traffickers post photos of victims in online advertisements, investigators can search the resulting database of images to try to determine the victim's location.[2]

Sex tourism that does not involve commercial sex is harder to define and is therefore less commonly recognized. This may be framed as seeking romance and/or a relationship rather than sex. This type of sex tourism is more often associated with women. Money may not be exchanged for the act of sex, but there still may be an economic motivation as tourists offer their partners drinks, meals, entertainment, and/or gifts during the course of their time together. The relationship that forms may continue after the tourists leave the destination; they may return later or pay to bring the partner to their home. The Caribbean, namely popular 3S destinations like Jamaica and the Dominican Republic, became associated with this phenomenon, where typically white female tourists were involved with black "beach boys." It became so common that women traveling without a male companion at these destinations were assumed to be sex tourists.

Romance, and sex, between tourists also occurs. The atmosphere of popular 3S destinations lends itself to this form of sex tourism, where the focus is on relaxation and pleasure (see box 3.1). Clothing may be minimal—perhaps even optional—and alcohol and/or recreational drugs may be present. Inhibitions may be lowered, and tourists may feel freer to have sexual relations with strangers than they would in their daily lives and home environments. However, this is not the only type of destination associated with such patterns. Far from the beach, the scenic and historic city of Lijiang in China's Yunnan Province has become known as a destination for *yanyu*—romantic encounters—among young Chinese tourists. Yanyu does not necessarily imply sexual relations but more generally the interactions that take place between male and female tourists.[3]

Perhaps not surprisingly, research has shown that, although many people have reported engaging in any one of the above behaviors, few would describe themselves as sex tourists.[4]

Box 3.1. Case Study: Mass S Tourism in the Mediterranean

> "You will either have the holiday of your life or a holiday from hell, all depending on your outlook on life."*

As this tour operator suggests, much of tourism experiences comes down to perspective. While some tourists avoid prototypical sun, sea, sand, sex, and spirits tourism, it clearly holds appeal for many tourists around the world, as evidenced by the tremendous popularity of resorts providing these experiences. In this case, the operator is describing Magaluf, one of the principal resorts on the Mediterranean island of Palma de Mallorca (map 3.1) and a destination often cited as having all of the excesses of S tourism in the region. Like many S destinations, Magaluf was once a small island fishing village. During the 1960s, the Mallorcan municipality of Calvià experienced significant investment in mass tourism infrastructure and high-rise resort development in both Magaluf and neighboring Palmanova. Today, the resort has little appearance of or connection to the rest of the island or the Spanish mainland. The municipality receives well over a million tourists annually, many of whom are foreign. In particular, Magaluf was developed specifically to cater to British tourists and has become a piece of Britain in the Mediterranean. English is widely spoken around the resort, and

* Islas Travel Guides, "Welcome to Magaluf," accessed July 14, 2016, http://www.majorca-mallorca.co.uk/magaluf.htm.

Map 3.1. Palma de Mallorca, Spain. Popular resorts at this Mediterranean destination like Palmanova and Magaluf are based on tourism's S's. *Source*: XNR Productions

hundreds of cafés and bars have British names, serve British foods and drinks, and even show British television programs.

Following a downturn in tourism during the 1990s, Mallorcan tourism officials developed a diversification strategy to get tourists involved in other activities throughout the destination. However, local tourism operators often discourage tourists from leaving the resorts as a result of possible inconveniences (e.g., crowded public transportation) or dangers (e.g., pickpockets). Most tourism promotions highlight Magaluf's beaches, with the promise of beautiful sand, clear water, and the relaxation of sunbathing during the day. While these S's may be the primary attraction for neighboring Palmanova, they are often only secondary considerations for Magaluf. Also known as "Shagaluf," this resort is better known for its other S's and the multitude of bars and nightclubs, cheap alcohol, the twenty-four-hour party atmosphere, and casual sex. In fact, tourism researcher Hazel Andrews found that there was an "expectation that sexual activity was a reason, if not *the* reason, for being there."[†] The principal tourist market for Magaluf tends to be young adult (from age 18 to the 30s) British working-class singles. Most arrive in groups on package tours, sometimes for stag and hen parties.

The atmosphere tends to be sexually charged, with references to and an abundance of naked bodies, including topless sunbathers, nudity in cabaret-type shows, exposure during bar crawl drinking games, and even in tourism-related imagery such as promotions and postcards. Tourists are warned about the noise levels of large quantities of inebriated tourists during the peak summer months. Females in particular are warned about unwanted attention, potential harassment. Although a subset of tourists return to the resort year after year, its negative reputation has been growing with muggings, rapes, and even accidental tourist deaths due to intoxication.

Mallorca's tourism stakeholders have become increasingly frustrated that Magaluf's reputation for 5S tourism predominates over its other natural, cultural heritage, and agricultural tourism products. The destination has initiated a process to rejuvenate, rebrand, and possibly even rename the resort. In addition, local officials sought to impose restrictions on the notorious pub crawls and to prohibit drinking in the streets during the 2015 tourist season.[‡] With the report of yet another tourist death attributed to a high level of alcohol in a young woman's bloodstream early in the 2016 season, it seems little has changed.

Discussion topic: Do you think Magaluf will be successful in its efforts to change its product and its image? What factors will influence the outcome?

Tourism on the web: Institut Balear del Turisme, "Mallorca, the Balearic Islands" at http://www.illesbalears.es/ing/majorca/home.jsp

---

[†] Hazel Andrews, "Feeling at Home: Embodying Britishness in a Spanish Charter Tourists Resort," *Tourist Studies* 5, no. 3 (2005): 251.

[‡] Tracy McVeigh, "Magaluf's Days of Drinking and Casual Sex Are Numbered—Or So Mallorca Hopes," *The Observer*, April 18, 2015, accessed June 17, 2016, http://www.theguardian.com/travel/2015/apr/18/vodka-sex-magaluf-tourists-spain-mallorca-shagaluf.

Nature Tourism

Nature tourism is a product that represents a diverse set of activities set in or based on the appreciation of natural attractions. These attractions may include unique natural features, landscape scenery, or the wildlife of a particular place. Such features may be protected as parks and preserves; in particular, the national park designation plays a role in the creation of opportunities for nature tourism, as both domestic and international tourists make a point to visit these places. Nature tourism may be the primary tourism product for a trip or one type of activity participated in during the course of a trip. For example, birding is a specialized nature tourism product that has been growing in recent years. The practice of bird watching and listening is particularly popular among older, affluent tourists, traditionally from more developed countries such as the United States and the United Kingdom, who enjoy traveling to new places in search of opportunities to observe different species. Dedicated tour companies, such as Birding Africa, provide entire trips oriented around the practice.

Although nature tourism may be positioned as niche tourism in opposition to mass tourism such as 3S, this product can also provide a diversionary activity for mass tourists. In the case of the Caribbean, islands depend on sun, sea, and sand to attract tourists. However, these destinations also promote nature tourism as an activity tourists can participate in for a day, or part of a day, during their vacation. This is not the primary motivation for the trip, but it allows tourists to experience more of an island than simply resort areas on the coast. These products may be packaged as nature walks or hikes, in which guides highlight local flora and fauna.

As the global tourism industry has been growing, more destinations around the world have utilized their natural attractions and developed a nature tourism product. There are, of course, good examples of nature-based tourism activities in which tourists have the opportunity to experience unique environments and/or wildlife with few negative impacts. At the same time, there are bad examples of nature being exploited for the purpose of tourism. This has generated considerable debate about how nature tourism should take place and resulted in the evolution of the ecotourism concept (see box 3.2).

Box 3.2. In-Depth: The Ecotourism Concept

The term **ecotourism** is frequently used as a synonym for nature tourism. In theory, there is overlap between the two products; in practice, there may be little distinction between them. However, the concept of ecotourism is intended to go beyond activities in nature and/or appreciating nature. The International Ecotourism Society (TIES) defines ecotourism as "responsible travel to natural areas that conserves the environment and improves the well-being of local people, and involves interpretation and education."* The concept was intended to maximize the benefits of tourism and ensure both economic and environmental sustainability. An argument can be made for tourism if it is shown to be as profitable as other, more environmentally destructive activities, such as logging or mining—or in fact, more profitable in the long term. However, this depends on the preservation of environmental resources that provide the basis for tourism. At the same

* The International Ecotourism Society, "What Is Ecotourism?," accessed July 14, 2016, http://www.ecotourism.org/what-is-ecotourism.

time, local people must be part of tourism. These people should be involved in activities to ensure that the tourism developed fits within their values and lifestyles. They should directly benefit from tourism, not only to improve their quality of life but also to ensure that they have a stake in it and will provide the necessary support.

Destinations around the world have attempted to translate the ecotourism concept into a tourism product, with varying results. Places such as Costa Rica and Kenya have become associated with ecotourism, while others offer some type of experience called ecotourism. As such, researchers argue that it may be useful to make a distinction between hard and soft variations of ecotourism that exist in practice.‡ In this model, "hard ecotourism" is a niche product involving small numbers of tourists who are explicitly interested in wilderness experiences as well as ensuring the sustainability of their actions. Typically categorized toward the drifter end of the spectrum, these tourists visit more remote destinations where there are few other tourists and tourist services. Ecotourism is the primary focus of the trip, which may be physically and/or mentally demanding. This may be done as part of a specialized tour package through a company such as Gap Adventures, but it may also be undertaken independently with the use of informal local resources.

"Soft ecotourism" has been criticized as just a different label for nature tourism, with little of the concept behind ecotourism. This variation provides mass tourists with an opportunity to have an "ecotourism" experience as part of their larger trip. These tourists have a more superficial interest in environmental issues. Their experiences are shorter and may be even just a day trip to a natural area that is relatively close to the principal destination region and has the appropriate infrastructure (e.g., paths, bathrooms, refreshments, etc.) to accommodate a large number of tourists. Consequently, interactions with nature are facilitated by a guide and tend to be more superficial. These hard and soft positions are, of course, two ends of a spectrum, and there are many examples of experiences that fall somewhere in between.

Thus, while ecotourism was intended to provide a sustainable framework for nature tourism, it has, to some extent, become just another buzzword to generate interest in tourism. This has led to the development of certification programs to help ensure that products being labeled ecotourism are, in fact, environmentally sustainable. For example, Ecotourism Australia is one of the most long-standing ecotourism accreditation systems. This organization developed a set of guidelines for various levels of environmentally sustainable tourism, from nature tourism that uses specific measures to minimize the impact of tourists' activities on the environment to a comprehensive form of ecotourism in which operators strive for the highest levels of sustainability. Businesses can then use this certification to support their claims of sustainability to knowledgeable tourists.

Discussion topic: Search for an ecotourism product on the Internet. Do you think that the experience/activity described should be considered nature tourism or ecotourism? Why?

Tourism on the web: Ecotourism Australia, "Welcome to Ecotourism Australia," at http://www.ecotourism.org.au/

‡ David A. Fennell, *Ecotourism*, 4th ed. (London: Routledge, 2015), 12.

Sport, Adventure, and Adrenaline Tourism

Sport tourism, adventure tourism, and adrenaline tourism are related products centered on physical activity—with varying degrees of intensity. In these products, sports—and other, more extreme physical activities—may be the primary motivation for a trip or simply one of the activities the tourist participates in during the course of a trip. These products encompass a range of activities, seasons, environments, and infrastructural requirements. The activity may be one that the tourist is involved in at home during his or her leisure time. Avid golfers often plan trips where they travel to play different courses, including famous ones associated with major professional golf tournaments such as Augusta National in Georgia. The activity may also be something that the tourist has limited opportunity to enjoy in their home environment. In the sub-tropical states of the American South, people will have to plan a trip to a resort in Wyoming or Colorado, to fulfill their demand for winter sports.

Alpine skiing, cross-country skiing, and snowboarding are popular winter recreation activities that provide the basis for winter sport tourism. Some ski resorts in Europe date back to the late nineteenth century, while the oldest resorts in North America date back to the early twentieth century. Summer sport tourism, or warm weather sport, includes a much more diverse set of activities, ranging from golf to bicycling to horseback riding. Water sport tourism is also immensely popular at coastal destinations and includes activities such as swimming, snorkeling, scuba diving, surfing, wind surfing, jet skiing, water skiing, sailing, fishing, and more. While these activities may take place at any number of coastal destinations, some have particularly been associated with water-sport tourism. For example, the island of Tortola in the British Virgin Islands is known as a sailing destination, while Bonaire, in the Netherlands Antilles, is known as a scuba destination.

Adventure tourism is typically a physical activity that tourists would not participate in at home and is more dependent on the natural resources of a place. These activities may require specialized equipment and training or skill, and there is some degree of excitement and/or perceived risk. Examples of adventure tourism might include zip-lining in rain forest canopies, kite boarding, whitewater rafting or kayaking, and mountain biking (figure 3.2) or trekking. Soft adventure tourism may be an activity for mass tourists close to well-developed destination areas, while hard adventure tourism transitions into adrenaline or extreme tourism with activities such as rock climbing, spelunking, bungee jumping, and skydiving. This product is more likely to be the focus of a trip that takes place in remote, possibly even dangerous, locations. The activities are intense and the risk greater, but the adrenaline rush is part of the attraction.

Rural and Urban Tourism

Rural tourism and urban tourism offer distinctive experiences based on their particular sets of resources. While there is some overlap between rural tourism and nature tourism, the former is more specifically associated with the general sensibility of "rural" that pertains to life in the countryside. Rural recreation activities such as scenic drives, country picnics, hunting, or fishing may be included in this product.

Figure 3.2. Adventure tourism often involves physical activities in spectacular natural environments. Mountain biking provides adrenaline-inducing excitement, while the scenery enhances the experience. *Source:* Scott Jeffcote

Farm tourism or agricultural tourism (agritourism/agrotourism) is a specialized product that has evolved out of rural tourism. Activities within this product vary widely. Tourists may participate in activities set in the farm environment, such as horseback riding or hiking farm trails. They may consume farm produce, such as eating at a farm restaurant, purchasing items at a farm market, or even doing the "pick-your-own" option. Tourists may also participate in farm activities. Some activities are simulated, such as tourist cattle drives and cowboy cookouts hosted by dude ranch resorts, but there are also working farms and ranches on which the tourist learns about and assists in daily chores or harvests. For these activities, tourists will stay on the farm in facilities converted to serve as a bed and breakfast or specially constructed facilities such as cottages. Farm tourism has a particularly strong tradition in Europe.

Likewise, urban tourism is an overarching product that may encompass cultural and heritage tourism or event tourism (below). Towns and cities have an array of attractions upon which urban tourism may be based, and some of the largest tourism destinations in the world are major cities, such as Paris, New York City, Bangkok, and Dubai. In many cases, attractions will be spatially concentrated in particular parts of the city, such as historic neighborhoods, shopping and entertainment districts, or waterfront developments.

Attractions include historic buildings, from ancient ruins to churches and cathedrals or castles and palaces. Although these places were not originally intended for tourism, they may be preserved as attractions. In other cases, places are transformed to become attractions. For example, the Meatpacking District in Manhattan developed based on the spatial concentration of slaughterhouses and packing plants. In recent years, this area has been redeveloped into one of New York's most fashionable neighborhoods with diverse restaurants, trendy clubs, stylish boutiques, and luxury hotels. This, combined with historic district designations, has given rise to walking and guided tours of the area. Other urban amenities, such as upscale or boutique shopping, museums, art galleries, theaters, concert venues, sports arenas, restaurants, bars, nightclubs, and more, serve the dual purpose of providing a desirable living environment for residents and an attractive environment for tourist visits.

Cultural and Heritage Tourism

Cultural tourism is arguably one of the oldest tourism products, as many of the earliest tourists traveled for the experience of other cultures and cultural attractions. As the tourism industry began to develop specialized products, cultural tourism was considered a niche tourism product oriented toward a small subset of affluent and educated tourists interested in authentic experiences of other cultures. Today, however, cultural tourism is recognized as one of the broadest tourism products, which encompasses a vast range of attractions and activities and has extensive overlap with other products. Some reports suggest that as much as 70 percent of international tourists today participate in cultural tourism.[5]

Cultural tourism is based on human attractions, and in particular, elements of a society's culture. For the most part, cultural tourism pertains to the unique cultural patterns that have evolved in a specific place over time to serve a purpose for that group of people, not to attract tourists. This includes the patterns of lifestyles, cuisine, clothing, art, music, folklore, religious practices, and more that make a place distinct. In the modern world, with globalization and the perception of uniform contemporary lifestyles, there is an interest in and a demand for experiences of different cultures. Although many of these patterns no longer have a role in the daily lives of these people, this demand means that elements of traditional culture may be maintained when they might otherwise be lost.

There is considerable overlap between this brand of cultural tourism and heritage tourism. Heritage tourism only recently emerged as a distinct tourism product and is considered one of the fastest growing. Heritage tourism encompasses travel to varied sites of historic (and often cultural) importance such as the Parthenon or the Taj Mahal, sites where important historic events occurred such as Independence Hall in Philadelphia or the D-Day landing beaches in Normandy, and sites that represent the stories and people of the past such as the National Museum of Cultural History in Pretoria, South Africa or Mozart's Birthplace museum in Salzburg, Austria. Prominent heritage tourism attractions are designated United Nations Educational, Scientific, and Cultural Organization (UNESCO)World Heritage Sites. Using a specific set of selection criteria, the organization identifies places with outstanding universal value.[6] While these places would likely serve as tourism attractions regardless, the World Heritage Site designation often raises the site's profile and increases visits.

Contemporary culture also plays a role in tourism. Elements of contemporary culture may be used to attract tourists, but tourism may also be a side effect of cultural activities. This brand of cultural tourism may be associated with some aspect of high culture, such as artistic works and performances. England is particularly known for its literary destinations—places specifically associated with a well-known writer or referenced in a widely read work. This may also be associated with popular culture. For example, major film and television projects, such as *Harry Potter*, *The Lord of the Rings*, *Game of Thrones*, and *Downton Abbey*, have all generated significant tourism to film sites.

Although attractions developed specifically for the purpose of tourism do not fit traditional ideas of cultural tourism, they are nonetheless shaped by and represent aspects of contemporary culture. For example, Disneyland and Disney World are two of the most visited tourism destinations in the world and have been described as epitomizing American culture. Similarly the theme park–like environment of Las Vegas is distinctly part of popular American culture. In contrast with tourism related to elements of traditional culture that may be associated with rural areas, these activities are more often associated with cities and overlap with urban tourism.

Food and Beverage Tourism

The food and drink of a place is a fundamental tourism resource. In the past, these items played a supporting role in tourism, providing tourists with sustenance while they were away from their home environment. However, in recent years, the popularity of food in the media and the expansion of ethnic restaurants have exposed more people to the foods of places they have not yet had the opportunity to visit.[7] This creates a demand for the experiences of those places and their foods.

This has given rise to a relatively new tourism product. Although the terms culinary tourism and gastronomic tourism are also used, often with more specific implications, food and beverage tourism provides a simple and comprehensive label. Food and beverage tourism describes travel motivated by, or involving an interest in, learning about and experiencing the food and drink of a place. This can include the experience of everyday foods served by street vendors and local restaurants; it can also include elaborate tasting menus served by restaurants with a Michelin star or featured on the S. Pellegrino and Acqua Panna World's 50 Best Restaurants list. For some, food has become the primary reason for traveling to a particular destination, but for many tourists today, enjoying the food of a destination is simply one part of the expectations they have for their trip.

This product is often considered a subset of cultural tourism in which tourists can observe, participate in, and gain an understanding of other cultures through food and

Figure 3.3. Wine tourism is a specialized tourism product that is part food and beverage tourism, with opportunities to learn about and sample wines, and part rural tourism, set in attractive rural landscapes such as this scene from Oregon's Willamette Valley. *Source:* Scott Jeffcote

eating experiences.[8] Food is a part of the culture of a place. Some destinations, like France and Italy, have a well-known reputation for quality food and eating experiences. Countless others around the world are currently engaged in the process of defining, or redefining, a food culture that can increase its pull factor for potential tourists. Peru is one of the most successful examples. The country has won World's Leading Culinary Destination from the World Travel Awards every year since the category was created in 2012 (as of 2015).[9]

While food and beverage are often considered part of an overall package, beverage tourism is an increasingly popular special interest tourism product in its own right. Wine tourism—in the form of wine trails, vineyard and winery tours, and wine tastings—has been the most developed product that has often overlapped with rural and agricultural tourism (figure 3.3). However, interest in craft beers, ciders, and liquors has been growing. These products are often connected to specific places that aficionados might wish to visit (e.g., the Kentucky Bourbon Trail). Others signify local or regional connections and therefore become a part of the experience of those places (e.g., the Great Lakes Brewing Company in Cleveland, Ohio).

Drug Tourism

Drug tourism is typically defined by an individual's interest in a destination due to the ability to consume licit or illicit drugs in that place. However, tourists may also travel to a destination without prior knowledge or intention of consuming but may ultimately participate in drug tourism. Studies show different motivations for participating in drug tourism. For some tourists, this is part of the inversions discussed in chapter 2. In their daily lives, these people operate within the legal frameworks and established norms of society; on vacation, they seek to escape from routine and break free from these restrictions. For other tourists, this is part of the experience of place and an opportunity to participate in local culture.[10] Still others seek self-actualization with the aid of substances not available to them in their home environment.

Amsterdam is perhaps the most well-known example of a drug tourism destination. Cannabis is an illicit drug in the Netherlands, but an official policy of tolerance allows "coffee shops" to sell small quantities to patrons. This has been a significant attraction for the city. The Dutch government has made several attempts to ban tourists from coffee shops, but none have been turned into law. Still, other policies have made it more difficult to obtain the product, leading some to proclaim Barcelona as the "new Amsterdam."[11] In another example, drug tourists have visited Amazonian Peru and Brazil with the intention of trying ayahuasca. Indigenous people have traditionally used this mixture of psychedelic plants for ceremonial and healing purposes. Studies have shown that local dealers posing as shaman exploit tourists seeking enlightenment through religious ceremonies or simply an unusual experience.[12]

In 2014, Colorado and Washington became the first American states to legalize recreational sales of marijuana. However, there are still challenges for drug tourists. Vendors are prohibited from traditional advertising. In addition, smoking marijuana in public is still illegal, and the majority of the traditional hotel infrastructure is also smoke-free. Therefore, it may be difficult for visitors to know where to go to both purchase products and consume them. This has given rise to a drug tourism industry. In addition to helping tourists shop and find marijuana-friendly lodging, companies offer opportunities to visit growers and learn about products, take cannabis cooking classes, and receive cannabis-related spa treatments.[13]

While drug tourism can provide an attraction for a destination and make a significant economic contribution, there are nonetheless many potential issues depending on the nature of the product. Businesses involved in drug tourism need to ensure they are in compliance with the laws of the places in which they operate and maintain any necessary permits and insurance. Communities may be concerned about unacceptable public behavior, an increased burden on public health systems, and potentially even drug trafficking and drug-related crime. Tourists should be aware of any potential legal and/or health consequences to their activities. Each year, drug-related tourist deaths occur.

Dark Tourism

Dark tourism is the term used to describe a range of experiences whereby people visit sites of death or attractions/exhibits that represent death. Thanatourism is a lesser used term derived from Greek mythology in which Thanatos was the personification of death. Interest in death is by no means a modern phenomenon; some scholars argue that public consumption of suffering and death in the Roman gladiatorial games or medieval executions were precursors to dark tourism.[14] Nonetheless, sites of—and associated with—natural disasters (e.g., Hurricane Katrina), human-caused disasters (e.g., Pripyat, Ukraine, the city that was abandoned after the Chernobyl nuclear meltdown), human atrocities (e.g., the Nazi concentration camp at Auschwitz), and death (e.g., Hiroshima Peace Memorial Park, figure 3.4) have had an increasing presence in modern tourism. Tourists have a variety of motivations for visiting such sites, ranging from a fascination with death to a desire to connect with their identity or heritage.

Philip Stone, Executive Director of the Institute for Dark Tourism, argues that there is a spectrum of dark tourism.[15] Some sites sensationalize events to entertain visitors (e.g., Jack the Ripper tours in London's East End), while others seek to educate visitors and to engage them with persistent issues (e.g., the International Slavery Museum in Liverpool that not only examines issues associated with slavery but also the legacies of racism and discrimination). Some sites are accidental, in that they naturally attract curiosity because of the death that occurred there (e.g., the site of the car crash in which Princess Diana was killed). Other sites are purposefully constructed for commemoration (e.g., the National Sept. 11 Memorial and Museum).

Evidence, artifacts, and memorials constitute the tangible heritage, or markers, associated with disasters or atrocities. They can be used as tools to articulate and promote remembrance of the event, but they are not always sufficient to engender empathy among visitors. Thus, memories of the event, such as testimonials of survivors, constitute the intangible heritage and an equally important component of these

Figure 3.4. Dark tourism refers to a range of experiences whereby people visit sites associated with death. The Hiroshima Peace Memorial Park in Hiroshima, Japan is such a site that receives over one million visitors a year. The park is intended to recognize the site of the first city in the world to experience a nuclear attack, to remember the victims, and to promote peace. *Source:* Velvet Nelson

sites.[16] Still, geographers remind us that places of memory are selective, often contested or politicized, interpretations of the past, and must be viewed critically.[17] Consider, for example, the different viewpoints on memorial sites associated with civil wars around the world. While many such sites are intended to promote reconciliation and healing, narratives and counternarratives of events can be contentious and foster resentment.

Event and MICE Tourism

Event tourism is based on special events as a category of tourism attractions. Special events have long been an attraction for localized markets and have often generated day trips, for example, as people travel to neighboring towns to participate in local festivals. However, as more people are enabled to travel farther distances, this product has exploded. It now encompasses a diverse set of events that are global in nature or highly localized, always in the same location or at various locations, religious or secular, annual or one time only.

Kumbh Mela is a riverside Hindu gathering held in India every twelve years. It is considered to be one of the world's oldest religious festivals and the largest. In 2013, more than 120 million people participated in this event over the course of 55 days.[18] Secular festivals are based on a range of attractions, from popular culture to local heritage or even local produce. Oktoberfest, held in Munich over a seventeen-day period at the end of September and the beginning of October, is one of the most famous—and largest—of these events. In 2015, the event drew in 5.9 million visitors who consumed an estimated 7.7 million liters of beer.[19] The Sauerkraut Festival started as a small local festival in Waynesville, Ohio, a town with less than three thousand residents, and has grown into a major event attracting approximately 350,000 visitors from all over the country. More than thirty food vendors offer everything from traditional pork and sauerkraut dishes to sauerkraut ice cream.[20]

Major sporting events are some of the most widely known forms of event tourism. In some cases, these are annual events that occur in the same place. As a result, these destinations are often closely associated with the event. For example, the Kentucky Derby is a world-famous thoroughbred horse race held in May at Churchill Downs in Louisville, Kentucky, that attracts approximately 150,000 visitors each year.[21] In other cases, events take place at certain intervals and are hosted at different venues around the world. The Olympic Games are a global event held every two years, alternating between the summer and winter games. Major cities compete to host the games, not only to bring international attention to the city and country but also to bring investment and tourism. London, England, host to the 2012 Summer Olympic Games, received over 10,000 athletes, 21,000 media representatives, and up to 180,000 spectators attended each day of the event.[22]

Tourists are also increasingly traveling to places for personal events. This type of event tourism is related to the Visiting Friends and Relatives product discussed below. For example, with dispersed family networks, reunions may be held at vacation destinations to take advantage of the tourism infrastructure as well as maximize the limited time and money many families have available for travel. Likewise, the rise of destination weddings has generated a type of event tourism where the couple, as well as friends and family, travel to a tourism destination for the ceremony as well as related activities.

MICE tourism refers to MICE meetings, incentives, conventions, and exhibitions. These events have become so important that major cities all over the world have developed extensive convention facilities and actively compete to host organizations' events. These events range from just a few hundred participants to tens of thousands. For example, the 2016 Annual Meeting of the Association of American Geographers, held in San Francisco, California, saw more than nine thousand participants from eighty-seven countries.[23] The majority of these participants must travel to and stay at the event site, and they may bring partners or children with them. The primary financial gain is often not from the event itself but from expenditures at hotels, restaurants, evening entertainment activities, shopping, etc. In addition, events can help sustain tourism revenues by bringing in visitors during non-peak times of the year.[24]

Visiting Friends and Relatives (VFR) and Roots Tourism

Visiting friends and relatives (VFR) is one of the most common tourism products in which people participate. This product developed significantly in the second half of the twentieth century with increased mobilities. People in many parts of the world experienced greater abilities to move to new locations, within their own country or abroad, based on a variety of opportunities for school, work, or otherwise. This created dispersed networks of family and friends. At the same time, people experienced greater abilities to travel to visit these family and friends. VFR is often domestic tourism; however, this is also one of the largest tourism products for places with high rates of emigration. This form of international tourism links the place of origin, whether it is a small Caribbean island like Trinidad or a large country like India, with the places to which these people have migrated.

Still, VFR tourism is a product that generally receives little attention. Many people do not consider VFR trips to be tourism because it may be seen as an obligation rather than a vacation or simply "going home" rather than "going away." Moreover, VFR tourists are often seen as existing "outside" the normal tourism industry. Their patterns of behavior might be determined by different factors than those of other tourists. For example, their destination choice is based on where friends or relatives live as opposed to the attractions of a place. They may be less reliant on the tourism infrastructure. They may stay in and eat at people's homes as opposed to staying in hotels and eating at restaurants. They may rely on their hosts' personal cars rather than renting one or using local transportation, and their hosts may serve as their guides rather than hiring one or taking a tour. Although these tourists may contribute less to the tourism industry at a destination, they are nonetheless contributing to the local economy.

There may be overlap between this product and others in terms of motivations. For example, people may travel to their birthplace for events such as weddings or reunions. In addition, there will be overlap with other products as people participate in tourism activities during the course of their visit with family and friends. Because of the tremendous variations within this product, some researchers have argued for it to be broken down into smaller segments, including domestic and international, short-haul and long-haul, or visiting friends and visiting relatives.[25]

VFR has some overlap with roots tourism. For example, migrants may return to their places of origin to renew their sense of identity that may feel lost the longer they are away. Much of roots tourism involves the descendants of migrants who may travel as tourists to their ancestral lands to meet relatives or to discover the culture of their ancestral society. Roots tourists' itineraries are often personal as they visit the places with which their ancestors were connected as opposed to typical destinations. Some scholars have likened this type of tourism to a secular pilgrimage involving a physical journey to the ancestral place as well as an emotional journey with the search for identity.[26] Particularly for descendants of immigrants, such journeys may ultimately reinforce tourists' ideas of homeland as the place in which they were born and raised.[27]

Service Tourism

Service tourism involves traveling to another place to volunteer one's time providing aid, assisting with local development, contributing to conservation efforts, participating in research projects, and more (box 3.3). We may not think of service work that takes place outside of the home environment as tourism because of the traditional association of tourism with leisure activities and the pursuit of pleasure. However, interest in volunteering in other places—and companies or organizations providing opportunities to do so—has been growing in recent years. This trend has given rise to the term "voluntourism."

Typically, service tourism involves the movement of tourists from more developed countries to less developed ones. Service tourism destinations may also be impoverished areas of developed countries (e.g., parts of the Appalachian region) or those devastated by a natural disaster (e.g., the hurricane-ravaged Gulf Coast region). Although the potential market for service tourism includes everyone, experiences are often oriented toward young people, particularly students or recent college graduates. Service tourists still pay for their travel, accommodation, and other daily expenses in addition to donating their time and labor. They may actually pay more for their experiences—while receiving fewer services—than they would for a traditional holiday.[28] For-profit companies may additionally require a fee for organizing the experience.

This product is sometimes considered "alternative tourism" in that it presents tourists with an alternative to other products that appeal to traditional touristic motivations. For example, the "alternative spring break" is targeted at students looking for something other than the typical spring break trip that covers all of the S's. Suppliers market these experiences as opportunities for adventurous, sophisticated, and thoughtful travelers, in opposition to indulgent, self-absorbed, and insensitive mass tourists.[29] Thus, tourists interested in this product generally have characteristics in common with drifters and explorers: they wish to visit out-of-the-ordinary places, interact with local people, and have a deeper experience of place. Unlike typical drifters and explorers who create their own experiences, service tourists must rely on a company or organization to create opportunities for service and facilitate their experience. However, this may open more doors for service tourists, allowing them to more effectively penetrate the back regions of a place than even drifters. In particular, these tourists have the opportunity to experience life as it is lived in that place by staying, eating, and socializing with local people.

Proponents of service tourism claim that there are both short- and long-term benefits of such experiences, although most center on tourists (e.g., personal development) rather than local communities. However, arguments cite an increase in intercultural awareness as well as social, political, or environmental consciousness. This can have an impact that lasts beyond the trip if it leads to greater global citizenship and activism to continue to learn about key issues and be involved in promoting change. Critics claim that service tourism not only has few benefits for host communities, but it can actually harm them. At the least, tourists may not have the skills to complete the tasks they are assigned (e.g., construction), and the resulting products may be sub-par.[30] At the very worst, children are separated from their families and kept in poor conditions at orphanages to attract visitors who will pay fees and/or donate money.[31]

Categorizing service tourism as good or bad is too simplistic. Instead, scholars propose a spectrum of experiences, including the harmful, the egocentric (i.e., providing for the self-centered interests of typically affluent youth), the harmless (i.e., offering little impact, positive or negative, for the host community), the helpful (i.e., offering constructive assistance), the educational (i.e., providing opportunities for greater cross-cultural understanding), and social action (i.e., leading to long-term involvement).[32] In addition to critically reflecting on ones' own motivations, potential service tourists should fully research the organization with which they seek to travel and to consider the potential consequences of their involvement in that place.

Box 3.3. Experience: Service Tourism and Life Changes in Lesotho

In recent years, service tourism has come under scrutiny, in both the academic literature and popular media, for a variety of reasons. Increasingly, complaints focus on the egocentric motivations of "voluntourists." For example, individuals may seek to make social media friends and followers envious because they traveled to off the-beaten-path places and had out-of-the-ordinary experiences, or they may seek praise for their perceived nobility or sacrifice. However, as Kayla shows below, many tourists who volunteer their time or participate in service projects have a genuine interest in the places they visit. This interest does not end with the trip. Instead, such experiences often create an attachment to place and strengthen the desire to be involved.

When I was in the fourth grade, I was given an assignment to research a country other than my own. As inquisitive as I was, I turned to the nearest map and picked the most foreign country that I could find. I landed on Lesotho, Africa. I fell in love with everything about this country: the land, the people, the family-based culture, and the customs. Ten years later, as a college student, I was amazed to find out that one of my professors offered a three-week long service trip in Lesotho. I was overwhelmed with excitement; my childhood dream of going to Lesotho was about to come true.

In the winter of 2015, my professor, his wife, their two children, fifteen other students, and I embarked on our eighteen-hour flight to Johannesburg, South Africa, followed by an eight-hour bus ride into Lesotho. With a naïve mind, I looked around at the beautiful landscape filled with enormous mountains and acres of dry grassland and thought about how I was going to make an impact on the lives of the people of Lesotho, the Basotho. I did not realize that the Basotho would have an even greater impact on my life.

For the first two weeks, we served in a town called Ramabanta. Day in and day out, we worked on projects that included repainting a grade school, constructing a tire playground for a kindergarten, and building a two-bedroom Habitat for Humanity house for an orphaned family of four boys, all under the age of eighteen. We constructed this house with the help of the oldest boy. He came at sunrise and left at sunset every day and was such a joy to be around. Despite the language barrier, he was able to show us how thankful he was for our help.

In the time that we were not climbing on make-shift landings, laying cinder blocks, or hand-mixing cement for these projects, we had the opportunity to play with the local children that ranged from infants to young teens. After the first day, the children began to meet us at the border of their town to walk to our work site. We danced, colored, played games, sang, and even did each other's hair. Friendships were formed, and we learned about each other's culture, with minimal conversation. It is amazing how much I could communicate with others simply by hand gestures, actions, smiles and hugs (figure 3.5). At the end of the day, these children would walk us home, not wanting us to leave, waving and yelling for us until we were completely out of sight.

Each day the same children come to play with us, wearing the same dirty clothes as the day before and without shoes. I also knew that due to the ongoing drought, not many of the children were eating adequate nutritious meals. However, this did not stop the children from

Figure 3.5. Kayla's service trip to Lesotho provided her with greater opportunities to interact with local people than a typical leisure trip to Africa. *Source*: Kayla Winn

playing with us with bright smiles. All they wanted was to be held, given love, and make us smile. From these children, I learned that we should put others' needs before our own. I learned that we need to have an open heart and love fiercely. But above all what I learned from these children it is *rea tsoana kaofela*—we are all the same.

After serving in Ramabanta, our group traveled to the city of Roma. There, we worked vigorously to build a playground for another school and paint a warehouse for a popular teen dance and lyrical content company, despite record-breaking temperatures. While working on the warehouse, we were able to interact with the dancers, have dance battles, share poetry, and exchange personal stories about our respective lives. While the differences were distinct, the similarities were also astounding. We bonded over our love for social media, music, and, of course, dance. In the end, the dancers were so elated to have a beautiful space of their own, and many of them even cried as we left.

We toured Sentebale, a charity facility dedicated to help children who were victims of Lesotho's HIV/AIDS epidemic, and hosted a carnival for the children at Baylor Hospital, which provides health care to children with HIV/AIDS from all over the country. As soon as we set up, the children at the hospital came running out and were so excited to play with us. It was such a humbling experience to know that, for a moment, these children did not have to worry about their illnesses and could have fun. It was initially hard to not think about the fate of these children, but after seeing their smiles, it was easy to get lost in the moment and try to make this day special for them. After leaving Baylor, however, all I could do was hope that these children received adequate medical care to save their lives in a country that lacks the resources to reduce the HIV/AIDS epidemic. This made me feel helpless.

Even though we were in Lesotho to work, we still had plenty of opportunities to experience the country. Although the land was barren from the drought, I found Lesotho to be full of life in every corner. Known as the Mountain Kingdom, we were able to hike the vast mountains, including Thaba Bosiu, the mountain upon which Lesotho was founded. After summiting each peak, we examined the beautiful land below and were overcome by the heights that we had reached. Not only were we breaking ground, we were shattering all of the doubts that we had within ourselves about our ability to hike such huge mountains. We saw the Maletsunyane Falls in Semonkong, although, at the time, just a trickle ran down the waterfall. We even saw dinosaur footprints that were thousands of years old while watching the sun set atop a mountain. We went into the towns and explored their communities, grocery stores, and homes. We even tried the Basothos' staple dish, *Papa*, a corn-based food that is served with almost every meal. It had a texture similar to mashed potatoes, but I didn't think that it had much flavor. No matter where we went, we were always welcomed with open arms by the people and embraced into their society.

Despite the heartaches of missing our families and knowing we were surrounded by very sick children, the difficult working conditions, and being pushed to levels we never thought possible, the Basotho people were worth every drop of sweat. Through the experience of this service trip, I was able to make a small dent into these peoples' lives—while they completely changed mine. They made me realize what I can accomplish through small acts and how a smile can make all the difference. There is not a day that goes by that I do not yearn to be back in Lesotho to be embraced in their loving culture. As for now, I try to apply what I learned there to my daily life. Each day, I try to wake up with a positive attitude and spread happiness with a simple smile or friendly hello. I also try to not take life and material things for granted by living in the moment and making sure that I say thank you.

I have not forgotten the Basotho. Since returning home, I joined the Lesotho Nutrition Initiative with my university. We pack and send a rice-based meal substance to Lesotho to try to combat the malnutrition epidemic that especially affects children in the country. This is my way of staying connected and continuing to serve the people of Lesotho.

—*Kayla*

Conclusion

Tourism is constantly evolving with ever-more specialized products to meet the demands of tourists. This chapter introduces only a few of the products now offered by destinations around the world. Moreover, it should be evident from this discussion that each category could be further subdivided into tourism products that reflect a specific interest, activity, or experience (e.g., golf tourism as a part of summer sport tourism as a part of sport tourism), which would require a far more in-depth examination. Nonetheless, this overview contributes to our broad understanding of tourism. It also provides the necessary context for discussions of the geographic foundation for and the effects of tourism, since different types of tourism and tourism products rely on varied resources and have their own distinct effects.

Key Terms

- ecotourism
- tourism products

Notes

1. United Nations Office on Drugs and Crime, "United Nations Organizations Cooperate to Stamp Out Human Trafficking and Sex Tourism," April 2012, accessed July 14, 2016, http://www.unodc.org/unodc/en/frontpage/2012/April/united-nations-organizations-cooperate-to-stamp-out-human-trafficking-and-sex-tourism.html.
2. Exchange Initiative, "About TraffickCam," accessed June 30, 2016, http://traffickcam.org/about.
3. Honggang Xu and Tian Ye, "Tourist Experience in Lijiang—The Capital of Yanyu," *Journal of China Tourism Research* 12, no. 1 (2016): 8.
4. Martin Oppermann, "Sex Tourism," *Annals of Tourism Research* 26, no. 2 (1999): 256.
5. Bob McKercher and Hilary duCros, *Cultural Tourism: The Partnership between Tourism and Cultural Heritage Management* (New York: Haworth Hospitality Press, 2002), 1.
6. United Nations Educational, Scientific, and Cultural Organization, "The Criteria for Selection," accessed September 10, 2011, http://whc.unesco.org/en/criteria/.
7. Athena H.N. Mak, Margaret Lumbers, and Anita Eves, "Globalisation and Food Consumption in Tourism," *Annals of Tourism Research* 39, no. 1 (2012): 184.
8. Barbara Santich, "The Study of Gastronomy and Its Relevance to Hospitality Education and Training," *Hospitality Management* 23 (2004):20.
9. World Travel Awards, "World Winners," accessed June 16, 2016, https://www.worldtravelawards.com.
10. Natan Uriely and Yniv Belhassen, "Drugs and Tourists' Experiences," *Journal of Travel Research* 43 (2005): 239, 242–3.
11. Winston Ross, "Holland's New Marijuana Laws Are Changing Old Amsterdam," *Newsweek*, February 22, 2015, accessed June 16, 2016, http://www.newsweek.com/marijuana-and-old-amsterdam-308218.

12. Marlene Dobkin deRios, "Drug Tourism in the Amazon," *Anthropology of Consciousness* 5, no. 1 (1994): 18.

13. Julie Weed, "Book Your 'Bud and Breakfast', Marijuana Tourism Is Growing in Colorado and Washington, *Forbes,* March 17, 2015, accessed June 16, 2016, http://www.forbes.com/sites/julieweed/2015/03/17/book-your-bud-and-breakfast-marijuana-tourism-is-growing-in-colorado-and-washington/#4af5ed2c73cf.

14. Philip Stone and Richard Sharpley, "Consuming Dark Tourism: A Thanatological Perspective," *Annals of Tourism Research* 35, no. 2 (2008): 574.

15. Philip R. Stone, "A Dark Tourism Spectrum: Towards a Typology of Death and Macabre Related Tourist Sites, Attractions and Exhibitions," *Tourism* 54, no. 2 (2006): 145–60.

16. Myriam Janse-Verbeke and Wanda George, "Reflections on the Great War Centenary: From Warscapes to Memoryscapes in 100 Years," in *Tourism and War*, ed. Richard Butler and Wantanee Suntikul (London: Routledge, 2013), 273–5.

17. Derek H. Alderman, "Surrogation and the Politics of Remembering Slavery in Savannah, Georgia (USA),"*Journal of Historical Geography*, 36 (2010): 90.

18. Kumbh Mela 2013, "Official Website of Kumbh Mela 2013," accessed June 17, 2016, http://kumbhmelaallahabad.gov.in/english/index.html.

19. Oktoberfest, "The Oktoberfest 2015 Roundup," accessed June 17, 2016, http://www.oktoberfest.de/en/article/Oktoberfest+2016/About+the+Oktoberfest/The+Oktoberfest+2015+roundup/4408/.

20. Ohio Sauerkraut Festival, "Festival History," accessed July 14, 2016, http://www.sauerkrautfestival.com/.

21. Churchill Downs Incorporated, "What to Expect at the Kentucky Derby," accessed June 17, 2016, https://www.kentuckyderby.com/visit/what-to-expect.

22. The International Olympic Committee, "London 2012 Summer Olympics," accessed June 17, 2016, https://www.olympic.org/london-2012.

23. American Association of Geographers, *"Highlights from the 2016 Annual Meeting in San Francisco,"* accessed June 17, 2016, http://www.aag.org/cs/annualmeeting/annual_meeting_archives/2016_san_francisco/2016_san_francisco_highlights.

24. Calvin Jones and ShiNa Li, "The Economic Importance of Meetings and Conferences: A Satellite Account Approach," *Annals of Tourism Research* 52 (2015): 117.

25. Gianna Moscardo, Philip Pearce, Alastair Morrison, David Green, and Joseph T. O'Leary, "Developing a Typology for Understanding Visiting Friends and Relatives Markets," *Journal of Travel Research* 38, no. 3 (2000): 251–9.

26. Paul Basu, "Route Metaphors of 'Roots-Tourism' in the Scottish Highland Diaspora," in *Reframing Pilgrimage: Cultures in Motion*, eds. Simon Coleman and John Eade (London: Routledge, 2004), 155–58.

27. Naho Maruyama and Amanda Stronza, "Roots Tourism of Chinese Americans," Ethnology, 49, 1 (2010): 24.

28. Stephen Wearing, *Volunteer Tourism: Experiences That Make a Difference* (Wallingford, UK: CABI, 2001), 2.

29. Keri Vacanti Brondo, "The Spectacle of Saving: Conservation Voluntourism and the New Neoliberal Economy on Utila, Honduras," *Journal of Sustainable Tourism* 23, no. 10 (2015):1407.

30. Brondo, "The Spectacle of Saving," 1407.

31. UNICEF, "Orphanage Voluntourism in Nepal: What You Should Know," accessed June 16, 2016, http://unicef.org.np/media-centre/reports-and-publications/2015/05/10/orphanage-voluntourism-in-nepal-what-you-should-know.

32. Regina Scheyvens, *Tourism and Poverty* (New York: Routledge, 2011), 98.

Part II

THE GEOGRAPHIC FOUNDATION OF TOURISM

Tourism is not simply a product of the modern world. People have been traveling for various reasons since ancient times. However, patterns of tourism have historically been spatially concentrated. Certain places became destinations for tourists, because those places were accessible based on the transportation systems available at the time and because they possessed the physical and/or cultural resources that were valued by people in the principal tourist-generating societies during that period. Today, we see the dynamic reshaping of tourism patterns. Places all over the world are more accessible than ever, and new tourist markets are demanding different types of experiences. Trying to understand these ever-changing patterns in places with widely varied circumstances can seem overwhelming. Yet the framework of geography—with its diverse set of topical branches across both physical and human geography—provides us with the means of exploring all of these issues.

This section begins our examination of tourism through the topical branches of geography. In particular, these chapters consider the geographic foundation of tourism. Chapter 4 discusses the historical geography of tourism. While we cannot make a comprehensive study of the historical geography of tourism in this introductory text, this chapter examines some of the geographic factors that contributed to the development of the modern tourism industry and some of the factors that presented challenges in this process. Chapter 5 discusses the transport geography of tourism. It examines the components of the transport system and some of the fundamental issues in tourism transport. Chapter 6 discusses the physical geography of tourism. This chapter illustrates how we can use the tools and concepts of branches in physical geography to understand the physical resources that provide the basis for tourism at destinations around the world and the physical factors that present a barrier to tourism. Finally, Chapter 7 discusses the human geography of tourism. It draws on the tools and concepts in some of human geography's most prominent branches to understand the human resources used in tourism, as well as human-created barriers.

CHAPTER 4

The Historical Geography of Tourism

Some of us may daydream about hopping in our car and driving across the West, cruising the Caribbean for four days and three nights, or getting on a plane and flying to Australia. Of course, we may never take these trips, for any number of reasons. But it is easy for us to imagine because we *can* do these things, which is something we often take for granted. If we look back, say, two hundred years, it is an entirely different story. At this time, the first expeditions across the American West had only just been completed by explorers like Lewis and Clark. The round trip journey took over two years of difficult travel on foot, on horseback, and in small boats. A sail to and around the Caribbean would take many months, dependent on wind and weather conditions. It could be a dangerous journey, with threats of hurricanes, tropical diseases, slave revolts, pirate attacks, or naval battles. Traveling to Australia would have been virtually unthinkable for an American.

Yet, as hard as it may be for us to comprehend, tourism was already a well-established phenomenon in the world. Clearly there are significant differences between this early tourism and today, but there are surprising parallels as well. Thinking about the past is not just a matter of idle curiosity; it is essential if we are to truly understand modern tourism in all of its complexity.

Historical geography is a topical branch of geography. Like the other branches, historical geography uses the framework of geography to examine topics in and contribute to the study of a particular field—in this case, history (see figure 1.1). Therefore, we can consider historical geography to be the study of the geography and geographic conditions of past periods. Yet, historical geography has another vital role to play. Geography is a means of understanding the world. However, as we examine current patterns and circumstances of places, we cannot truly understand them if we do not understand how they came to be. Thus, historical geography can also be used to examine the processes of change that have taken place over time so that we might better understand the geography of the present. Perhaps taken a step further, if we understand this evolution over time, we might be able to project the geography of the future. As such, historical geography has a part to play in all geography, including the geography of tourism.

Research in tourism studies has often focused on contemporary issues, and tourism has typically been a neglected topic in historical research. Likewise, there has been little relationship between tourism geography and historical geography, despite the fact that the latter framework clearly has potential to contribute to both the study of tourism in past periods and the evolution of tourism over time. Consequently, historical tourism research has been somewhat uneven, focusing on tourism during specific time periods or in particular places. Yet, the evolution of tourism as a mass phenomenon is considered one of the most significant social developments in recent history. The importance of studying the past to understand the factors that allowed the development of tourism, and the origins of many of the patterns we see today, should be clear.

This chapter continues to lay the foundation for our discussion of the geography of tourism. It provides an overview of tourism in key past periods and the development

Figure 4.1. We can examine patterns and trends of tourism in past periods to help us understand the circumstances of modern tourism. ***Source:*** **Nancy Collins**

of the modern tourism industry (figure 4.1). The framework of geography can help us understand the factors that allowed this development to take place. We will consider some of these factors here so that we may begin to develop an appreciation for broad patterns and trends in tourism; however, we will continue to examine these issues as they pertain to the topical branches discussed in the following chapters.

Pre-modern Travel

Travel has taken place throughout human history, leading some scholars to argue that we can trace this history back to the Sumerians some six thousand years ago.[1] Most scholars do not go back so far, as little is known about this period. However, some scholars point to evidence found in the ancient Mediterranean world among the Greek and Roman civilizations as some of the earliest examples of tourism. Others consider health-related travel and religious pilgrimages as important predecessors of tourism. There are often few sources of data available to us that could afford us insight into patterns of travel and tourism in these pre-modern time periods. Additionally, little research has focused on these patterns in non-European contexts.[2] For example, there is comparatively little English-language literature on travel within Asian cultures, including China, Japan, and India, although these cultures have a rich legacy of travel writing. Given this context, the following discussion is necessarily selective.

ROMAN TOURISTS

Although examples of travel for health, culture, or even pleasure may be found in other ancient civilizations, such as Greece, the Romans may be considered the first true tourists based on a number of parallels with later—even modern—eras of tourism. We do not have the benefit of historical sources, such as letters and diaries, to provide in-depth perspectives on tourists and their activities during this era; however, information can be obtained from archaeological evidence and the writings of scholars and social commentators that have survived the passage of time.

There were several key factors that laid the foundation for tourism in the Roman Empire. One of the most important was the two-hundred-year long period of peace and stability that the empire enjoyed (called the Pax Romana—from the end of the first century BCE to the end of the second century AD), which is typically a precondition for tourism. This helped create a prosperous society that was able to develop an interest in traveling to other places for health or pleasure without fear of having to cross hostile territory.[3]

At the same time, the Roman Empire had a well-developed transportation infrastructure. This extensive network of paved roads was originally built for military purposes and to connect the empire's vast land area, as well as providing the basis for commercial trade. Increased patterns of movement within the empire also generated new developments in public transportation, with organized relays of horses at five or six mile intervals, by which a person could travel up to 100 miles per day. Likewise, inns were established along the roads to accommodate traveling government officials and merchants. This infrastructure also facilitated travel for pleasure.[4]

The Romans had various motivations for travel, many of which had a distinctly practical basis. For example, one motivation was military tourism. Soldiers had explicit reasons for traveling, but these expeditions could also be combined with pleasure. Women and children might be allowed to travel with their husbands and fathers, and families could visit attractions along the way. The Romans had developed an appreciation for leisure and entertainment activities. This meant that at least part of the population had free time outside of work and necessary daily chores and that they enjoyed celebrations that were distinct from religious rituals or ceremonies.

Health tourism was also widely practiced among the Romans. Some invalids traveled to places with distinct physical properties, such as mineral waters or hot springs, that would be beneficial to those with certain health conditions. Perhaps more significantly, people traveled to escape places with conditions that would be detrimental to their health. All but the poorest citizens left Rome during the summer due to extreme heat and the rampant spread of disease among the crowded urban population. These middle- and upper-class citizens would retreat to the surrounding countryside in lower altitudes of mountainous regions, where temperatures would be lower and the air fresher.[5] Similarly, seaside resorts in the coastal region between Rome and Naples became popular destinations for those seeking to get away from the city. These fashionable resorts replicated the best parts of social life from Rome and offered entertainments including baths, dining, concerts and theater performances, and even gladiator games.[6]

Only a few privileged groups had the time and resources to be able to travel farther afield in the Mediterranean region. This included the most affluent families, high-ranking government officials, and young men from the upper class in the process of completing their education. Cultural attractions such as temples and ancient monuments formed the basis of many destinations. Most cities had temples that not only represented a god or goddess but also served the function of museum with collections of statues, paintings, and artifacts. The list of the Seven Wonders of the World created some of the most sought-after destinations and formed the basis for an early version of the Grand Tour. This was a tourist itinerary, typically through Greece, Asia Minor, and Egypt, comprised of the most important sights. Egypt, in particular, boasted of wonders such as the Pyramids of Giza and the Lighthouse of Alexandria, as well as landscapes and a culture that would have seemed different and exotic to Roman tourists. In contrast, the mountainous landscapes of the Alps were generally avoided, as they were considered barriers to travel rather than attractions.[7]

Roman tourism has been described as being "typically modern" and having "nearly all of the trappings of its late-twentieth century counterpart."[8] Roman tourists visited many of the same sites popular among tourists today. They had the benefit of guidebooks to instruct them on what they were to see; however, they had to read about the sites before their travels because the books were expensive, large, heavy, leather-bound volumes of papyrus sheets. During the course of their travels, Roman tourists would hire guides. To remember their experiences, they would sketch the scenes they saw or purchase souvenirs, such as paintings, artifacts, or miniature replicas of statues or monuments.[9]

PILGRIMS

Despite the apparent familiarity of the type of tourism seen during the Pax Romana, it was not to last. The collapse of the Roman Empire brought an end to these patterns. The transportation infrastructure fell into disrepair, and traveling became a dangerous proposition with the poor condition of roads, closed inns, and various threats of wild animals, thieves, and hostile territories. As such, there was little thought of traveling for pleasure. Only the most adventurous, the most determined, or those who absolutely had to would risk travel.

One of the most common forms of travel in Europe during the Middle Ages (from the fifth to the fifteenth centuries) was undertaken by devout individuals with strongly held spiritual beliefs. Some of the best-known and frequently visited shrines included Santiago de Compostela in Spain (as early as the ninth century) and Canterbury in England (from the twelfth century). Pilgrimages were also undertaken to the Holy Land, although this was a much more difficult, time-consuming, expensive, and dangerous journey for those traveling from Europe.[10] At this time, travel for health reasons became intertwined with religious pilgrimages. The Roman Church had an extremely powerful influence over life during the Middle Ages, and people increasingly turned to faith healing. They traveled to shrines with the express purpose of appealing to the patron saint for miraculous cures. Given the generally poor living conditions during this period, with high rates of malnutrition and disease, this became a relatively common practice.[11]

Pilgrimages have taken place in Asian societies over an even greater scope of time than in Europe.[12] Buddhist monks would often travel to learn from renowned teachers. Particularly between the fifth and eighth centuries, there was significant religious traffic between India and China. Pilgrimages were long, typically spanning a period of years, and overland journeys involved crossing much difficult terrain.[13] For example, over the course of eighteen years (627–645), the Buddhist pilgrim Xuanzang traveled approximately 16,000 miles through modern-day Kyrgyzstan, Kazakhstan, Uzbekistan, Afghanistan, Pakistan, Kashmir, and India and returned to China with over 600 religious texts.[14] Likewise, Indian and subsequently Japanese monks traveled to China.

In the seventh century, the Hajj was established as one of the five pillars of Islam. Caravan routes to Mecca were established from several origins, including Kufa (Iraq), Damascus (Syria), and Cairo (Egypt). Starting in the eighth century, the ruling Abbasids, who made multiple pilgrimages themselves, subsidized improvements to the 900-mile pilgrim route from Kufa. These included wells, rest stations, milestones, fire beacons, and forts. Although the journey became easier, pilgrims were still vulnerable to predatory nomads.[15] Al-Abdari, an educated, religious scholar and poet, left a narrative of his pilgrimage to Mecca and Medina (1289–1290) that describes his journey through North Africa into the Arabian Peninsula.[16]

EXPLORERS AND TRAVELERS

Exploration provided another significant motivation for travel during this era. Explorers' written accounts and detailed descriptions of peoples and places encountered offer some of the most significant records of pre-modern travel. These writings range from Gerald of Wales' *Topographia Hibernica*, first published in 1188 based on the author's journey to Ireland, to what is thought to be the sole surviving anonymously written journal from Vasco da Gama's journey to India (1497–1499).

Exploration has often been associated with the European age of expansion and colonization, but people all over the world have had a curiosity about and a desire to experience other places. Exploration flourished during the Ming dynasty (1368–1644) in China. In the early years of the dynasty, Zheng He made seven large-scale, officially sponsored overseas journeys that extended as far as the eastern coast of Africa. Although these costly expeditions were short-lived, exploration within China continued. Its size and diversity in both landscapes and people offered ample opportunities for journeys, and explorers sought evermore remote regions. Many of these explorers were officials who held posts throughout the country, but others came from wealthy backgrounds and were able to travel without the need for an official position.[17] For example, Xu Xiake is considered one of the greatest Chinese travelers and a prolific travel writer. From 1609 to 1636, he traveled extensively throughout China, primarily on foot, with a particular interest in physical geography research.[18]

One of the best-known African travelers was ibn Battuta. He was a North African Muslim scholar who began his travels with a pilgrimage to Mecca at the age of twenty-one. After that, he continued to both visit famous places and to hold official positions. From 1325 to 1354, ibn Battuta traveled through forty-five modern-day nations in Africa and Asia. Upon his return to Morocco, his dictations of his experiences produced more than twenty-five manuscripts.[19] Iraqi priest Ilyas Hannaal-Mawsuli began his journey to Europe in 1668. He then obtained a permit from the Spanish king to travel to South America. Although little is known about who he was, his manner of travel (e.g., a private cabin while crossing the Atlantic, litters and coaches while traveling in Peru, an entourage of slaves, servants, and dogs, etc.) indicates that he was wealthy or well sponsored. Based on his account, al-Mawsuli followed the same general route and visited the same "attractions" as other travelers to the region at that time.[20]

The Evolution of Modern Tourism

The origin of tourism is a subject of debate among scholars. This is partially attributed to the lack of a clear definition and what motivations and/or activities should (or should not) be considered tourism. The beginning of modern tourism development is commonly placed in eighteenth-century Western Europe. The verb *tour* had come into usage in the English language in the seventeenth century, and by the eighteenth century, the noun *tourist* had developed to describe those who traveled, typically for pleasure or culture.[21] At this time, tourism became a popular activity among the elite upper classes who had sufficient disposable income and leisure time. In particular,

Britain is cited as not only one of the first nations to develop tourism but also one of the largest sources of tourists during this early era. With new innovations in transportation, tourism was increasingly expanded to the middle classes as well. As a result, the greatest quantity of research has focused on the emergence and expansion of tourism that started in Europe in the eighteenth century and accelerated throughout the nineteenth and twentieth centuries.

THE GRAND TOUR

The European Grand Tour represents a key component in the evolution of tourism. A variation of the Grand Tour took place as early as the Elizabethan era in the sixteenth century and evolved into the traditional Grand Tour era from the mid-seventeenth century through the eighteenth century. This was originally intended to provide young British men from the aristocratic class with a classical education. Often traveling with tutors, they would visit the cultural centers of Renaissance Europe and sites of classic antiquity. Italy, above all, was the focal point of such a tour, with destinations such as Venice, Florence, Rome, and Naples.[22]

The average length of the Grand Tour was forty months, and the journey often followed a designated route through France, Italy, Germany, Switzerland, and/or the Low Countries (modern Belgium, the Netherlands, and Luxembourg). Few tourists strayed from this route into other areas. Particularly early in this era, traveling conditions were difficult, so the route was distinctly shaped by geographic conditions and available transportation technologies. As with Roman tourism, the Alps were considered a barrier to be crossed en route to the highlighted destinations rather than an attraction in themselves. A widespread, efficient network of transportation that met the needs of these tourists was slow to develop. Likewise, there were few accommodations. Although some of the main cities on the tour developed hotels, these Grand Tourists generally had to use the same inns, hostels, and post houses as other travelers.[23]

Toward the end of the eighteenth century, the Grand Tour began to experience a number of changes. The demographics of the Grand Tourists steadily expanded to include aristocrats from other Northern European countries as well as the sons of the growing class of affluent but not titled British families. The territory of the tour expanded, as tourists searched for newer and more exclusive destinations, such as Greece, Portugal, and Turkey. The focus of the Grand Tour also began to shift. Education continued to play a role, but sightseeing gained in importance. Tourists visited archaeological sites, museums, and art galleries, and they attended concerts and theater performances. Socialization and the development of social contact with others in the same class at assemblies and balls also came to be a part of the Grand Tour. Given the increasing importance of these latter activities, some critics argued that the Grand Tour had become nothing more than the pursuit of pleasure.[24]

The onset of the French Revolution in 1789, followed by the conflict surrounding the Napoleonic Wars, effectively halted Continental travel. While this interval brought a boost in British domestic travel, it also created a pent-up demand for experiences abroad. Napoleon's defeat at Waterloo and the Second Treaty of Paris in 1815 created a host of new opportunities for international travel. Many of the changes to the Grand Tour that had begun before the Napoleonic Wars continued after travel resumed once more. This effectively ended the Grand Tour era and ushered in a new era of international tourism in the nineteenth century in which more people participated than ever before.

In this era, more adults and families began to travel. This expanded females' participation in travel, a trend that would continue with an increasingly organized tourism infrastructure. Members of the middle class also began to participate, which generated further changes in the nature of the experience. Middle-class tourists did not have the advantage of invitations from local nobility, so they had to rely on the developing tourism infrastructure (e.g., hotels). They were less likely to travel with servants and household staff, which created a demand for local serving staff at the places of destination. These tourists had less time and money available to travel, so before the middle

Box 4.1. In-Depth: The Rise of Organized Mass Tourism

Thomas Cook is often described as the father of modern mass tourism because of the role he and his company played in organizing tourism services that made tourism easier and more accessible to more people.* Cook was a bookseller and a Baptist preacher who got his start by organizing a train trip for 570 people to attend a temperance meeting in 1841 England. With the success of this trip, he began to organize excursions for other groups, which quickly evolved into organizing low-cost pleasure trips primarily utilizing rail transport and the growing accommodation industry. Within a few years, Cook's Tours had opened up new opportunities for tourism among the working classes, as well as for females traveling without male companions. Travel by rail was quick, cheap, and generally considered safe. The company preplanned all aspects of the trip. The tourists did not have to know anyone at the destination, and they did not have to worry about whether the accommodations would be suitable.

Based on the existing popularity of seaside resorts, they were one of the key destinations for Cook's Tours. However, preferences were changing, and demands for new experiences were arising. Cook's Tours were responsive to these demands and consistently offered trips to destinations that were becoming popular, such as England's Lake District. In addition, his tours helped create new destinations not only in England but also in Wales, Scotland, and Ireland. These destinations were quickly followed by Continental tours. By the middle of the 1850s, Cook was organizing tours in France and Germany, followed by Switzerland and Italy. Thus, a new generation and a new class of tourists could experience many of the same places as the Grand Tour, albeit on a far more compressed time frame and with ever more of the comforts of home.†

Thomas Cook was the innovator, but his company was soon joined by others providing similar types of experiences. However, a century later, Thomas Cook & Sons Ltd. remained at the forefront of organized mass tourism when the company launched a new type of packaged

* Freya Higgins-Desbiolles, "More than an 'Industry': The Forgotten Power of Tourism as a Social Force," *Tourism Management* 27 (2006): 1193.

† Orvar Löfgren, *On Holiday: A History of Vacationing* (Berkeley: University of California Press, 1999): 163; Jack Simmons, "Railways, Hotels, and Tourism in Great Britain 1839-1914," *Journal of Contemporary History* 19 (1984): 208.

trip: instead of traveling by chartered rail transport, this trip was based on chartered air transport. The company began with trips from Britain to Corsica and continued to expand into new destinations.‡

Discussion topic: What do you think was the most significant development of Cook's Tours in the evolution of modern tourism, and why?

‡ Gareth Shaw and Allan M. Williams, *Critical Issues in Tourism: A Geographical Perspective* (Malden, MA: Blackwell Publishing, 2002), 227.

of the nineteenth century, the average length of a European tour had been reduced to four months. As tourists had less time to spend at the destination, seeing the sights took precedence over learning about them. Many continued to follow the same route and visit the same cities for their well-known attractions.[25]

During this same period in the nineteenth century, European tourists also began to extend their reach into new regions. Explorers and travelers had already been in Africa, Asia, and the Americas, but the new generations of tourists visited these places for pleasure. Transatlantic travel had become safer following the end of the Napoleonic Wars, as well as easier and faster with the development of steamships. Lingering concerns, however, focused on the hazards of tropical storms and the fear of diseases such as yellow fever and malaria. Scholars have argued that this wave of European tourists arriving in other parts of the world was a new form of colonialism. Soon after the arrival of the first tourists at a destination, their numbers steadily increased. On one hand, this had the positive effect of creating a demand for new businesses to cater to the needs of these tourists. On the other hand, local residents and even other tourist groups complained about negative effects ranging from increased use of European languages to higher costs of living.

European Grand Tour travel is significant in that it is the first era of tourism in which there is considerable source material for analysis.[26] In that era and since, a tremendous number of documents—including tourists' personal diaries, letters, and published narratives, as well as travel company literature and promotions—were produced that provide us with insights into why people traveled, where they went, how they got there, and what their experiences were. We do not always have access to this same type of data about travel in other periods or places. Even the information that we do have is limited in perspective. Most of the sources from the Grand Tour era come from tourists. Considerably less data are available from a supply side perspective, including from individuals providing services to these tourists.

RESORT TOURISM

The development of spas and resorts also played a role in the evolution of modern tourism. Health had long been a primary motivator for travel. Physicians put forth many theories about which environments possessed the best curative properties for various conditions, most notably tuberculosis. Spas—places usually possessing mineral springs—had been used intermittently over time as destinations for invalids seeking cures for different ailments. The role of faith healing during the medieval era led to a decline in early spas, but by the seventeenth century these places experienced a resurgence with visits from members of royal and noble families. There was a growing interest in balneotherapy, or water therapy, and physicians widely promoted cures from either drinking or bathing in mineral waters. Thus, spas had the dual benefit of possessing health-giving properties and providing an escape from the poor environmental conditions of the increasingly polluted industrial cities. As a result, by the eighteenth century, English spas such as Bath and Tunbridge Wells had become immensely popular.[27]

Although spas were initially developed for those seeking cures, and in some cases prevention, soon they became known as fashionable and exclusive resorts. As the socialization function became more important, resorts increasingly built promenades and assembly rooms and offered theater performances, concerts, dances, receptions, card parties, and gambling. Eventually, "seasons" developed in which the upper classes would converge on spa towns for the entertainment and to both see and be seen.[28]

The earliest English resorts were located around mineral springs in areas that were inland and relatively accessible to London. At this time, the coast and the sea were seen as dangerous places to be avoided if possible. It was a wild landscape full of hazards, from unpredictable weather to pirates and smugglers. However, by the late eighteenth century, several factors contributed to a change in attitudes and allowed new spa resorts to emerge. First, a new appreciation began to develop for rugged natural scenery and the forces of nature that had formerly generated fear. Second, physicians began to advocate the health advantages of the seaside, including taking brisk walks along the beach, drinking seawater, and even sea bathing. Sea bathing was a carefully regulated activity, typically undertaken with the aid of bathing machines. These wooden structures allowed the bather to be gradually immersed in the water safely and privately, the latter being especially important for ladies.[29]

Initially, seaside spas provided a complement to inland resorts. Visits to the seaside would take place at different times of the year than the social season at the fashionable inland destinations. However, the seaside spas were increasingly developed into resorts with the same comforts and entertainments and thus started to compete with inland resorts for status and clientele. As with the inland resorts, the most successful spas, such as Brighton, were those that were relatively accessible from London. Transportation by stagecoach often made farther resorts impractical because this mode was expensive, and poor roads made travel both slow and uncomfortable.[30]

The development of seaside spas changed the nature of the coastline, which had once been characterized by scattered fishing villages.[31] These resort towns began to be connected by new modes of transportation, including steamships and passenger trains. Such innovations shortened travel time and reduced the expense of travel, allowing more people from the middle classes to make the trip. This brought further changes in the nature of the resorts. The earlier, upper-class tourists rented houses for the season and established a temporary residence complete with their own serving staff.[32] The increase in middle-class tourists, who spent shorter amounts of time at the destination, created a demand for accommodation facilities such as hotels and boarding houses.

By the second half of the nineteenth century, faster and more reliable rail service allowed day trips to the seaside. This meant that even the working classes, who were not able to get away for extended periods of time or did not have the money to stay in a hotel, were also able to enjoy the resorts. By this time, less emphasis was placed on curing illnesses and more on promoting well-being. Sea bathing with the use of expensive bathing machines fell out of favor, and tourists were encouraged to get out and enjoy the fresh sea air. Perhaps the most important component of a seaside holiday was the pursuit of pleasure, as these tourists sought to emulate the life of leisure displayed by the upper classes—at least for a short time.[33]

Once these resorts were seen as less exclusive, the upper class, followed by the middle class, began looking for new destinations, often abroad. The same transportation innovations that made resorts at home more accessible also helped open up new resorts across Europe. These tourists particularly looked to the new winter resorts developing in the Mediterranean region, such as the Côte d'Azur in France. As with coastal resorts in England, these areas were previously underutilized for tourism. However, with the development of spa tourism, the region's mild climate was highly desirable among northern tourists and was popularized by the British royal family. Likewise, members of the Austrian royal family made other resorts fashionable, particularly within their own empire, such as Opatija on the Istrian Peninsula (Croatia).

Although these resorts provided relief from the cold, damp northern winters, they were generally to be avoided during the summer. In the Victorian era, tanned skin was highly unfashionable and considered a sign of the working classes. In addition, clothing styles were tight and made from heavy materials that would have been unsuitable for the Mediterranean summer heat.[34] By the early twentieth century, however, physicians began recommending heliotherapy, based on exposure to sunlight. Clothing styles became less restrictive and hot, which allowed people to spend more time outside. As more people swam freely in the ocean, swimwear was also needed. Suntans became fashionable, as the upper classes had time to spend at resort destinations in the sun, while the working classes were stuck inside in factories.

Thus, a new tourism product, based on the combination of sun and sea, became enormously popular. The Mediterranean was at the heart of this new trend. Developments in air transport and relatively inexpensive foreign package vacations made the Mediterranean more accessible. At the same time, new and exotic resorts were developed around the world, including the Caribbean basin and Southeast Asia. Interestingly, the original coastal resorts in England experienced a decline. Upper- and middle-class tourists had the opportunity to visit new resorts, which left the old resorts to day-trippers and lower income tourists who could not afford to travel abroad. Moreover, with little new investment, the infrastructure became outdated. For example, at some of the early resorts where tourists had arrived by train, there were few parking facilities to accommodate those now arriving by car.

Spa and seaside resorts were among the first modern tourism destinations to emerge in many parts of the world. For example, English beliefs about the curative properties of mineral waters carried over to American society; therefore, some of the earliest destinations here were also spas. Ballston Spa in Saratoga County, New York, was one of the first resorts around the turn of the nineteenth century. It boasted the first hotel in the country built outside of one of the major cities. As travel opened up in the nineteenth century, these spas began to follow a similar path to development as earlier British spas. New accommodations and entertainments drew ever more tourists, who came for leisure and socialization rather than health. Then, as these spas became overrun with visitors, the original and wealthy tourists sought newer, more exclusive resorts, and they were replaced at existing resorts by the middle classes.[35] Seaside resorts also gained in popularity. Atlantic City in New Jersey dates back to the mid-nineteenth century. By the early 1900s, it was widely known and boasted reputations such as "The Queen of the Jersey Shore" and "The World's Playground."

Beach culture in Sydney, Australia, began to develop in the mid-nineteenth century. Entrepreneurs initially sought to replicate British and American resort models, but local circumstances proved different. Beaches afforded day-trip leisure opportunities initially for middle-class urban residents before eventually opening up to the working class as well. Especially after daylight surf bathing was permitted in the early twentieth century, visitors were primarily interested in the beach as opposed to amusement parks and other entertainment facilities.[36]

Changes and Challenges in Tourism

We could examine many other significant eras and developments that facilitated the evolution of tourism. In the late eighteenth and early nineteenth centuries, changing attitudes toward nature helped create a demand for new types of tourism experiences in new destinations. For example, new types of resorts were developed in the Alps, whereas earlier generations of Grand Tourists had viewed them as a barrier to their destinations. To immerse themselves in nature, more people took to walking, hiking, and roving the countryside. New modes of transport brought further changes to both destinations and experiences. With increased automobile ownership and highway construction in the interwar years, "autotouring" became popular, especially in the United States. In the years after World War II, air travel was opened up for mass passenger transport. Particularly as the price of air travel came down, destinations around the world were suddenly far more accessible. This contributed to a surge in international tourist arrivals. In 1948, there were an estimated fourteen million international tourists. By 1965, that number had grown to 144 million (figure 4.2).[37]

That being said, tourism has not come easily to all places or equally to all peoples as a result of various barriers, often political or economic in nature. For example, we could look at the Cold War era to understand the political policies that restricted patterns of travel for people in countries behind Europe's "Iron Curtain" (box 4.2). Likewise, we could look at different periods of economic development for countries, such as China or Brazil, to understand when more segments of the population gained the disposable income and/or leisure time to travel for pleasure. One era that has gotten relatively little attention, however, is the Jim Crow era of racial segregation in the United States. This period reshaped patterns of travel and tourism for African Americans and, some scholars argue, left a legacy that can be seen today.

Figure 4.2. By the 1970s, international air transport was becoming increasingly accessible, allowing more people to travel than ever before. This group of international tourists is preparing to board a plane in Columbus, Ohio for a European tour in 1971. *Source:* Carolyn Nelson

TOURISM IN THE JIM CROW ERA

In the late nineteenth century, there was a relatively small (approximately 10 percent), affluent segment of the African American population that was able to travel for pleasure. These doctors, lawyers, entrepreneurs, and politicians visited newly developed resorts such Saratoga Springs (New York) and Atlantic City (New Jersey) along with white tourists.[38] However, this was relatively short-lived. Geographers Derek Alderman and Joshua Inwood state:

Box 4.2. Experience: Guiding Tours in Yugoslavia

Dr. Anton Gosar began his career in the early 1960s as a tour guide in Yugoslavia. In the late 1970s, he joined a research team to study Yugoslavian tourism. In 2016, he retired from his position as dean of the Faculty of Tourism Studies at the University of Primorska in Slovenia. Throughout his career, he made many contributions to tourism geography as well as political geography. In what follows, he recounts some of his early experiences as a guide during a period in which the country—and the tourism system—was undergoing significant changes.

In 1961, Yugoslavia opened its borders. Some two million Yugoslavs visited border towns on shopping sprees each month. Requiring a personal bank deposit of 500 dinars (equivalent to about US$50) or only permitting cars with odd or even numbers to cross the border did little to deter this movement. Likewise, no preventive measures implemented by the government could stop the import (and smuggling) of Levi's jeans, Minas coffee, margarine, toilet paper, detergents, and more. At the same time, foreign tourists also began to cross the border. Traveling to and from the coast, Sun, Sea, and Sand (3S) tourism was the primary motivation. My parents began to open up our house for bed-and-breakfast to these transit tourists during peak seasons. On weekends, the main road through my hometown could hardly be crossed due to the many Volkswagens (Germans) and increasingly Fiats (Italians) that were passing through. Inland, only urban centers and attractions along major transcontinental highways gained the attention of these transit tourists. Few looked to enrich their knowledge of the cultural diversity to be found in the multiethnic state and the natural diversity of the region.

I had been employed as a guide at the Cave of Postojna, but as patterns of tourism changed, I began to guide on the Adriatic coast. Initially, I would meet a group of foreign tourists at the border railway station in Jesenice, talk to them on the train, and take them to the port of Rijeka. From there, we embarked on a ship bound for Zadar, Split, and Dubrovnik, where I showed them the cultural sites. Then I was assigned to lead groups through several modern-day Southeast European nation states by bus. Many of these tours combined British and American—and sometimes even French and German—tourists on the same bus. It was expected the guide be trilingual.

These tourists were interested in the natural and cultural diversity of Yugoslavia, including its five major geographic landscapes, four major languages, three major religions, and so on. Perhaps the biggest attraction was experiencing adventure—that is, in a communistic nation state. In the 1960s, such a trip could indeed be an adventure. There were often long checks at border posts, local guides on each nation-state's territory, constant currency changes, buses and hotels with no air-conditioning, and minimal hygiene. Communistic oppression was carefully hidden from these tourists, particularly in Sarajevo where things like the open expression of religion was repressed to extremes. Tourist guides, like me, praised Yugoslavian"brotherhood and unity" (not directly communism) and achievements on the domestic level, like increased highway and railway networks (often a result of military construction activities). I never spoke to tourists about the discomfort the one-party system had caused my own family, in particular my mother's German side.

Later, I worked in the travel agency to accommodate the rare Czechoslovak tourists. Their group had to have individual visas, a trustworthy tour manager, and sufficient hard currency. Occasionally, I served as the tour guide for their two-day bus excursion to Venice. On two occasions, several young couples disappeared from the hotel and never returned to the group. The receptionist at the hotel told me they were picked up by relatives or friends. This disappearance, and defection, was most likely arranged in advance. I had to file a report about this, but my travel agency did not blame me. However, the Czech escort of the group was most likely held responsible for their disappearance; he never again appeared with the group of tourists in Yugoslavia.

In the 1970s, Yugoslavia was ranked in the top ten most-visited countries in the world; in the 1980s, it was in the top fifteen. However, in the 1990s, the Balkan wars dramatically reduced international visits, hindered tourism growth, and resulted into the nation-states of Bosnia and Herzegovina, Croatia, Kosovo, Macedonia, Montenegro, Serbia, and Slovenia. Tourism has changed much since that time. Today, tourism contributes more than 10 percent of the national gross domestic product for Croatia, Montenegro, and Slovenia. The majority of tourists are residents of EU countries, but an increasing number of visitors come from Russia, Israel, India, China, and Japan. They aren't just looking for Sun, Sea, and Sand anymore; other products from food and beverage tourism to adventure and adrenalin tourism have become popular. Today's tourists do not simply follow the footprints of others; they are making their own footprints and new paths to be discovered.

—Anton

> The term "Jim Crow" refers to a racial caste-like system that began as early as 1877 with the end of Reconstruction and operated primarily, but not exclusively, in the southeastern United States. While Jim Crow is often identified with rigid laws that marginalized and excluded African American, it actually represented a broad array of formal and informal social, economic, and political practices that segregated blacks and whites and justified rampant racism, intimidation, and violence toward African Americans.[39]

With the Jim Crow system of segregation, African American tourists, regardless of means or status, increasingly found these resorts and other facilities closed to them. Travel in general became more difficult and even perilous.

Because African Americans suffered from various expressions of discrimination using public transportation, many welcomed the freedom afforded by the automobile. Still, African Americans faced significant barriers to traveling beyond familiar places. Segregation practices could vary considerably from town to town, so travelers never knew what to expect. On the road, they faced constant apprehension about if they would be able to find a place to eat or to spend the night.[40] The latter was particularly serious, because they might encounter "sundown towns" that prohibited African Americans from being out after nightfall.[41] Defying such laws and practices, even if unknowingly, could lead to violent reprisals.[42] Even if they were able to obtain products and services while traveling, they often faced poor quality, rude treatment, and inflated prices. Essentially, travel for African Americans could be harder, longer and less direct, and more expensive than that for their white counterparts.[43]

Increasingly, African American–owned businesses provided the necessary services for African American travelers and tourists. In 1936, Victor Green, an African American postal worker in Harlem produced the first edition of *The Negro Motorist Green Book*. It began as a local guide for New York City, using his experiences and those of other postal workers, to provide a listing of businesses that welcomed African American customers. This often included private residences that lodged African American

Figure 4.3. Although automobiles offered African Americans freedom from the discrimination and harassment common on public transportation, they still faced many barriers to travel. *Source:* Al Stephens

Box 4.3. Case Study: Michigan's Black Eden

According to one historian, in the Jim Crow era "blacks could almost never achieve total relaxation, but . . . they came closest to doing so when there were no whites around. It is hardly surprising that successful blacks did all they could to *insulate* themselves, and particularly their children, from unpleasant confrontations with whites."* In the early twentieth century, entrepreneurs began to recognize the need to establish resorts for middle-class, urban African American professionals and business owners. Like their white counterparts, these peoples sought to escape the summer heat, poor environmental conditions, and social pressures associated with cities at the time. To avoid potential problems, these resorts needed

* Mark S. Foster, "In the Face of 'Jim Crow': Prosperous Blacks and Vacations, Travel and Outdoor Leisure, 1890–1945," *The Journal of Negro History*, 84, no. 2 (1999): 131, original emphasis.

to be developed in relatively remote places.[†] In rural Michigan, Idlewild offered a place that was a reasonable driving distance from Midwestern cities like Detroit, (Michigan), Cleveland (Ohio), Chicago (Illinois), Indianapolis (Indiana), and even St. Louis (Missouri), but far enough from the racism and discrimination present in these cities.[‡]

In 1912, four white couples purchased 2,700 acres of land in the northwestern part of Lower Michigan near Idlewild Lake and founded the Idlewild Resort Company. With the intention of appealing to the African American community, they organized train and bus tours to bring people from key cities to see the area and sell lots.[§] Development at the resort occurred slowly and facilities were initially limited, but vacationers enjoyed its simplicity and rusticity.[**] After World War I, development accelerated. Popular activities included hiking, horseback riding, swimming, boating, and fishing as well as relaxing, reading, and playing cards in the clubhouse. With the establishment of entertainment venues and jazz clubs, such as the Flamingo Club and the Paradise Club, the resort gained a reputation as a cultural mecca. In addition to regular acts that ranged from magicians to showgirls, many well-known and up-and-coming African American entertainers performed in Idlewild, including Duke Ellington, Louis Armstrong, B.B. King, Sarah Vaughn, Dinah Washington, Della Reese, the Four Tops, and the Temptations.[††]

In its heyday, Idlewild was known as the Black Eden of Michigan. When it reached its peak in the 1940s, the resort's summertime population exceeded twenty thousand. It was a place to see and be seen. In the post–World War II era, Idlewild began to attract African American vacationers from the working middle class; however, after the Civil Rights Act of 1964, the community began to decline. African Americans faced fewer restrictions in terms of where they could vacation and began to explore new destinations. Entertainers had the opportunity to perform at other resorts with bigger audiences and more money. Over time, little new investment was made in the tourism infrastructure, and the resort failed to remain competitive.[‡‡] The economy collapsed and the population plummeted.

Then a growing number of retirees chose to settle in the community. Many of these retirees had visited the area during its prime and wanted to see its history preserved. Revitalization efforts began in the 1990s, and the Idlewild Music Festival was established in the early 2000s to help connect modern tourists with the resort's heritage.[§§] In 2012, the community hosted its centenary. While no one expects to recreate the Idlewild of the past, its significance should not be forgotten. Idlewild is considered to be one of the oldest, most famous, and most memorable African American resort communities in the United States.[***] More than that, it was one of the few places of retreat for middle-class African Americans during the Jim Crow era, a place where they could relax and enjoy the same leisure opportunities available to white Americans.

Discussion topic: What aspects of place made Idlewild an attractive resort? Do you think it could be redeveloped into a modern tourism destination?

Tourism on the web: Idlewild Music Festival, "Honoring the Past, Standing in the Present; Looking Towards the Future," at http://www.idlewildmusicfestival.com

---

[†] Foster, "In the Face of 'Jim Crow'," 140.

[‡] Ronald J. Stephens, *Idlewild: The Black Eden of Michigan* (Charleston, SC: Arcadia, 2001), 10.

[§] Kathlyn Gay, *African-American Holidays, Festivals, and Celebrations: The History, Customs, and Symbols Associated with Both Traditional and Contemporary Religious and Secular Events Observed by Americans of African Descent* (Detroit, MI: Omnigraphics, 2007), 223.

[**] Myra B. Young Armstead, "Revisiting Hotels and Other Lodgings: American Tourist Spaces through the Lens of Black Pleasure-Travelers, 1880–1950," *The Journal of Decorative and Propaganda Arts* 25 (2005), 145.

[††] Foster, "In the Face of 'Jim Crow'," 139–40.

[‡‡] Stephens, *Idlewild*, 10, 119.

[§§] Gay, *African-American Holidays*, 224.

[***] Stephens, *Idlewild*, 8.

travelers in the absence of other available accommodations. The guide was eventually expanded to include cities in locations across the United States as well as several international destinations. In this process, he relied on an early form of user-generated content (UGC) by asking African American travelers to submit the names of suitable businesses they encountered.[44] Still, the majority of such businesses were spatially concentrated in the eastern United States; few facilities were available in the West.[45] African Americans also developed their own resorts where they could enjoy the same leisure activities as other vacationers without fear of discrimination (box 4.3).

In the early 1940s, one journalist concluded that African Americans would have an easier time traveling abroad than in the United States.[46] Foreign travel was still limited at this time, but those with sufficient means who were able to do so often reported that they experienced less discrimination. Still, they were conscious of their appearance and behavior. Some noted that they seemed to be curiosities in the places they visited. Others felt that, as they were the first African American many people had encountered, they bore the responsibility of shaping these peoples' opinions about the group as a whole.[47]

In the introduction to the 1949 edition of the *Green Book*, the author wrote:

> There will be a day sometime in the near future when this guide will not have to be published. That is when we as a race will have equal opportunities and privileges in the United States. It will be a great day for us to suspend this publication for then we can go wherever we please, and without embarrassment. But until that time comes we shall continue to publish this information for your convenience each year.[48]

Indeed, Green continued to publish his guide until the mid-1960s. The Civil Rights Act of 1964 outlawed discrimination and opened up greater opportunities for African American travel and tourism. Geographer Perry Carter argues that we need to consider this legacy of racial discrimination as we seek to understand present patterns of travel for African Americans. He notes that African Americans, in comparison to white Americans, are more likely to travel in large groups, visit destinations recommended by people they know, follow a pre-set itinerary, and avoid unfamiliar things. Although places are no longer segregated, they are still racialized. African Americans may still be perceived as out of place in "white" spaces, including popular tourism destinations, which could potentially lead to racist encounters. Thus, such places and experiences become more stressful than relaxing.[49]

Conclusion

Much research on tourism in past periods has focused on several key eras considered instrumental in the evolution of modern tourism. In particular, the Grand Tour is often cited as the origin of modern international tourism. In fact, some scholars argue that the Grand Tour lives on:

> The true descendants of . . . the Grand Tour tradition, however, consist of the young interrailers who roam the city in search of other interrailers and the groups of American and Japanese college students doing the modern version of the Grand Tour. Just as in the seventeenth century, they are here with the blessing of their parents. A season of interrailing or a European tour is still supposed to be a good investment in a middle class education.[50]

The historical geography of tourism is a fundamental component in our investigation of the geography of tourism. Historical geography provides the framework for examining the geographic patterns of tourism in past periods, and the changes that have taken place over time. This is the foundation for the patterns that we see today. Although it may be hard for us to imagine tourism in earlier periods, clearly many parallels may be seen. Moreover, starting from the early nineteenth century, we can trace the evolution of infrastructure, organization, experiences, and even many of the problems of tourism directly to the patterns that we see today and will be exploring in greater depth in the remaining chapters.

Key Terms

- historical geography

Notes

1. Charles R. Goeldner and J.R. Brent Ritchie, *Tourism: Principles, Practices, Philosophies*, 9th ed. (Hoboken, NJ: Wiley, 2006), 41.
2. John Towner, "What Is Tourism's History?" *Tourism Management* 16, no. 5 (1995): 340.
3. Maxine Feifer, *Tourism in History: From Imperial Rome to the Present* (New York: Stein and Day, 1986), 9.
4. Feifer, *Tourism in History*, 9–10; Bruce Prideaux, "The Role of the Transport System in Destination Development," *Tourism Management* 21 (2000): 53; Loykie Lomine, "Tourism in Augustan Society (44BC–AD69)," in *Histories of Tourism: Representation, Identity and Conflict*, ed. John Walton (Clevedon: Channel View Publications, 2005), 84; Goeldner and Ritchie, *Tourism*, 44.
5. Simon Kevan, "Quests for Cures: A History of Tourism for Climate and Health," *International Journal of Biometeorology* 37 (1993): 114.
6. Lomine, "Tourism in Augustan Society," 78.
7. Lomine, "Tourism in Augustan Society,", 73–7.
8. Lomine, "Tourism in Augustan Society,", 69; Feifer, *Tourism in History*, 8.
9. Feifer, *Tourism in History*, 16; Lomine, "Tourism in Augustan Society," 74; Goeldner and Ritchie, *Tourism*, 44.
10. Jack Simmons, "Railways, Hotels, and Tourism in Great Britain 1839-1914," *Journal of Contemporary History* 19 (1984): 207; Goeldner and Ritchie, *Tourism*, 50.
11. Kevan, "Quests for Cures," 115.
12. Goeldner and Ritchie, *Tourism*, 49.

13. Tabish Khair, Martin Leer, Justin D. Edwards, and Hanna Ziadeh, *Other Routes: 1500 Years of African and Asian Travel Writing* (Bloomington: Indiana University Press, 2005), 32–3.

14. Julian Ward, *Xu Xiake (1587-1641): The Art of Travel Writing* (London: Routledge, 2001),8.

15. F.E. Peters, *The Hajj: The Muslim Pilgrimage to Mecca and the Holy Places* (Princeton, NJ: Princeton University Press, 1994), 73–4.

16. Khair et al., *Other Routes*, 281–2.

17. Ward, *Xu Xiake*, 14–5, 20.

18. Khair et al., *Other Routes*, 184–5.

19. Khair et al., *Other Routes*, 289–91.

20. Khair et al., *Other Routes*, 299–300.

21. Simmons, "Railways, Hotels, and Tourism," 207; Marguerite Shaffer, *See America First: Tourism and National Identity, 1880-1940* (Washington, DC.: Smithsonian Institution Press, 2001), 11.

22. John Towner, "The Grand Tour: A Key Phase in the History of Tourism," *Annals of Tourism Research* 12 (1985): 301; Feifer, *Tourism in History*, 64; Kevan, "Quests for Cures," 116; Tom Baum, "Images of Tourism Past and Present," *International Journal of Contemporary Hospitality Management* 8 (1996): 25; Orvar Löfgren, *On Holiday: A History of Vacationing* (Berkeley: University of California Press, 1999), 157; Goeldner and Ritchie, *Tourism*, 51.

23. Towner, "The Grand Tour," 321–2.

24. Feifer, *Tourism in History*, 98; Goeldner and Ritchie, *Tourism*, 51; Löfgren, *On Holiday*, 161; Towner, "The Grand Tour," 301; Stephen Williams, *Tourism Geography* (London, Routledge, 1998), 44.

25. Towner, "The Grand Tour," 301, 316–17; Baum, "Images of Tourism," 25, 28; Rudy Koshar, "'What Ought to Be Seen': Tourists' Guidebooks and National Identities in Modern Germany and Europe," *Journal of Contemporary History* 33 (1998): 326; Williams, *Tourism Geography*, 44; Löfgren, *On Holiday*, 163.

26. Towner, "The Grand Tour," 298.

27. John Beckerson and John K. Walton, "Selling Air: Marketing the Intangible at British Resorts," in *Histories of Tourism: Representation, Identity and Conflict*, ed. John Walton (Clevedon: Channel View Publications, 2005), 55; Williams, *Tourism Geography*, 23.

28. Goeldner and Ritchie, *Tourism*, 54; Löfgren, *On Holiday*, 160; Williams, *Tourism Geography*, 23.

29. Kevan, "Quests for Cures," 116; Löfgren, *On Holiday*, 113–16; Williams, *Tourism Geography*, 23.

30. Beckerson and Walton, "Selling Air," 55; Löfgren, *On Holiday*, 112; Williams, *Tourism Geography*, 23–4.

31. John K. Walton, "Prospects in Tourism History: Evolution, State of Play and Future Development," *Tourism Management* 30 (2009): 787.

32. Baum, "Images of Tourism," 27–8.

33. Beckerson and Walton, "Selling Air," 55–6; Feifer, *Tourism in History*, 205; Goeldner and Ritchie, *Tourism*, 54; Kevan, "Quests for Cures," 118; Löfgren, *On Holiday*, 120; Williams, *Tourism Geography*, 26.

34. Kevan, "Quests for Cures," 118–19; Löfgren, *On Holiday*, 163; Williams, *Tourism Geography*, 44–5.

35. Richard H. Gassan, *The Birth of American Tourism: New York, the Hudson Valley, and American Culture, 1790-1830* (Amherst: University of Massachusetts Press, 2008), 5, 13–14.

36. Caroline Ford, "A Summer Fling: The Rise and Fall of Aquariums and Fun Parks on Sydney's Ocean Coast 1885-1920," *Journal of Tourism History* 1, no. 2 (2009): 96.

37. Gareth Shaw and Allan M. Williams, *Critical Issues in Tourism: A Geographical Perspective*, 2nd ed. (Malden, MA: Blackwell, 2002), 30.

38. Myra B. Young Armstead, "Revisiting Hotels and Other Lodgings: American Tourist Spaces through the Lens of Black Pleasure-Travelers, 1880-1950," *The Journal of Decorative and Propaganda Arts* 25 (2005): 137–8.

39. Derek H. Alderman and Joshua Inwood, "Toward a Pedagogy of Jim Crow: A Geographic Reading of *The Green Book*, in Teaching Ethnic Geography in the 21^{st}Century," in *Teaching Ethnic Geography in the 21^{st}Century*, ed. Lawrence E. Estaville, Edris J. Montalvo, and Fenda A. Akiwumi (Washington, DC.: National Council for Geographic Education, 2014), 68.

40. Mark S. Foster, "In the Face of 'Jim Crow': Prosperous Blacks and Vacations, Travel and Outdoor Leisure, 1890-1945," *The Journal of Negro History* 84, no. 2 (1999): 140–1.

41. Alderman and Inwood, "Toward a Pedagogy of Jim Crow," 70.

42. Armstead, "Revisiting Hotels and Other Lodgings," 140.

43. Alderman and Inwood, "Toward a Pedagogy of Jim Crow," 73; Foster, "In the Face of 'Jim Crow'," 136.

44. Jacinda Townsend, "How the Green Book Helped African-American Tourists Navigate a Segregated Nation," *Smithsonian Magazine*, April 2016, accessed October 10, 2016, http://www.smithsonianmag.com/smithsonian-institution/history-green-book-african-american-travelers-180958506/?no-ist.

45. Foster, "In the Face of 'Jim Crow'," 137.

46. Armstead, "Revisiting Hotels and Other Lodgings," 140.

47. Foster, "In the Face of 'Jim Crow'," 132.

48. Victor H. Green, *The Negro Motorist Green Book, 1949 Edition* (New York: Victor H. Green& Co., Publishers, 1949), 1.

49. Perry L. Carter, "Coloured Places and Pigmented Holidays: Racialized Leisure Travel," *Tourism Geographies* 10, no. 3 (2008): 266–7, 278, 281.

50. Löfgren, *On Holiday*, 160.

CHAPTER 5

The Transport Geography of Tourism

Although it is easy to daydream about the places we would like to visit, our travel decisions are often based on far more practical logistical issues. Few of us have the luxury of traveling without constraints. We may be limited by our travel budget, the amount of vacation time available, even our personal preferences or biases. Consequently, we have to consider how accessible the places are that we want to go by asking questions like: How would I get there? How long would it take? How much would it cost? Then we have to negotiate between where we *want* to go and where we *can* (reasonably) go. Much of this is contingent upon transportation.

While the transportation infrastructure may shape *where* we travel today, in the early eras of travel discussed in the previous chapter, it determined whether people could travel at all. The development and improvement of transportation was one of the most important factors in allowing modern tourism to develop on a large scale and become a regular part of the lives of billions of people around the world. Technological advances provided the basis for the exponential expansion of local, regional, and global transportation networks and made travel faster, easier, and cheaper. This not only created new tourist-generating and tourist-receiving regions but also prompted a host of other changes in the tourism infrastructure, such as accommodations. As a result, the availability of transportation infrastructure and services has been considered a fundamental precondition for tourism.[1]

Transport geography is a topical branch of geography that evolved out of economic geography. Like tourism, transportation is, of course, inherently geographic because it connects places and facilitates the movement of goods and people from one place to another. Transport geography fundamentally depends on some of the basic geographic concepts introduced in chapter 1, such as location or scale. For example, location shapes patterns of movement, including whether movement is possible from and/or to a given location, and how that movement might occur. Transportation networks exist at local and regional scales and, in the modern world, are increasingly being connected into a global system. In addition, there are many geographic factors of places—both physical and human—that either allow or constrain transportation.

There is a distinct and reciprocal relationship between tourism and transport. Tourists constitute an important demand for transportation services and therefore play a role in the study of transport geography. At the same time, transportation is a fundamental component of tourism and thus is important for our purposes in the geography of tourism. We need to understand the means of connection between the people who demand tourism experiences and the places that are able to supply those experiences. As tourism is based on the temporary movements of people across space, the transportation that facilitates these movements is key in converting suppressed demand to effective demand. Beyond getting tourists to a destination, transportation also facilitates their experience of that destination. Given the extent of interconnection between tourism and transport, a recent "progress report" on the state of transport geography argued that there should be a closer relationship between this topical branch of geography and the geography of tourism.[2]

This chapter further develops the geographic foundation of tourism by examining transportation as a fundamental component of tourism through the concepts of transport geography. This topical branch provides us with the framework to examine the transport system, particularly the role of different transportation modes in tourism, the geographic factors that facilitate movement, and the spatial patterns of movement in tourism. In addition, we will look at some of the ways researchers are examining the intersection between tourism and transport.

Studying Tourism Transport

It is widely recognized that transport is a vital element in tourism, but its role is not always well developed in the literature. In fact, there are some inherent difficulties in trying to understand tourism transport. It can be difficult to even identify what would be considered "tourism transport." Tourists use a multitude of different types of transportation in varied contexts for a wide array of purposes. In some cases, these tourists may be able to use a personal form of transport. In others, they must pay for the services of government-subsidized public transport or those provided by a private company. Although there are several examples of dedicated tourism transport at a destination or specific attraction, tourists are typically only one group of users of transportation facilities and services. Employees of the tourism industry, local residents, and other transit passengers may also use the same transportation. Generally, no distinction is made between these different passengers, which provides us with little data on how tourists are using transportation. Moreover, tourism and transport are typically managed by different governmental agencies with little in the way of collaboration.

The approach to studying issues in tourism transport is not always clear either. Tourism studies and transport studies are interdisciplinary fields that have drawn upon concepts and theories from different perspectives.[3] Geography has been one of the common areas. Spatial concepts in geography have been applied to understand the role of transport in facilitating tourists' movements, both from tourist-generating regions to receiving regions and within tourist-receiving regions.

Yet, the application of geographic models to tourism has not always been successful. For example, urban transportation models for commuting patterns are based on the assumption that the majority of people will take the most efficient route possible from home to work. This is predicated on commuters' knowledge of the situation, including the type of transportation available to them, potential routes, traffic, and congestion patterns.[4] However, these assumptions cannot be made in the context of tourism. The decision-making process of tourists about where to go and how to get there is not always rational. They frequently have little or no knowledge about the place they are visiting. They rely on guidebooks, maps, tourist information services, hotel personnel, and random strangers to provide them with information about how to get to a particular destination. Those tourists who try to take the most efficient route may get lost several times along the way, and many others will voluntarily choose a route that is less than direct because they consider exploration to be part of the experience. Even when tourist routes can be modeled for a specific destination, there is typically little transferability to other contexts.

Nonetheless, geographers argue that an understanding of the distribution of accommodations and attractions at a destination, as well as the transportation network that connects these places, can be extremely valuable. This type of data should allow the destination to efficiently plan and manage the transportation system to meet the needs of tourists and better provide them with the information they need. In particular, geographers cite the potential for geospatial technologies to more effectively understand the patterns of and opportunities for tourism transport in the context of specific destinations.

The Evolution of Transportation and Tourism

From the historical geography of tourism, we can begin to appreciate how vital transportation has been to the development of tourism. For example, the ancient Romans were among the earliest societies to travel, and an extensive road network—combined with an organized system of horse-and-cart transport—was one of the key factors in this development. Likewise, the deterioration of these roads after the collapse of the Roman Empire was one of the issues that brought all nonessential travel to a halt.

Over time, new transportation systems developed throughout Europe that allowed greater opportunities for travel. From the mid to late Middle Ages, while roads remained poor, water transport provided some means for travel. Major river systems such as the Rhine, Danube, and Loire, as well as canal networks, formed the basis for transportation within the region and provided regular passenger services. New options for travel over land also gradually developed and expanded across the region. In the fifteenth century, the post system was developed in France, where travelers could change horses at relay stations established at regular intervals. This evolved into a widespread network of coach services by the middle of the eighteenth century.[5]

The innovations with the greatest impact on tourism came at the beginning of the nineteenth century with the development of commercially successful steam locomotion. Regular steamboat service offered faster, more reliable, and increasingly comfortable transportation.[6] Steamboats operated along river systems and supplanted earlier, slower, riskier oceanic sailing vessels. Steam packets traveling regular transatlantic routes were the most efficient means of travel throughout much of the nineteenth century. Originally intended for transporting the mail, they also began carrying cargo and passengers. Then, as rail service developed and expanded, it trumped all previous means of transportation. Although railways were originally intended for carrying heavy freight, like coal, they also proved extremely successful for passenger travel. Not only were railroads faster and more efficient than other available modes, they could also routinely carry ten times the number of passengers as a horse-drawn coach.[7] In addition, the typical charge of 1 penny per mile for rail travel was substantially lower than coach fares.[8]

Both forms of steam locomotion reshaped patterns of tourism in a myriad of ways. Due to decreased travel time and cost, more people from the middle and lower classes were enabled to participate. This increase in tourism raised concerns among the earlier generations of tourists. In some cases, these earlier tourists sought new destinations in previously distant or inaccessible places. In other cases, they fought to limit the changes that were taking place at existing destinations. For example, prominent English poet William Wordsworth strongly objected to the proposed rail development in the Lake District on the grounds that it would destroy the natural beauty of the area that he and other visitors came there for. Although this line was not built, rail stations on the periphery of the area nonetheless brought substantial numbers of tourists, who traveled into the area on foot or by coach.[9]

The invention of the sleeping car provided greater opportunities for long, uninterrupted train trips. This idea evolved into the Pullman car—luxury sleeping cars that effectively served as a hotel on wheels and allowed the upper classes to travel longer distances in comfort. However, rail travel created new challenges as well. For example, where it formerly took weeks for a tourist to travel from locations in Northern Europe to destinations in Southern Europe, trains reduced the trip from London to Nice to just one-and-a-half days. Prominent physicians claimed that this was not enough time for passengers, particularly those traveling for health reasons, to adjust to changing environmental conditions. To avoid potentially serious health complications, these physicians argued that travelers should break the journey down into intermittent stages.[10]

Steam-based transportation also changed the ways people experienced places. Tourists had the opportunity to see different landscapes in locations farther afield and to see them in a new way. Instead of stopping at strategic vantage points, tourists viewed the landscape through glass windows on scenic cruises and railroads, and they had to learn to focus on a moving landscape. These tourists were also somewhat restricted in what they saw along transportation corridors, whether it was a river, canal, or rail line.[11] Additional modes of transport were necessary for further exploration. Tourists would have to take horse-drawn carriages or buses to nature sites and scenic vistas. Secondary transport was even needed to reach downtown centers, as rail stations were typically located outside of town. Consequently, enterprising innkeepers invested in shuttle services and opened inns near the station to capture the in-transit market.

At the beginning of the twentieth century, the automobile further reshaped and expanded tourism. Widespread personal car ownership is considered to be key in the development of modern mass tourism. In the United States, this—combined with the expansion of the interstate highway system—allowed tourists increasing freedom to visit multiple destinations during the course of a single trip and to explore new areas of the country. New attractions and destination regions emerged, leading to the development of new types of accommodation, such as the motor hotel (motel), to meet tourists' needs.[12]

Air transportation created ever more opportunities for tourism—especially mass international tourism—in the second half of the twentieth century. Air travel had been made available to a select group of affluent tourists following the end of World War I, but it was greatly expanded in the years following World War II. Innovations in air transportation, such as the jet engine and wide-bodied passenger jets, allowed planes to increase both the distance traveled and the numbers of passengers carried. As such, all parts of the world have been opened up to tourism, including many destinations that are almost entirely dependent on air transportation for international tourist arrivals.

While these innovations in transportation were not driven by tourism, the tourism industry directly benefited from improvements in safety and efficiency as well as reductions in cost. The framework of historical geography helps us examine these changes that have taken place over time to better understand the interconnections between transport and tourism. This gives us valuable insight because the transport industry continues to evolve. For example, regulations and/or prohibitive emissions taxes could limit where we travel in the future and how often we travel. In contrast, technological developments may open up far greater opportunities.

Interest in space tourism has existed since the space race. In 1989, Pan American Airways reportedly had a wait list of over 93,000 people for its first passenger flight to the moon. The company folded in 1991. Ten years later, the world saw its first space tourist. American Dennis Tito traveled on a Russian spacecraft to the International Space Station. Over the next decade, six additional tourists paid an estimated US$20–35 million for an orbital trip.[13] While the Russians continue to be willing to offer tourist flights, there is presently a constraint on supply as the Soyuz spacecraft only holds three people.[14] Virgin Galactic initially planned to start offering suborbital space trips in 2009, but the project has experienced numerous delays. In the latest setback, the company's *VSS Enterprise* was destroyed during a test flight in 2014. Seven hundred seats had already been reserved for its first commercial flight scheduled for the following year. Tickets cost US$200,000.[15]

The idea of space tourism was initially dismissed by agencies such as NASA due to safety hazards as well as the amount of time and resources involved; however, with recent developments, it is now regarded as a possibility to generate revenues for space agencies. Still there are many areas of research on the potential for space tourism. For example, actual demand for the experience needs to be determined. While surveys conducted over the years show a high level of interest in space tourism, the idea is still considered hypothetical for most due to technological and/or financial barriers. One study found that at least one-quarter of respondents would be unlikely to participate in space tourism, even if cost were not a consideration. Perceived risk was the primary reason given. Other issues pertain to health and training of participants; regulation, liability, and insurance; socioeconomic impacts of space tourism at spaceport locations; and environmental impact and carbon footprint.[16]

Space tourism is a captivating idea. According to one set of scholars, "Space appears to be the next natural step in satisfying people's need for exploration, adventure and new recreation activities."[17] It will be exciting to see how this intersection of transportation and tourism continues to evolve in the future.

The Transport System

Transport geography recognizes transport as a system that involves networks, nodes, and modes and is based on demand. For tourism, the primary function of this transport system is to facilitate the movements of passengers to and from destinations. Secondary functions include getting tourists to the transport terminal and supporting the movements of tourists within the destination. Since tourism is typically considered to be nonessential travel, transport services must be safe, relatively convenient and comfortable, and competitively priced to support tourism. However, the networks, nodes, and modes of this system will not be solely used for the purposes of tourism. Instead, the tourism industry generally takes advantage of existing transport systems, with the exception of new destinations that were specifically planned for the purpose of tourism. Yet, even in this case, when the transport infrastructure must be developed to facilitate tourism, the network will serve other transportation needs as well.

A **transportation network** is the spatial structure and organization of the infrastructure that supports, and to some extent determines, patterns of movement. The transportation infrastructure has been expanding at both the local and global scales, becoming an ever-more complex web of interconnections. At the same time, the relative cost of transportation has declined. These factors have allowed more movement to take place than ever before.[18] The nature of the network can encourage people to travel along one route or discourage them from traveling along another. These networks may be highly dependent on geography. For example, the physical geography of a place will affect patterns of transportation, whether it is physical features like mountains, river systems, and ground stability; atmospheric conditions such as wind directions; or oceanic conditions such as currents. The human geography of a place, such as the circumstances of political geography, can also have an effect on transport. National boundaries may affect the ability to create a transportation network and efficiently connect places, either on the ground or in the air through no-fly zones.

Transportation nodes are the access points to the network. These nodes may be **terminals**, where transport flows begin or end, or **interchanges** within a network. Population geography often plays the most significant role in determining the location of nodes. In general, nodes are likely to be situated in areas with high population densities, and terminals in particular will be located in or near major cities. **Transportation modes** are the means of movement or the type of transportation. Broadly, there are three categories of modes based on where this movement takes place—over land (surface), water, or air—with different types within each category[19] (see table 5.1).

Table 5.1. Summary of the Advantages and Disadvantages Associated with the Modes and Types of Tourism Transport

Mode	*Tourist Considerations*		*Destination Considerations*	
Surface				
Walk	Free Flexibility Required in traffic-free zones	Used only for short distances Dependent on physical condition	Reduces traffic congestion and pollution Increases access to businesses	Requires investment in some infrastructure (e.g., paths, sidewalks, signs)
Rail	Well suited for short-to medium-haul travel Ease of navigation versus driving May facilitate tourism experiences (e.g., historic or scenic trains)	Constrained by schedules and routes Costs vary by destination	Schedules and routes can be modified based on demand May serve as an attraction	May require investment in infrastructure, new technology, and/or maintenance
Personal car	Flexibility Privacy Well suited for short- to medium-haul travel	Associated with varied costs (e.g., fuel, tolls, parking) May be difficult to navigate unfamiliar roads	Increases accessibility of destinations not served by mass transit	May require investment in roads and parking facilities Brings fewer visitors per vehicle May generate congestion and pollution
Scheduled bus	Low cost Ease of navigation versus driving May be used for short- or medium-haul travel	Constrained by schedules and routes Lack of privacy, personal space, and security	Requires little additional investment Schedules and routes can be modified based on demand	May be subject to significant perceptual constraints (e.g., location of stations, prevalence of crime)
Water				
Ferry	Provides access to small destinations Offers the ability to take personal cars	Slow Not used for long-haul travel	Increases access to small destinations	Requires docking and terminal facilities
Cruise	Allows travel to multiple destinations Provides a vacation experience	Requires travel to port Slow Not well suited for long-haul travel	Increases access to island/ coastal destinations Has the potential to bring large quantities of tourists	Requires a deepwater harbor or ferry service, docking and terminal facilities
Air				
Scheduled	Provides access to more destinations Well suited for long-haul travel	May have high costs May contribute to increased stress levels May have health risks (e.g., jet lag, deep-vein thrombosis)	Increases access to remote or hard-to-reach destinations Has the potential to bring large quantities of tourists	Requires space for/investment in runway and terminal facilities Must be regulated Generates pollution (e.g., air, noise)
Sightseeing	Offers a unique experience	May have high costs	Facilitates access to large-scale attractions	May be disruptive

Figure 5.1. Bicycle rentals are an increasingly popular option for easily getting around and experiencing a destination. *Source:* Velvet Nelson

In comparison with the other primary categories of mode, surface transport is more dependent on geography because the development of a network is restricted by land area, infrastructure, and possibly even national boundaries. Yet, surface transport continues to be the most widely used mode of tourism transport during the movement stage to reach destinations and to move around during the experience stage. Self-powered surface transport, such as walking or bicycling (figure 5.1), is used to get around a destination or as a form of transportation-as-experience along with other forms of modern personal transport (e.g., Segway tours). Additionally, surface transport can be broken down into rail and road transport.

Today, the use of rail transport in tourism is highly uneven. Countries like the United States that have placed an emphasis on expansion of road networks have seen some of the greatest declines in passenger rail transport. Few developing countries have invested in creating rail networks. However, throughout Europe and parts of East Asia, extensively developed rail networks continue to be used on a wide scale. Some countries have even made new investments in infrastructure and technological advances to improve rail transport and provide a competitive means of getting to or from a destination. For example, France's Train à Grande Vitesse (TGV) is an intercity high-speed rail service. Based on the success of the TGV, other countries in the region such as Germany, Spain, and Italy developed their own high-speed rail services. High-speed rail service is relatively new in China, but the country now has the longest network in the world and the most riders.

Personal vehicles are now the dominant mode of tourism transport, accounting for approximately 77 percent of all trips.[20] Still, road transport has not been as significant in developing regions of the world where fewer people have access to private cars. Recreational vehicles (RVs) are a subcategory of personal vehicles that offer a unique form of tourism contingent upon transport. Scheduled bus services account for a small amount of tourism transport to and from destinations, while local bus services and taxis may be used to reach a major transport node or for travel within a destination. Charter bus services are used in package tours and excursions from a resort area, and specialized tourism transport, such as sightseeing and hop-on hop-off buses that stop at major attractions, provide an additional option at major destinations.

Developments in other modes, particularly air, have changed the role of water transport in tourism. Ferries may be used as a means of reaching a destination, particularly those areas that are not large enough to support an airport. For example, ferries and hydrofoils are used to transport tourists to some of the islands off the coast of southern Europe, such as the Greek Islands. Yet, in other cases, ferries have been replaced with "island hopper" flights, which is a common occurrence in places such as the Hawaiian Islands.

Nonetheless, water transport continues to play an important role in tourism with the cruise ship industry, where transportation constitutes the tourism experience rather than serving as a means to an end. The Caribbean continues to be the most popular cruising region, followed by the Mediterranean; however, new routes are being developed all over the world. Cruises were once marketed to older age groups, but many lines have expanded their target markets by offering various price options, attractions for younger markets, such as families with small children, and theme cruises for special interest groups. In addition, water transport can provide the means of participating in activities at the destination, such as sunset cruises, whale watching, and snorkeling or scuba-diving expeditions.

Air transport is the most recently developed mode and has primarily been used in tourism as a means of reaching the destination. In fact, this mode has been vital in increasing the accessibility of remote and poorly connected destinations (figure 5.2). Due to high costs, air travel has been unavailable for much of the world's population. However, recent developments such as the introduction of low-cost carriers have started to make changes in the way airlines do business. In fact, cheap regional flights have begun to compete with both rail and road transport. Air transport can also be an integral part of a tourism experience, such as a panoramic helicopter flight over large scenic attractions like the Grand Canyon or Iguazu Falls on the border between Argentina and Brazil.

Figure 5.2. Air transport provides the means of reaching destinations that are isolated by poor surface transportation networks—in this case, Dangriga, Belize. ***Source*****: Tom Nelson**

Box 5.1. Experience: Tourism on the Job

In this chapter, we consider the relationship between transport and tourism. For individuals working in transport industries, there is a certain tension in this relationship. As a corporate aviation pilot, Jeremy has had the opportunity to experience more places than the majority of people who have to pay commercial airfares. However, participating in tourism on the job comes with constraints.

I always wanted to fly. My dad was a pilot, and I flew with him as a child. I got my first pilot's license when I was seventeen, and flew my own plane across the country when I was twenty-one. I planned my route so that I would stop in places where there were interesting things to see and do. One such amazing day included flying from St. George, Utah, crossing Lake Powell, dodging the sandstone buttes of Monument Valley, and passing the Four Corners to finally stop in Durango, Colorado, for the night: an incredible experience, especially from 3,000 feet and 90 mph. With my feet on the ground, I was then able to spend a day visiting Mesa Verde, a truly amazing archeological marvel. Since then, I have traveled as often as I could, even when I didn't have much money. Some of my best trips have been the ones where I got to spend three to four weeks in a place. Not only did that give me the opportunity to have amazing experiences in countries like Peru and South Africa, but it also allowed me to get to know the places better. Although I got a degree in a non-aviation field, I kept flying and was able to get a job flying for a commercial airline. People often think airline pilots get to see the world, but it seemed that all I got to see were hotel rooms. After a few years, I moved to corporate aviation.

In this business, we cater to the needs of various types of private clients. For famous actors or athletes, their primary concern might be privacy. We would take them to a small, exclusive airport at their destination rather than the big, commercial hub. For business people, they might need to get to a place as quickly as possible. We would take them to whatever airport was closest to their final destination. For wealthy individuals, they might want to go to a specific place, such as where they have a vacation house. We could take them directly to their destination, unlike several legs on a commercial airline. The specific type of client often depends on the location. In my area, our clients tend to come from the oil and gas industries, but picking up passengers from all over the country, we certainly see all types.

We fly our clients to wherever they need to go. Most travel to destinations in the United States as well as to Canada, Mexico, and the Caribbean, although anywhere in the world is possible. Then we wait until they are ready to return. We don't have an unlimited budget, but when we are at a tourist destination, we often do the same things tourists do. In Colorado, I can go skiing; in New York City, I may see a Broadway show; and of course I get some beach time and watersports when in the Islands. When we are in unfamiliar places with business travelers, we usually rely on local advice regarding where to go and what to do. These trips often become food-centric. Of course, getting something to eat is one of the first things we are going to do after we arrive, but food is also something that can help differentiate places when one small town seems much like another.

We often fly to the same places regularly. For example, I fly one client to Asheville, North Carolina, for a few days every month. The more time I spend there, the more my experience changes. I am more familiar with the area, and I have moved beyond the types of things a typical tourist would see and do. Now when I go, I don't visit the Biltmore Estate; instead I might visit my favorite coffee shop, meet up with friends from previous trips, or maybe take a drive on the Blue Ridge Parkway to enjoy the fresh air. In essence, we become "locals" in a town far from home—a comforting change from sitting in another chain hotel, waiting for the next flight.

It sounds very glamorous to get to be a tourist while I'm working, but ultimately I am still working. I am on call all day, every day. I don't choose my destinations, which could be anywhere from Seattle, Washington, to Bainbridge, Georgia. I often have very little notice of a trip, so I never have time to plan where I want to go or what I want to see. When I am at a destination, I have to make sure that I can be reached at any time and that I never get too far away in case the clients' plans change. I'm not with my family or friends in these places; I'm with a coworker. And, if you do something enough times, it becomes work. All that being said, the ability to travel somewhere unknown and unplanned holds a degree of excitement. I go to places that I may never have chosen to go in my personal life, see things that I may never have known existed and certainly get off the beaten track. Often this leads me to find places to which I would like to return, in my own time, armed with a basic knowledge of a place and eager to find more. If nothing else, I can tell most anyone who asks the best dining spots in a vast number of cities, and as the cliché goes, *I have the best office view in the world*!

—*Jeremy*

These modes serve different roles in tourism, and trips frequently require the use of multiple modes. For example, personal cars, taxis and ride-share services, or inner-city train systems may be used to get to and from the terminal node (i.e., a train station, airport, or seaport). Likewise, tourists may rent a car, take taxis and tour buses, or use public transportation systems to reach and get around their destination. As such, it is important that a destination develop a comprehensive transportation system in which the networks of the different modes used in tourism are integrated. This will allow tourists to change from one mode to another as seamlessly as possible.

As tourists, we typically evaluate our options instinctively, with little reflection; therefore, these issues often seem self-evident. However, there are many factors that determine the appropriate mode(s) of transportation for a trip. There are both practical considerations involved (i.e., what modes are available for a trip) and a variety of perceptual considerations (i.e., personal preferences and constraints).[21]

Distance is one of the most important practical considerations that may automatically eliminate one or more modes. Greater distances require longer travel times and/or higher transportation costs. Monetary cost is one of the most important perceptual considerations. Transportation accounts for some of the largest expenditures in tourism. For many trips, the experience stage begins when the tourists reach the destination; thus, they are interested in reaching their destination as quickly and efficiently as possible. However, for tourists with a limited budget for a trip, it may come down to a choice: spend more money on transportation or on the experience at the destination. For example, a direct flight may be the option that requires the shortest amount of travel time, and, by extension, often the least amount of hassles and potential problems in the form of lines, security screenings, delays, lost luggage, and so on. This allows the maximum amount of time spent at the destination. Yet, a flight with multiple connections or even other modes of transport, such as a personal car, may be lower-cost options that are longer or less convenient but could allow the tourists to spend an additional day at the destination, participate in a particular tourism activity, or make other expenditures. Likewise, if transportation cost is a significant factor, tourists may need to consider a second- or third-choice destination as an alternative.

Personal goals and preferences also play a role in transportation decisions. As we saw above, we cannot assume that tourists will take the most efficient route from the point of origin to the destination based on shortest distance or travel time. Rather than being primarily concerned with getting from point A to point B in the fastest manner possible, tourists might be interested in seeing things along the way, where the movement stage is as important as the experience stage in the tourism process. Consequently, they might choose to drive a personal car to get a better view of the landscape through which they are passing, as well as to have the freedom to make stops or take intentional detours along the way. Additionally, tourists who are afraid to fly will take an alternative mode of transportation to get to their destination, regardless of whether air transport is the quickest, easiest, or cheapest mode.

This choice of mode affects the level of interaction tourists will have with both people and places. The development of new modes of transportation changed the ways in which people experienced places. The faster the mode, the less of the passing landscape is seen. For air transportation in particular, observing the landscape is generally not considered a part of travel; therefore, tourists only have the opportunity to experience the place of the destination as opposed to the places of travel as well. The choice of mode also has the potential to create opportunities for interaction with other people or limit it. Personal cars tend to isolate tourists, both from locals and from other tourists. Specialized tourism transport, such as charter and sightseeing buses, fosters interaction with other tourists, while walking and taking public transportation often allows tourists the greatest opportunities to interact with local people.

Patterns of Movement in Tourism

Transport geography also considers spatial patterns of movement. Movement is a fundamental part of the tourism process. Depending on the type of **tourism itinerary**, or the planned route or journey for a trip (figure 5.3), movement may comprise two distinct stages in the tourism process: travel to and travel from the destination. While perhaps just a means to an end, travel to the destination may hold a certain measure of excitement and anticipation. This is less likely the case for travel home. After the experience at the destination, tourists may be tired and ready to just get home. The length of the trip, and any potential hassles that arise during the course of travel, can shape the post-trip stage, in which people remember their trip. Yet, movement may also be an integral part of the trip and encompass the entire experience stage.

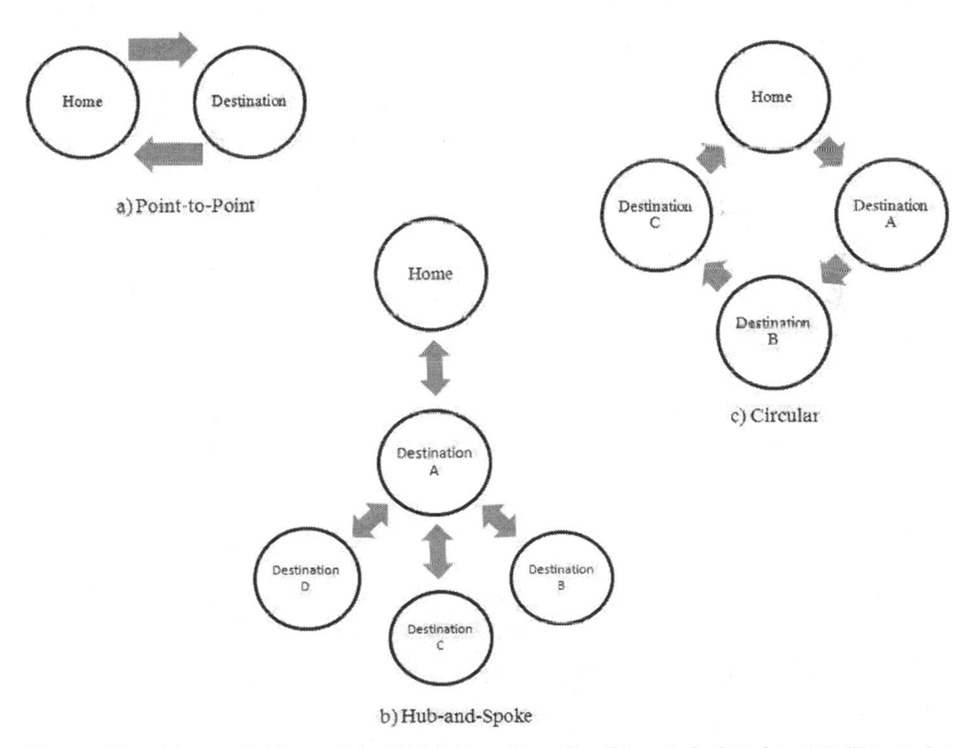

Figure 5.3. The point-to-point (a), hub-and-spoke (b), and circular (c) itineraries provide three examples of tourism itineraries. There are many variations on these patterns that are, to some extent, dependent on the available transport system for the destination.

Tourism scholars have identified many different types of tourism itineraries. The most basic type is **point-to-point**. In this itinerary, people travel from their home to a destination where they stay for the duration of their vacation and then return home. The selected mode for such a trip will depend on the transportation infrastructure and relative locations of the tourist's origin and destination. Typically, movement will be a means to an end with this type of itinerary. For example, tourists traveling from New York City to an all-inclusive resort in the Dominican Republic will follow a point-to-point itinerary using air transport to reach the destination, at which point the experience stage will begin. However, the movement stage can nonetheless be incorporated into the experience stage. For tourists using road transport to drive from New York City to the Finger Lakes region in Upstate New York, the scenery witnessed en route may be considered part of the experience, in addition to the time spent at the destination.

A variation on the point-to-point is the **hub-and-spoke itinerary.** This involves travel from home to a destination. That destination then becomes a base for visiting other destinations, each time returning to the first destination before returning home. For example, tourists to Oregon might use the Portland area as their hub and take day-trips (the spokes) to participate in a multitude of activities at sites around Mount Hood, in the Columbia River Gorge, in the Willamette Valley wine region, and possibly even at the coast depending on how much time they have. In this case, movement is interspersed with experience.

The **circular itinerary** is the pattern most dependent on transportation, as travel is distinctly part of the tourism experience. In this itinerary, tourists travel from home to one destination and then another (and perhaps another) for varying lengths of time before returning home. At one end of the tourist spectrum, drifters and explorers such as backpackers may follow a circular itinerary as they utilize public transportation systems like trains or buses to travel from destination to destination. At the other end of the spectrum, a common pattern for organized mass tourists is to follow a circular itinerary on a charter bus tour of multiple destinations, possibly even in multiple countries. Cruises are the ultimate transportation-as-tourism experience. However, it must be noted that for tourists who do not live near a deepwater port, their tourism process still involves movement to and from the terminal port.

These models are, of course, simplifications. In reality, the possibilities are endless. Itineraries are typically shaped by factors such as the distance traveled and the mode of transport used. However, the specific itinerary often comes from individual choices about visiting a single destination or spending time at several, taking a direct route or making detours and side trips, or following the same route to and from the destination or returning by a different path.

Transportation and Destination Development

Practically, transport geography can be used to provide valuable information about specific patterns for destinations. New destinations seeking to establish tourism, as well as existing destinations looking to expand their industry, need to consider issues such as where their potential tourist markets are located and how accessible the destination is to those markets. This is entirely contingent upon the transport system.

Potential destinations must analyze the existing system to determine whether the appropriate framework already exists and can be utilized for the purposes of tourism. If not, they will have to invest in the development of new networks, nodes, and/or modes. In most cases, at least part of the existing system can be used for tourism purposes. For example, a nation may have an international airport located in the capital city that can accommodate international tourists, but a regional airport or surface transport may need to be developed to connect these tourists with the country's destination regions. In the case of Switzerland, Zürich is the largest city and a prominent international business center, and the Zürich airport is considered to be the country's international gateway. Yet, many of the country's international tourists continue on to other destinations, such as Lucerne, Interlaken, or Grindewald, via its well-developed transportation network (box 5.2).

Box 5.2. Case Study: The Swiss Travel System

Switzerland is widely recognized for having one of the most efficient public transport systems in the world and one of the densest networks. The Swiss Travel System has over 16,000 miles of public transport.* While this system serves local and domestic needs, it also provides an incredible foundation for tourism. Few countries have the same potential to offer tourists such easy access to their destinations through both comprehensive connections between places and regular timetables. Although air transport provides access to the country for regional and global tourists, the public transport system within the country spans both water and surface transport. Water transport consists of boat routes on rivers and lakes, while surface transport includes a road network used by cars, intercity buses, and metro buses, as well as the internationally renowned rail network that provides the foundation for the entire system.

In an effort to both increase use of this network and promote international tourism, the Swiss Federal Railways (SBB) created the Swiss Travel System. This system provides a range of ticket options for foreign travelers and increases the ease of travel within the country by providing access to each of the train, bus, and boat routes within a network that spans some 12,500 miles. For example, the Swiss Pass (the most comprehensive option) allows unlimited travel on any of the standard transport modes for a specified number of days and discounted fares on specialized transport (i.e., mountain railway services such as cog trains or aerial cable cars) in destination areas. In addition, the pass offers free entry to nearly five hundred museums across the country and discounts at some hotels. Discounted fares are also provided to youth (defined as under the age of twenty-six) and free fares for children (defined as under the age of sixteen) when accompanied by a parent.

While most routes in Switzerland could arguably be described as scenic, there are specifically designated scenic rail and water routes, most of which are included with the Swiss Pass. On scenic routes, panoramic trains allow passengers to view the landscape through large viewing windows. For example, the Glacier Express is advertised as "The slowest express train

* SBB, "Visit Switzerland," accessed July 22, 2016, https://www.sbb.ch/en/leisure-holidays/holidays--short-breaks-in-switzerland/swisstravelsystem.html

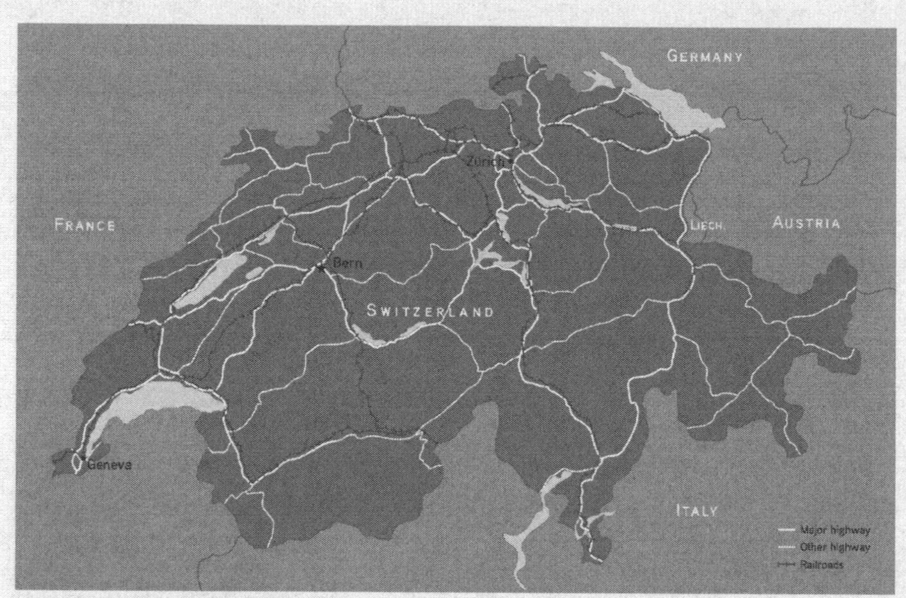

Map 5.1. Switzerland. This popular European tourism destination is well connected by a comprehensive and efficient transportation network. The major surface transport links (road and railroad) shown here are further connected by numerous minor ones. *Source:* XNR Productions

in the world."† Traveling from Zermatt to St. Moritz, the journey begins in the shadow of the world-famous Matterhorn, crosses distinctive natural and cultural landscapes, and travels part of the UNESCO World Heritage Albula Railway. Additionally, restored "nostalgia" trains have been developed for the use of tourists. Some railways, such as the Jungfraujoch, which takes visitors to the "Top of Europe," Europe's highest railway station, were constructed especially for the purpose of tourism.

The Swiss Travel System network provides an excellent example of the interrelationship between tourism and transport. In addition to simply facilitating travel from one place to another, this network can also be considered a distinct part of the tourism experience. Moreover, the Swiss Pass minimizes the potential hassles and stress for international tourists as they navigate an unfamiliar transportation network. This, in itself, can be a distinct incentive for choosing Switzerland as a destination.

Discussion topic: What role do you think intra-destination transport (such as the Swiss Travel System) plays in tourism decisions?

Tourism on the web: Switzerland Tourism, "Vacation, Holiday, Travel, Meetings," at http://www.myswitzerland.com; Swiss Travel System; "Home," at http://www.swisstravelsystem.com/en/home.html

† Swiss Travel System, "Glacier Express," accessed July 22, 2016, http://www.swisstravelsystem.com/en/highlights-en/train-bus-boat/railway/glacier-express-st-moritz-zermatt.html

Similarly, destinations targeting new markets must consider their distance from that market and the available transportation modes between the two places, with the associated travel time and cost factors. The geographic concept of **distance decay** indicates that demand for a product or service decreases as the distance traveled to obtain that product or service increases. In other words, if a consumer perceives that two products or services are comparable, he or she will choose the one that is easier to obtain. In theory, then, the increased time and cost of greater distances will decrease the desirability of a destination. However, tourism represents a special case that requires some modification of the model. Tourism demand does not "decay" immediately. In fact, demand is highest at a certain distance from the tourists' home.[22] After that point is reached, increasing distance will result in decreasing demand, unless there is a high degree of **complementarity**. Complementarity refers to the relationship between people in one place who have the desire for certain travel experiences and a place that has the ability to satisfy that desire.

It is also important that destinations analyze tourists' patterns of movement and itineraries, preferences in modes, cost thresholds, and any factors that might influence their decision-making process. For example, drifters and other categories of tourists interested in niche tourism products like nature and adventure tourism may specifically seek out undeveloped destinations and therefore have lower expectations from the transportation infrastructure (e.g., unpaved roads). However, mass tourists from the major generating regions, specifically Europe and North America, are used to a modern and efficient transportation infrastructure where they can get from one place to another quickly, comfortably, and safely. If these tourists perceive that transportation at a destination is unreliable, slow, difficult, and possibly even dangerous, they are likely to choose an alternative destination. As such, destination stakeholders may need to coordinate their efforts to increase investment in the infrastructure and regulate the provision of transportation services to create a better environment for tourism.

Data on these patterns can identify transport flows and places where traffic is concentrated at a destination. This information can aid destination planners to formulate effective strategies to eliminate potential bottlenecks or alleviate problems with congestion and overcrowding during peak seasons. These solutions may include designating alternative routes, directing tourists to alternate attractions, or creating new policies such as the establishment of restricted or traffic-free zones. This will help reduce the negative impacts on the destination created by traffic and pollution and improve the visitor experience.[23] Management strategies such as these will be discussed in greater detail in chapter 11.

In addition, destination stakeholders need to consider usage patterns, profitability, and competition for limited mass transportation resources due to seasonality. For most destinations, tourism is a seasonal industry with highs and lows. During the high season, mass transportation may run at capacity or, in fact, over capacity. If the supply of transport services is not expanded to meet the increased demand, systems can become congested. Thus, all users experience decreased access to and quality of transportation services. For example, tourists arriving at major urban destinations are often reliant on public transportation systems due to prohibitive costs (e.g., car rental, insurance, parking, congestion charges for driving in the city center, and the like) or unfamiliarity with roads and traffic patterns. Tourists' usage can complement commuters' usage when they visit on weekends and holidays. However, during peak tourism seasons, typically during the summer months, tourists compete with residents.[24] For residents, this can generate animosity toward tourists. For tourists, this can have a negative impact on their ability to visit the desired attractions and their overall satisfaction with the destination.

Conversely, mass transportation is a perishable product because on regularly scheduled routes, unsold seats "expire"—in other words, those seats are not available to be sold at another time or on a later date. Transportation systems may experience losses during the low or off tourism season. As a result, operators may need to reduce the number of scheduled routes to the destination. They may also advertise discounted fares or work with other stakeholders to offer package deals (e.g., combining transport with accommodation or activities) in an effort to increase off-season visitors.

Directions in Research

Clearly it is important that we continue to try to understand the evolving patterns of where tourists are going and how they are getting there. Additionally, researchers have been exploring the intersection between tourism and transport in other ways. This research takes into consideration the implications of current trends in tourism on transportation (box 5.3) and the role of tourist transport in issues of global human and environmental significance.

Box 5.3. In-Depth: Low-Cost Carriers

Low-cost carriers (LCCs), also known as discount, budget or no-frills airlines, concentrate on reducing operating costs to offer lower ticket prices to passengers. These carriers run out of secondary airports that not only charge lower airport fees but also allow quicker operations. They offer point-to-point services (i.e., limited or no connections to other places) on shorter routes to maximize the number of trips in both directions, and they maintain fast turnaround times to increase aircraft utilization.[*] Carriers benefit from little competition for specific routes. For example, over 90 percent of flight routes in Europe are serviced by a single LCC. They operate with minimal personnel, and staff tend to work the maximum hours legally allowed with relatively low pay and few benefits. These factors allow a cost advantage over full-service carriers (FSC) of up to 60 percent.[†] Most LCCs have only one seating class, no seat assignments, and limited personal space to maximize the number of seats. They typically offer little in the way of customer service.[‡]

The term ultra low-cost carrier has been applied to some of the newest entrants to the low-cost field (e.g., Spirit Airlines) in addition to some LCCs looking for a new competitive

* Juan L. Eugenio-Martin and Federico Inchuasti-Sintes, "Low-Cost Travel and Tourism Expenditures," *Annals of Tourism Research* 57 (2016): 142.

† Sven Gross and Louisa Klemmer, *Introduction to Tourism Transport* (Oxfordshire: CABI, 2014): 34.

‡ John F. O'Connell and George Williams, "Passengers' Perceptions of Low Cost Airlines and Full Service Carriers: A Case Study Involving Ryanair, Aer Lingus, Air Asia and Malaysia Airlines," *Journal of Air Transport Management* 11 (2005): 260.

advantage (e.g., Frontier Airlines). These carriers offer the lowest fares but have a high level of additional fees, from assorted luggage fees to boarding pass fees, seat assignment fees, and credit card fees (even when credit card is the only payment method accepted). This makes calculating the actual cost of travel more complicated. As such, many travelers question whether the final price is truly lower than what they could get through other carriers and whether the additional hassle is worth potential savings. The ultra low-cost carriers have seen an increase in passenger volume in the United States, but they also have some of the worst on-time records, the worst overall rankings, and the worst records for passenger complaints.

Obviously, price is the primary consideration for passengers traveling on LCCs. These passengers are willing to accept fewer and lower-quality services as a trade-off for these low prices, and they are willing to travel farther to access airports serviced by LCCs. For the most part, LCC passengers have generally been younger. While passengers traveling on FSCs are also concerned about price, they are more likely to consider other factors such as the reliability of the company, safety records, quality of customer service, flight schedules and connections, comfort, complimentary products and services, loyalty programs, and others.[§]

Research on the impact of LCCs on air travel is inconclusive. Some studies indicate that LCCs generate new demand; in other words, passengers on LCCs would not have been able to afford to take that particular trip if they had to pay the higher fare on a FSC. Other studies show that LCCs simply take business away from the FSCs and thus total demand remains the same. Likewise, research on the impact of LCCs on destinations is unclear. Studies suggest that LCCs can help increase tourists to a destination, especially during low-season periods of the year, which has the potential to increase revenues.[**] LCCs also have the potential to generate an increase in tourist expenditures. For example, if tourists with a set budget for a trip travel on a LCC, they may have more money to spend on accommodations, food and drink, or activities at the destination. However, if tourists are using LCCs to travel more frequently or to travel on a lower budget, they may choose to stay fewer nights, look for low-cost accommodation alternatives, and otherwise spend less money at the destination.

Southwest Airlines, operating in the United States, Central America, and the Caribbean, is the largest LCC in the world. As of 2012, Europe accounted for the highest passenger volume on LCCs with approximately 37 percent of seat capacity on scheduled domestic service. LCCs accounted for approximately 31 percent in the United States and 23 percent in Asia.[††] To remain competitive, several FSCs such as Delta and United attempted to launch corporate spinoffs or subsidiaries with a low-cost model, but these projects were ultimately unsuccessful.[‡‡] Other changes in the FSCs also appear to be moving these providers closer to the low-cost model, whether it is adding fees for services that were once complimentary or using temporary, hourly personnel instead of a permanent staff vested in the company. These changes have the potential to undermine the reasons many travelers continue to pay more for the FSCs. The evolution of the airline industry will remain a topic of interest for tourism geographers for years to come.

Discussion topic: Do you think LCCs are a positive development for the tourism system? Explain.

Tourism on the web: Southwest Airlines, "About Southwest," https://www.southwest.com/html/about-southwest/index.html?clk=GFOOTER-ABOUT-ABOUT

[§] O'Connell and Williams, "Passengers' Perceptions of Low Cost Airlines," 264.

[**] Jin Young Chung and Taehee Whang, "The Impact of Low Cost Carriers on Korean Island Tourism," *Journal of Transport Geography* 19 (2011): 1335–6.

[††] International Civil Aviation Organization, "Air Transport Policy and Regulation," accessed July 22, 2016, http://www.icao.int/sustainability/Pages/Low-Cost-Carriers.aspx

[§§] Chung and Whang, "The Impact of Low Cost Carriers," 1337.

For example, growth in some of the tourism products discussed in chapter 3 has created new opportunities and challenges with regard to transport. In particular, both heritage and event tourism have been cited as bringing the relationship between tourism and transport even closer. There is a significant niche market of "transport enthusiasts." Interest in appreciating historic transport has generated a demand for themed museums, exhibitions, and car shows. Interest in experiencing these modes has generated a demand for tourist trips on horse-drawn carriages, historic trains, vintage cars, gondolas, paddle steamers, and others (figure 5.4).[25] At the same time, interest in advances in transport technology has generated a demand for opportunities to experience the newest planes, very fast trains, and possibly even those suborbital spacecraft discussed above.

Conversely, the rise of products like nature, rural, and sport tourism has at times strained the relationship between tourism and transport. These products typically involve the transfer of tourists from an urban market or a centrally located terminal node (e.g., an airport) to remote locations that are not well served by public transportation. In addition, tourists participating in various sports activities may be carrying heavy or bulky equipment, such as golf clubs, skis, surfboards, and bicycles.[26] These items are difficult to take on public transportation, if in fact they are permitted at all. As such, these tourism products are heavily reliant on private

Figure 5.4. The Sea to Sky Gondola in Squamish, British Columbia (Canada), transports visitors to just under 3,000 feet above sea level in a 10-minute ride. At the summit, visitors have panoramic views and access to hiking, rock climbing, and other outdoor recreation activities. *Source*: Velvet Nelson

cars. Tourists must either be able to reach the destination in their own vehicles or rent one upon their arrival at the terminal, and destination stakeholders must plan accordingly to manage vehicles in an area that is perhaps unaccustomed to high volumes of traffic.

Another relatively recent direction in research considers tourism, transport, and health. Although early health concerns about the faster speeds of rail travel proved to be unfounded, the increase in long-distance air transport associated with tourism has generated new risks. This can range from the comparatively mild effects of jet lag to traveler's thrombosis, which, in the most serious cases, can result in a potentially fatal pulmonary embolism. In addition, new attention is being given to the increased levels of stress that can result from anxiety about flying, the threat of terrorism, missed flights, lost baggage, and more.[27] While this can lead to occasional (and well-publicized) incidents of aggression or violence, it also can exacerbate existing medical conditions, resulting in in-flight emergencies. Geographers have also begun to study the rapid diffusion of infectious diseases, like severe acute respiratory syndrome (SARS), by air travel.[28]

Other researchers are considering issues of access to tourism transport and inequality. Tourism is dependent on transport to facilitate experiences. As such, tourism becomes unavailable to people without ready access to or the ability to pay for transportation, such as lower-income groups in inner-city areas, populations living in remote rural areas, or large segments of the population in developing countries. Tourism transport can become a symbol of inequality and a means of segregating tourists from residents. For example, the modes used for tourists are typically modern, safe, and comfortable, whereas the public transportation used by residents may be old, deficient, and overcrowded. Tourist transport does not generally serve the needs of residents and may, in fact, be off-limits to them.

Finally, a significant direction in current research focuses on the environmental impacts of tourism transport, management policies, the adaptations to and potential effects of regulations, and attempts to change patterns of tourist behavior. Some of these issues will be discussed further in chapter 10, "The Environmental Geography of Tourism."

Conclusion

Transport is a fundamental component in tourism, as it facilitates the movement of tourists from their place of origin to their destination. Transportation systems were a precondition for tourism, and new innovations helped usher in several key eras in tourism. In particular, transport was one of the factors in the development of modern mass tourism, which allows more people—in more parts of the world—to travel than ever before. At the same time, transportation can distinctly shape the tourism experience. As transport geography provides the means of exploring the spatial patterns of movement and the geographic factors that allow or constrain this movement, we can apply the concepts of this topical branch to contribute to our understanding of the role transport plays in tourism.

Key Terms

- circular itinerary
- complementarity
- distance decay
- hub-and-spoke itinerary
- interchange
- point-to-point itinerary
- terminal
- tourism itinerary
- transport geography
- transportation mode
- transportation network
- transportation node

Notes

1. Bruce Prideaux, "The Role of the Transport System in Destination Development," *Tourism Management* 21 (2000): 54.
2. David J. Keeling, "Transportation Geography: New Directions on Well-Worn Trails," *Progress in Human Geography* 31 (2007): 221.
3. Stephen Page, *Transport and Tourism: Global Perspectives*, 2nd ed. (Harlow, UK: Pearson Prentice Hall, 2005), 34.
4. Alan Lew and Bob McKercher, "Modeling Tourist Movements: A Local Destination Analysis," *Annals of Tourism Research* 33 (2006): 405.
5. John Towner, "The Grand Tour: A Key Phase in the History of Tourism," *Annals of Tourism Research* 12 (1985): 322.
6. Marguerite Shaffer, *See America First: Tourism and National Identity, 1880–1940* (Washington, D.C.: Smithsonian Institution Press, 2001), 13.
7. Jack Simmons, "Railways, Hotels, and Tourism in Great Britain, 1839–1914," *Journal of Contemporary History* 19 (1984): 207.
8. Charles R. Goeldner and J.R. Brent Ritchie, *Tourism: Principles, Practices, Philosophies*, 9th ed. (Hoboken, NJ: Wiley, 2006), 56.
9. Simmons, "Railways, Hotels, and Tourism," 212.
10. Simon M. Kevan, "Quests for Cures: A History of Tourism for Climate and Health," *International Journal of Biometeorology* 37 (1993): 118.
11. Orvar Löfgren, *On Holiday: A History of Vacationing* (Berkeley: University of California Press, 1999). 43.
12. Gareth Shaw and Allan M. Williams, *Critical Issues in Tourism: A Geographical Perspective*. 2nd ed. (Malden, MA: Blackwell, 2002), 216–17.
13. Sam Cole, "Space Tourism: Prospects, Positioning, and Planning," *Journal of Tourism Futures* 1, no. 2 (2015): 132.
14. Derek Webber, "Space Tourism: Its History, Future and Importance," *Acta Astronautica* 92 (2013): 141.
15. Cole, "Space Tourism," 133.
16. Maharaj Vijay Reddy, Mirela Nica, and Keith Wilkes, "Space Tourism: Research Recommendations for the Future of the Industry and Perspectives of Potential Participants," *Tourism Management* 33 (2012): 1095–7.
17. Reddy, Nica, and Wilkes, "Space Tourism," 1101.
18. Jean-Paul Rodrigue, Claude Comtois, and Brian Slack, *The Geography of Transport Systems*, 2nd ed. (New York: Routledge, 2009), 5, accessed February 10, 2011, http://people.hofstra.edu/geotrans

19. Rodrigue, Comtois, and Slack, *The Geography of Transport Systems*.
20. Rodrigue, Comtois, and Slack, *The Geography of Transport Systems*.
21. Lew and McKercher, "Modeling Tourist Movements," 407.
22. Bob McKercher and Alan A. Lew, "Tourist Flows and the Spatial Distribution of Tourists," in *A Companion to Tourism*, ed. Alan A. Lew, C. Michael Hall, and Allan M. Williams (Malden, MA: Blackwell, 2004), 40–42.
23. Lew and McKercher, "Modeling Tourist Movements," 420.
24. Daniel Albalate, and Germà Bel, "Tourism and Urban Public Transport: Holding Demand Pressure under Supply Constraints," *Tourism Management* 31 (2010): 432.
25. Derek R. Hall, "Conceptualising Tourism Transport: Inequality and Externality Issues," *Journal of Transport Geography* 7 (1999): 182.
26. Jennifer Reilly, Peter Williams, and Wolfgang Haider, "Moving Towards More Eco-Efficient Tourist Transportation to a Resort Destination: The Case of Whistler, British Columbia," *Research in Transportation Economics* 26 (2010): 71.
27. Stephen Page and Joanne Connell, "Transport and Tourism," in *The Wiley Blackwell Companion to Tourism*, ed. Alan A. Lew, C. Michael Hall, and Allan M. Williams (Malden: Wiley Blackwell, 2014), 163.
28. Les Lumsdon and Stephen J. Page, "Progress in Transport and Tourism Research: Reformulating the Transport-Tourism Interface and Future Research Agendas," in *Tourism and Transport: Issues and Agenda for the New Millennium*, ed. Les Lumsdon and Stephen J. Page (Amsterdam: Elsevier, 2004), 11–12.

CHAPTER 6

The Physical Geography of Tourism: Resources and Barriers

Tourists routinely evaluate the physical geography of potential destinations. For most tourists, this is not something that is done scientifically, systematically, or even consciously; they simply want to know if a place has the physical setting they are looking for, if its physical conditions will provide them with the opportunity to participate in the activities they want, or if its conditions might keep them from doing those things. Some rely on stereotypes about places, but those tourists who do a little research into the physical geography of a place are able to make informed decisions about where to go as well as when. This type of research is going to become more important in the future with changing environmental conditions that will make it more difficult to rely on assumptions.

Likewise, tourism stakeholders also consider the physical geography of their destinations. In the planning stage, stakeholders must assess the physical resources that will provide the foundation for tourism in that place as well as the physical barriers to tourism. Then, they must devise strategies that will allow them to use resources sustainably and to manage the challenges presented by barriers. These stakeholders, consisting of various community members, business owners, government officials, and/or development workers, may not have the knowledge or expertise they need to do this on their own. The work of physical scientists and environmental consultants, many of whom come from a background in geography, can provide stakeholders with the information they need to develop a tourism destination, manage it, and plan for future environmental changes.

Because tourism is a human phenomenon, greater emphasis has been placed on examining tourism through the topical branches of human geography. Nonetheless, it is clearly important to consider the physical side of geography as well. Physical geography is the subdivision of geography that studies the earth's physical systems. As in human geography, physical geography is further organized into topical branches, such as meteorology and climatology, hydrology and oceanography, geomorphology, and biogeography. This chapter introduces each of these topical branches and examines how the elements in the earth's physical system *affect* (see box 1.3) tourism, either as a resource that provides the basis for tourism or as a barrier that prevents tourism. In addition, it considers how global environmental change is also affecting patterns of tourism. First, however, we will discuss the concept of resources as applied in the context of tourism.

Resources, Barriers, and the Tourism Resource Audit

In general terms, resources refer to some type of product that is perceived to have value and may be used to satisfy human needs and/or wants. Geographic research on resources recognizes that these products are relative and subjective. This means that what is considered a resource depends on the cultural, political, economic, and/or technological circumstances of a society at a given point in time.[1] Consequently, something that might be considered a resource for one group of people might not be for another due to different cultural values, political priorities, economic conditions, or levels of technology. Likewise, what is considered a resource in one time period might not be in another due to changes in all of these factors. While resources may be human or cultural, we typically think of physical or natural resources that are elements in the earth system. The availability of these resources is dependent on physical processes but also human efforts.

Applied to the context of tourism geography, **tourism resources** are those components of a destination's environment (physical or human) that have the potential to facilitate tourism or provide the basis for tourism attractions. Physical tourism resources are considered to be "an invaluable tourism asset and . . . fundamental to the development of tourism for virtually all destinations. They tend to be the foundation from which other resources are developed, and thus often play both a principal and key supporting role in tourism."[2] Moreover, tourism activities are contingent on not one but a combination of resources. These resources may be readily available tangible features in the geography of a place, but for many resources, destinations must still develop them to be used in tourism. This is based on the goals and values of the target tourist market to meet their demands and create that complementarity between places discussed in chapter 5.

Whereas the presence of resources can allow a destination to develop, the presence of barriers can prevent it. A barrier refers to something material in the environment that constitutes a physical impediment or something immaterial that creates a logistical or perceptual impediment. As with resources, what is considered a barrier—and the extent to which it functions as a one—varies with different cultural norms, political policies, economic circumstances, or technological advancement. Elements in the earth–ocean–atmosphere system can present distinct physical barriers, but they also have the potential to become perceptual barriers as well.

In tourism, both physical and perceptual barriers may prevent tourists from visiting certain destinations. Additionally, these barriers have the potential to shape the ways in which destinations develop. Thus, destination stakeholders need to evaluate its physical geography not only for potential resources but also for any barriers and to find ways of overcoming them—whether it is grading the landscape, installing artificial snowmakers, or convincing potential tourists that the weather is really not as bad as they think it is going to be.

Tourism stakeholders, especially those at emerging destinations, frequently fail to fully understand the conditions of their own resource base. With economic benefit as the goal, stakeholders may take shortcuts in the development process. They may choose to model their industry on that of a successful destination, even though circumstances are different for each place. They may conduct only a superficial analysis of the area's resources, or they may simply assume that they already have all of the information they need. Yet, it is hardly ever that simple. Some resources are attractions in themselves; these are the ones that are often easy to spot (e.g., Half Dome, Yosemite National Park). Others, however, simply provide the framework that allows for tourism. It can be much more difficult to understand how the quality, quantity, distribution, accessibility, seasonality, and so forth of these resources are going to affect tourism in that place.

The **tourism resource audit** (TRA) is a tool that can be used by destination stakeholders to systematically identify, classify, and assess all of the features of a place that will impact the supply of tourism. Because resources are subjective, however, this can be tricky. Typically, a range of stakeholders, coming from different perspectives, should be involved to create the most comprehensive and appropriate dataset. This will include experts to provide scientific data and analysis, community members to contribute local knowledge, industry analysts to assess market potential, and even tourists to offer the demand-side perspective. A variety of strategies can be used to create an exhaustive list of resources that are critically evaluated to understand how they might affect tourism. Geographic information systems (GIS) are used to manage the often large datasets created by a TRA. Analysis of this data allows stakeholders to determine the strengths and weaknesses of tourism at the destination, improvements that need to be made, and strategies that should be put in place for both immediate and long-term development.

Although this process is, perhaps, less exciting than other aspects of tourism development and promotion, it is fundamental. According to the authors of *The Tourism Development Handbook*, "The effort put in at this stage should be well rewarded later on with the development of a more successful and sustainable tourism destination."[3] Still, a TRA only captures the condition of resources at a given time. Resources, and what are considered resources, are not static. Consequently, the TRA database should be updated regularly, and tourism strategies reevaluated accordingly.

The Physical System, Physical Geography, and Tourism

A system is defined as an interrelated set of things that are linked by flows of energy and matter and are distinct from that which is outside the system. This is an important organizing concept in physical geography, as the earth is made up of interrelated physical systems, including the abiotic systems (i.e., the overlapping, nonliving systems consisting of the atmosphere, hydrosphere, and lithosphere) that provide the basis for the biotic system (i.e., the living system made up of the biosphere). Specifically, the atmosphere is the thin, gaseous layer surrounding the earth's surface. The hydrosphere encompasses the waters that exist in the atmosphere, on the earth's surface, and in the crust near the surface. The lithosphere includes the solid part of the earth. Finally, these three spheres form the basis for the biosphere, which is the area where living organisms can exist.

Each of these spheres can be studied through different but ultimately interrelated topical branches in geography, including meteorology and climatology (atmosphere), hydrology and oceanography (hydrosphere), geomorphology (lithosphere), and biogeography (biosphere). Table 6.1 provides a summary of the resources and barriers associated with each of these branches of physical geography, and the issues are discussed below. In the past, there have been fewer links between tourism and these topical branches of geography, in comparison with those on the human side of geography. Nonetheless, physical geography plays a crucial role in our understanding of the earth and our place in it, and geographers recognize that these physical systems have distinct impacts on all aspects of human life. Thus, there is clear potential for greater research connecting physical geography and tourism geography.

METEOROLOGY, CLIMATOLOGY, AND TOURISM

While it may seem like the atmosphere is beyond the scope of geography, it is still an integral part in the earth system. Not only do atmospheric processes affect what happens in other spheres, these phenomena also affect human life every day. Geographers are interested in both weather and climate to understand how patterns vary from place to place, how they shape those places, and how they affect human activities on the earth's surface. Meteorology and climatology are interrelated atmospheric sciences. **Meteorology** is the study of weather, which refers to the atmospheric conditions (e.g., air temperature and pressure, humidity, precipitation, wind speed and direction, cloud cover and type, etc.) for a given place and time. Because these conditions are dynamic, in an almost constant state of change, there is a distinct focus on short-term patterns. **Climatology** is the study of climate, which refers to the aggregate of weather conditions for a given place over time. Climatology expands upon meteorology by considering longer-term trends, making generalizations about average weather conditions, and identifying variations or extremes.

In one introduction to physical geography, the distinction between weather and climate is bluntly put in this way: the idea of a place's climate is what attracts people to that place, but it is the reality of day-to-day weather conditions that makes them leave.[4] While overly simplistic, this does raise an important consideration for the demand perspective in the geography of tourism. Tourists depend on information about the climate of a destination to try to make an informed decision about whether or not that place generally has the right conditions for the desired tourism activities at the time of year in which they intend to visit. Yet, climate data do not predict specific weather conditions. Forecasts become increasingly unreliable beyond just a few days, and most trips will be planned well in advance of that. Consequently, tourists may find that the actual weather conditions at the destination during their vacation are not what they expected. This can be simply an inconvenience or prompt small changes in their plans, but it can also fundamentally alter or even cancel a trip.

One tourism scholar notes: "It is generally accepted that climate is an important part of the region's tourism resource base, but the role of climate in determining the suitability of a region for tourism or outdoor recreation is often assumed to be self-evident and therefore to require no elaboration."[5] Another argues that tourism planning rarely considers anything more than "simple, general descriptions of the climate, which are often unconnected to the needs of tourism."[6] When we consider all of the ways in which weather and climate impact tourism, we should begin to realize that this cannot be taken for granted. While the intersection between the geography of tourism and climatology has been explored in greater depth than the other branches of physical geography, the literature still clearly argues for more work to be done.

Weather and Climate as a Resource and a Barrier for Tourism

It is said that weather and climate have a greater influence over what can and cannot be done in a given place than any other physical feature, and this applies to the development of tourism. These elements determine the time and length of the tourism season, the products that can be developed, the location of activities and infrastructure, and more. Generally speaking, climate is the feature that a destination is least able to manipulate to provide the desired conditions for tourism. There are exceptions; for example, winter sport destinations use snowmakers to ensure that tourists have the experience they came for, even though natural conditions (i.e., a day of sun with 5 to 10 inches of fresh powder) would still be preferred.

Whether it is nature-based or in an urban area, much of tourism takes place outside. As such, elements of weather and climate can be a resource that does not generate tourism but provides the conditions that allow for tourism activities to take place. Tourism is voluntary; tourists will only participate in an activity if the conditions allow it to be done safely and relatively comfortably. Consequently, there is an important correlation between weather and tourism revenues, either directly (e.g., financial losses due to poor or unexpected weather conditions) or indirectly (e.g., financial gains in secondary tourism activities that are less sensitive to the weather).

These elements of weather and climate can also be the resource on which tourism depends. Obviously, sun is a vital resource for sun, sea, and sand tourism. For these elements of the physical system, though, what is considered a resource for or a barrier to tourism is variable, depending on the activity and perceptions. This means that the same feature can, in fact, be both. For example, in the case of Tarifa, Spain, located between the popular 3S resorts of Costa del Sol and Costa de la Luz, the presence of high winds was a barrier to the development of sun, sea, and sand tourism. However, stakeholders turned this feature into a tourism resource by promoting the destination as the "capital of wind" and developing niche tourism activities like windsurfing.[7] Likewise, fog and mist might present a barrier to viewing scenic landscapes, but it could also add to the mystique of the place or increase the feeling of wonder as a scene suddenly presents itself to the viewer (figure 6.1)

Destinations seek to reassure potential tourists in target markets of their conditions, such as in the case of Barbados and their "perfect weather" (box 2.3). However, even destinations with notoriously poor weather conditions for tourism activities try to make the most of it. For example, Scotland's National Tourism Organization website, Visit Scotland, reads: "We've all heard plenty of jokes about the Scottish weather—but most of them aren't true! Scotland's climate is actually quite moderate and very changeable, although on occasion we get really hot or really cold weather. As the old Scottish saying goes, 'there's no such thing as bad weather, only the wrong clothes!' "[8] Moreover, rather than shying away from their bad weather, they make light of it while highlighting other attractions. The site playfully itemizes their reasons why rain is actually a good thing, from providing the foundation for the country's lush natural vegetation and waterfalls to creating opportunities for visiting museums and drinking whiskey. More than just attracting tourists, studies have also shown that warmer temperatures have the potential to increase tourism expenditures.[9]

Table 6.1. Summary of How Features in Each of the Topical Branches of Physical Geography Can Become Resources for or Barriers to Tourism

Branch	*Resources*	*Barriers*
Meteorology and climatology	**Attraction** In general, good weather conditions Perceptual depending on individual and cultural preferences and desired activities	**Detraction** Perceptual depending on individual and cultural preferences and desired activities
	Basis for activities Moderate temperatures Sun (e.g., sunbathing) Precipitation (e.g., skiing)/lack of precipitation (e.g., most outdoor activities) Wind (e.g., windsurfing)/lack of wind (e.g., swimming)	**Disrupt activities** Extreme temperatures Precipitation/lack of precipitation Wind/lack of wind Natural hazards (e.g., thunderstorms, hurricanes, blizzards)
Hydrology and oceanography	**Attraction** Unique water features (e.g., waterfalls, geysers) Specific characteristics (e.g., meandering rivers for floating, rapids for whitewater rafting and kayaking) Distinct properties (e.g., thermal or mineral springs for medical treatments) Foundation for attractive tourism landscapes (e.g., green golf courses, landscaped resorts, decorative fountains)	**Detraction** Perceptual (e.g., lack of available water to create attractive tourism landscapes) Physical (e.g., poor water quality)
	Basis for activities Swimming and bathing Boating and rafting Watersports Fishing	**Disrupt activities** Lack of available water to participate in tourism activities Health risks from poor water quality Natural hazards (e.g., flooding, tidal surges, tsunamis)
	Necessary quantity and quality Drinking and bathing Cooking and cleaning	

Branch	Resources	Barriers
Geomorphology	**Attraction** Unique landforms (e.g., islands, mountains, canyons, caves) Cultural values (e.g., sacred landscapes) Landform processes (e.g., erupting volcanoes)	**Detraction** Perceptual (e.g., cultural and personal perceptions of uninteresting or ugly landscapes)
	Location for resorts High-altitude summer retreats and health resorts	**Prevent accessibility** Physical (e.g., landforms that cut a destination off from major markets and/or make transportation difficult)
	Basis for activities Mountain hiking/climbing Winter sports	**Disrupt activities** Natural hazards (e.g., earthquakes, volcanic eruptions)
Biogeography	**Attraction** Distinct biomes (e.g., tropical rainforest, temperate rain forest, desert) Attractive vegetation (e.g., flowering plants, fall colors) Unique, rare, or endangered plant and animal species	**Detraction** Lack of expected vegetation (e.g., barren instead of lush) Deforested landscapes Diminished wildlife populations due to habitat loss, overhunting, and poaching
	Basis for activities Nature hikes, canopy tours Fruit picking, truffle hunting Bird watching, wildlife safaris	**Disrupt activities** Natural hazards (e.g., wildfires) Outbreaks of animal diseases (e.g., foot-and-mouth disease)

Figure 6.1. One travel writer describes the Castle of Lousã, set in the mountains of Portugal, as "veiled in mystery and myth."* The mountain mist creates a fitting setting for such a place. *Source:* Tom Nelson

Travel 2.0 presents a challenge to stakeholder efforts to maintain a positive reputation. In addition to permitting or preventing tourist activities, weather conditions affect tourists' comfort and mood.[10] The tourists who were wet, cold, and miserable while visiting a destination or participating in an activity will have negative associations with that experience. This may have nothing to do with the characteristics of the site the tourists visited or the quality of the activity in which they participated. Their memories of the experience will be tied to their emotions. If they choose to share their experiences in various online forums, their posts, reviews, or blogs will reflect these emotions. Other potential tourists who read these accounts may be deterred from the destination, even though they might have a different set of conditions and therefore a completely different experience.

Finally, extreme weather events such as hurricanes or blizzards present a barrier to tourism. As a perceptual barrier, tourists may avoid destinations when and where there is the potential for a hazard to occur (e.g., the low tourism season for destinations in the Caribbean and the Pacific corresponds to the hurricane season). As a physical barrier, these events have the potential to prevent tourists from reaching a destination or participating in the desired activities at a particular time. In addition, the damage and destruction caused by an extreme weather event has a long-term effect on the destination. It will face not only the cost of repairs but also the lost revenues while it is partially or completely closed to tourists. Additionally, the destination may have to work to recover those tourists who went elsewhere for the duration by advertising that they are open again or by offering discount specials.

* "The Castle of Lousã and the Lost Village," *Portugal Adventures*, May 31, 2015, accessed August 8, 2016, http://www.azores-adventures.com/2015/05/the-castle-of-lousã-.html

Climate Change Impacts

Over the past decade, the relationship between climate and tourism has gotten increased attention due to climate change. Part of this new research agenda considers the ways in which the tourism industry is contributing to climate change. This aspect will be discussed in chapter 10. Additional research considers the ways in which climate change is affecting patterns of tourism and will continue to do so in the future. The tourism industry is considered to be highly sensitive to changing climatic conditions, and some of the world's most popular tourism destinations are considered among the most vulnerable places (e.g., islands, other coastal areas, and mountains). Even protected UNESCO World Heritage Sites—ranging from Australia's Great Barrier Reef to the Glacier-Waterton International Peace Park on the border between the United States and Canada—are considered threatened by the effects of climate change.

These effects—direct and indirect—have the potential to dramatically reshape patterns of tourism. The direct effects of climate change may create new opportunities, new challenges, or simply changing conditions with little net gain or loss for destinations. For example, warmer temperatures may present an opportunity for places at higher latitudes or in higher elevations to develop or expand their summer tourism offerings. This could represent a shift in destination regions. As existing summer resorts in tropical and subtropical locations experience even hotter temperatures and heat waves that are uncomfortable at best and deadly at worst, they will become less desirable. At the same time, tourists from the significant tourist-generating regions in Northern Europe, North America, and parts of East Asia may find places closer to home more attractive.[11] This type of shift would have a significant economic impact on highly dependent tourism regions like the Caribbean and the Mediterranean.

While there are real implications for tourists' comfort and well-being associated with high temperatures, this may largely be a perceptual barrier based on media representations of climate change and heat waves. For instance, one much-cited article in the online edition of *The Guardian* refers to a study predicting that Mediterranean summers could become too hot for tourists after the year 2020.[12] However, there is little indication of what, exactly, is deemed "too hot" for tourists. In reality, thresholds for heat are both personal (i.e., what the individual is accustomed to and prefers) and contextual (i.e., what the same temperature feels like in different locations).[13]

Warmer temperatures could also lead to a shift in tourism seasons. The expansion of warm-weather tourism activities in traditionally cold-weather destinations may be offset by declines in winter tourism. Winter sport tourism is particularly vulnerable to changing climatic conditions. Many popular ski resorts, such as those in New England, are facing increased average temperatures, shorter winter seasons, and increasingly unreliable snowfalls. If these resorts are unable to supply snow artificially due to water or energy constraints, winter sport tourism may become unfeasible in the future. As stakeholders at these destinations refocus on summer offerings (e.g., hiking trails, mountain bike trails, zip lines, ropes courses, Frisbee golf courses, etc.), their efforts are to maintain tourism in that place as opposed to expanding it for greater benefit.

Climate change also has the potential to magnify existing barriers. Warmer temperatures are projected to contribute to an increase in the frequency and severity of extreme weather events. Destinations affected by these events will experience an increase in operating costs from higher insurance premiums and investments in emergency infrastructure such as backup water and power systems.[14] As tourism is disrupted more regularly, however, these destinations will also suffer from lost income, not only during and after each event but also over the long term as tourists begin to avoid places perceived to be risky.

HYDROLOGY, OCEANOGRAPHY, AND TOURISM

The hydrosphere includes the surface water in oceans, lakes, and rivers; subsurface water; frozen water; and even water vapor in the atmosphere. As a result, there is significant overlap between this sphere and the others. Broadly, **hydrology** is the science of water and considers the properties, distribution, and circulation of water in the hydrosphere. However, modern hydrology is specifically concerned with fresh water.[15] Fresh water is incredibly important in shaping human activities; consequently, the study of hydrology provides us with the means of understanding the availability of fresh water so that this fundamental resource can be appropriately managed to provide people with both the quality and quantity of water that they need. At its most basic, oceanography is the study of processes in oceans and seas and is therefore concerned with saline water. The global ocean is the most extensive feature of the hydrosphere. Covering 71 percent of the earth's surface, oceans make up approximately 97 percent of the earth's surface water.[16]

Water as a Resource and a Barrier for Tourism

Water is a tremendously significant resource for tourism. Combined with the environments surrounding it, this feature provides the basis for countless tourism attractions and activities around the world. Features such as waterfalls (figure 6.2) and geysers are often scenic attractions, while thermal and mineral springs have long provided the basis for health resorts. Rivers and lakes (both natural and artificial) allow for recreational activities, such as boating, fishing, rafting, kayaking, wildlife viewing, and more. Today, some of the most significant destinations are located in coastal areas. The beach, in particular, is considered to have a powerful appeal to the physical senses. For many societies, it is considered to be an aesthetically pleasing place that provides the potential for recreation from sunbathing to water sports.

Knowledge about these environments is important for stakeholders in the development and maintenance of a destination. The characteristics of a coast can shape the attractiveness of the area for tourism as well as its potential for tourist activities. Depositional coastlines characterized by beaches are common mass 3S destinations, while the more rugged erosional coastlines can be a resource for scenic tourism. White sand is often perceived to be the most desirable for beach tourism, although volcanic black sand can be found on beaches in the Caribbean and Hawaiian islands, and some popular beaches in the Mediterranean are composed of rocks and pebbles. The calm waters of sheltered coves may be an important resource for mass tourism but not for niche tourism based on adventure and sport. Stakeholders also need to be aware of the physical processes at work along coastlines that can affect these resources and other infrastructure. Tourism destinations may need to periodically undertake beach nourishment to artificially replace lost sand or improve sand quality. For example, nearly US$70 billion have been spent in efforts to maintain, rebuild, and replenish beaches

Figure 6.2. Waterfalls, such as the iconic Multnomah Falls in Oregon's Columbia River Gorge, have long been popular tourist attractions. *Source:* Scott Jeffcote

in Miami-Dade County, Florida.[17] Finally, stakeholders need information about these resources to provide a safe and suitable environment for tourism activities. Data about tides, currents, and waves should be used to identify the optimal times to participate in water sports (e.g., swimming, snorkeling, scuba diving, surfing, etc.) and to provide tourists with warnings about potentially hazardous conditions.

A lack of water—in terms of appropriate quality or quantity—can present a tangible barrier to tourism development; however, this can be overcome. Water is a necessary precondition for tourism because it is a fundamental human resource. A destination needs to ensure adequate levels of water quality for both tourism resources (e.g., quality of surface water for aesthetic purposes and tourism activities) and human resources (e.g., quality of water for drinking and bathing). At the same time, stakeholders must understand the constraints of water supply at the destination and consumption patterns to balance the needs of local economic activities, the resident population, and tourists. Las Vegas, a desert destination that received over forty-two million tourists in 2015,[18] is unable to provide enough water to support this demand from locally available surface and groundwater reservoirs and must import water with expensive diversion systems. Small island destinations (e.g., Curaçao, Cyprus, and Mauritius), as well as dry coastal destinations (e.g., Australia, Dubai, southern California), are increasingly looking to desalination of seawater (box 6.1) to meet their needs.

Box 6.1. Case Study: Desalination to Meet Increased Water Demand in Egypt's Red Sea Resorts

Over the past few decades, Egypt experienced tremendous population and economic growth. This put significant pressure on its water resources, and the country is now facing increasing problems with water scarcity. This is especially an issue for the Red Sea coast and the Sinai Peninsula, where there are extremely limited fresh water resources, and the distance from the Nile River makes the cost of transporting potable water via pipeline prohibitive.[*] The peninsula's traditional population was nomadic Bedouin tribes, who migrated based on the availability of water and pastures. The remainder of the historically small population was concentrated in the northern part of the peninsula where there is greater access to fresh water. However, the development of tourism over the past few decades brought many changes to the peninsula.

Following the return of the Sinai Peninsula to Egypt in 1982, the southern and coastal regions were targeted for economic growth and development, fueled by foreign investment. These areas were considered to have a good tourism resource base, including a consistently warm and dry climate year-round, beaches, and renowned coral reefs for scuba diving and snorkeling. While the lack of fresh water presented the greatest barrier to tourism development, this was something that the destination was able to overcome. In particular, desalination was identified as the most appropriate means of meeting the need for water to accommodate tourism.[†]

Between 1980 and 2001, the number of tourist resorts in this area grew from 50 to 630.[‡] At the southern tip of the Peninsula, Sharm El Sheikh developed into one of Egypt's most popular destinations. It expanded beyond niche tourism based on scuba diving into one of the world's significant mass tourism destinations. Hotel construction continued until the destination was nearly at capacity, which prompted the development of other resorts in the region. At its peak, prior to the Arab Uprisings, Sharm El Sheikh was receiving over three million tourists annually, with little seasonal fluctuation. The city's resident population also grew, as more people were attracted to the area by the influx of jobs in tourism and related industries. This growth put ever-greater pressure on limited fresh water resources. These demands quickly outstripped the state-owned facilities' ability to supply enough water. Private companies also began to build desalination facilities in the region and sell water to resorts at a relatively high cost.

Despite the vital need for desalination to meet the water needs of this region, there are some concerns about the process and its outcomes. Some studies have shown that the desalination process is highly efficient at removing the salts from the water. However, the process does not necessarily produce bacteriologically safe water for drinking, which can create a health risk for international tourists.[§] Other studies have focused on the environmental effects of the process. Desalination yields approximately 30 percent drinking water and 70 percent brine, which contains all of the salt. This brine cannot be discharged back into the sea because

[*] M. Shehata, M. Mahgoub, and R. Hinklemann, "High Resolution 3D Model of Desalination Brine Spreading: Test Cases and Field Case El-Gouna, Egypt," *E-Proceedings of the 36th IAHR World Congress*, June 28-July 3 (2015): 1.

[†] Magdy Abou Rayan, Berge Djebedjian, and Ibrahim Khaled, "Water Supply and Demand and a Desalination Option for Sinai, Egypt," *Desalination* 136 (2001): 81.

[‡] Shehata, Mahgoub, and Hinklemann, "High Resolution 3D Model," 1.

[§] Atef M. Diab, "Bacteriological Studies on the Potability, Efficacy, and EIA of Desalination Operations at Sharm El-Sheikh Region, Egypt," *Egyptian Journal of Biology* 3 (2001): 63.

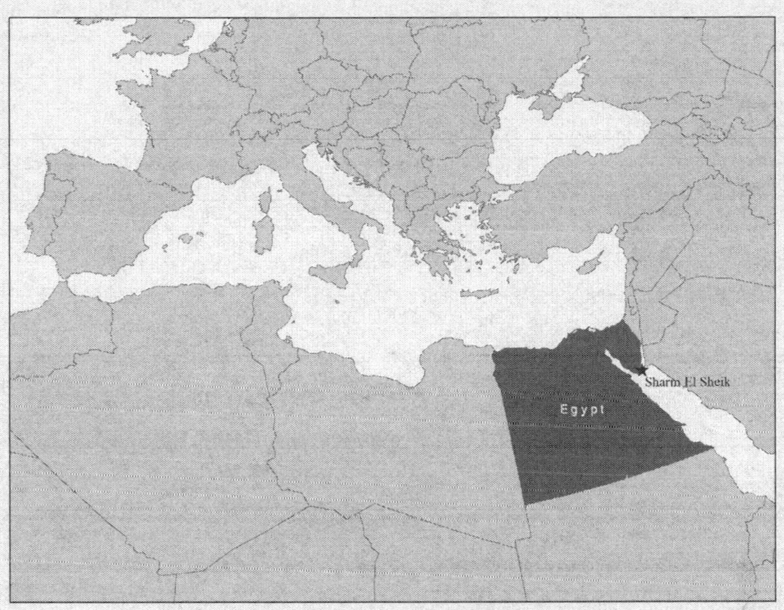

Map 6.1. Sharm El Sheikh, Egypt. For this destination located on the Red Sea, water is both a resource for, and a barrier to, tourism. *Source:* Gang Gong

of the effect it would have on the highly diverse coral reef ecosystem that is one of the most significant tourism resources for the destination. Instead, the brine is discharged back into the aquifer by injection wells, which increase the groundwater salinity over time.** This causes a decrease in the efficiency of desalination plants and an increase in costs.

The significant amount of energy required by the process has been another key concern for Red Sea resorts. Isolated resorts and communities may not have sufficient energy capacity to support desalination, and rising energy costs will make the process more expensive. Recent research has investigated the potential for small-scale desalination systems powered by renewable energy sources to provide fresh water to these areas without increasing local air pollution or contributing to global climate change.[††]

Amidst heightened security concerns in the northern Sinai with the activities of extremist groups linked to the Islamic State of Iraq and the Levant (ISIL), travel to Sharm El Sheikh has dropped significantly. The southern resort weathered a series of terrorist attacks in 2005. Tourists will likely return as conditions stabilize. As they do, the lack of fresh water resources will continue to be an issue. While Sharm El Sheikh is certainly not the only destination to face this problem, it does present an interesting case study because tourism has been the primary driver of economic development and population growth in this region.

Discussion topic: Give another example of a physical feature that can be both a resource for and a barrier to tourism in a specific place. How does one weigh against the other?

Tourism on the web: Egypt Tourism, "Sharm El Sheikh: The Classic Red Sea Destination," at http://en.egypt.travel/city/index/sharm-el-sheikh

---

** Ramadan A. Awwad, T.N. Olsthoorn, Y. Zhou, Stefan Uhlenbrook, and Ebel Smidt. "Optimum Pumping-Injection System for Saline Groundwater Desalination in Sharm El Sheikh," *WaterMill Working Paper Series* 11 (2008): 8.

[††] Faten Hosney Fahmy, Ninet Mohamed Ahmed, and Hanaa Mohamed Farghally, "Optimization of Renewable Energy Power System for Small Scale Brackish Reverse Osmosis Desalination Unit and a Tourism Motel in Egypt," *Smart Grid and Renewable Energy* 3 (2012): 43–4.

Water also presents a barrier to tourism in the form of hazards, although this is tied to meteorological or geomorphic hazards. For example, coastal destinations are affected by tidal surges caused by hurricanes or tsunamis as a result of earthquakes. One of the most devastating disasters in recent times was the 2004 Indian Ocean tsunami that killed approximately 230,000 people across fourteen countries, including an estimated 9,000 tourists enjoying a beach vacation. In the case of Sri Lanka, this disaster was estimated to have cost the tourism industry US$250 million and 27,000 tourism-related jobs. Across the region, countless small and medium-size tourism businesses did not have insurance that would allow them to rebuild.[19] In addition, many tourists stayed away from these popular beach destinations out of fear due to the traumatic and highly publicized nature of the event. Interestingly, the tsunami that struck the northeast Japan in 2011, and the resulting emergency at the Fukushima nuclear power plant, had less of an impact on the country's tourism industry than might have been expected. Immediately after the event, Japan registered a 62 percent drop in international tourist arrivals, but the World Travel & Tourism Council determined that the country's tourism industry had recovered by the end of the year.[20]

Climate Change Impacts

The impacts of climate change go beyond the direct effects on patterns of weather and climate. Climate change will affect other physical resources for tourism and exacerbate conditions that constitute barriers. Unlike the potential opportunities created by the direct effects of climate change, these indirect effects of climate-induced environmental change are anticipated to create significant challenges for destinations.[21]

Sea level rise poses a threat to various coastal tourism resources. Destinations are projected to see increased beach and coastal erosion as well as damage to, or destruction of, coastal infrastructure and beachfront real estate. Stakeholders at such destinations will have to make investments in projects to both minimize and repair the damage to maintain its attractiveness to tourists and investors. These destinations may experience a heightened risk of coastal flooding and potentially the intrusion of saline water into fresh water aquifers.[22] This can negatively impact a destination's ability to meet local and tourist water requirements. In addition, tropical destinations are projected to experience increased problems with coral bleaching, which will lead to the loss of distinctive marine ecosystems and a decrease in tourism demand for the experience of these ecosystems. In the worst-case scenarios, sea-level rise could submerge low-lying coastal areas and islands (figure 6.3).

Additionally, dry regions are projected to experience longer periods of drought, and desert and semi-desert areas are projected to expand. This has the potential to create new, or exacerbate existing, problems of water shortages. This directly affects the tourism industry in terms of water availability, and the impacts are magnified by the fact that the high tourist season often corresponds with the dry season. It also affects related industries, such as agriculture, which provides not only consumable products but also experiences for tourists (e.g., agritourism). As competition for scarce water resources intensifies, this has the potential to generate resentment and hostility toward the tourism industry.[23] At the same time, temperate regions are projected to experience wetter conditions. These conditions could allow an increase in the risk and geographic

Figure 6.3. Sea-level rise is threatening tourism resources and infrastructure at low-lying destinations such as the Maldives in the Indian Ocean. *Source:* Nick Wise

spread of vector borne infectious diseases, such as malaria, West Nile virus, and Zika virus. More tourists may need to seek the advice of medical professionals regarding their risks associated with travel to these areas and use prophylactic pharmaceuticals. Some tourists may choose not to travel to these places altogether.

GEOMORPHOLOGY AND TOURISM

Geomorphology is the study of landforms, which refers to the shapes of the earth's surface. This includes identifiable forms such as mountains and hills, valleys and plains. In particular, the study of geomorphology considers the characteristics and spatial distribution patterns of landforms, as well as the internal and external geographic processes that create and shape them. Landforms are changing constantly as a result of a variety of forces in the atmosphere, hydrosphere, and biosphere that are continuously at work on the surface of the lithosphere. The internal forces are generally constructive in nature, meaning that they increase the relief (the changes in elevation and slope) of the earth's surface, while the external forces are more likely to be destructive in that they wear features down and decrease the relief of the surface. Over time, the action of these forces has created the landforms that provide the basis of destinations we know today—for example, the Hawaiian Islands (formed by a hotspot or magma plume), the Alps (formed by compressional folding and faulting), or Arches National Park (formed by erosion).

Landforms as a Resource and a Barrier for Tourism

As a resource, landforms and landform processes can be a natural tourism attraction (box 6.2). One of the most well-known examples of a landform-based natural attraction is the Grand Canyon, but others include Uluru (also known as Ayers Rock) in Australia, the Rock of Gibraltar on the Iberian Peninsula, or the fairy chimney rocks at Göreme, Turkey. Cultural values are often attributed to these landforms, and they are visited for that reason. For example, Mount Emei is one of the Four Sacred Buddhist Mountains in China. Places where we can safely see the physical (internal) forces of landscape formation at work can also become tourism attractions. Arenal Volcano became one of Costa Rica's most popular sites, where tourists witnessed the almost constant effusive eruptions—with ash plumes and lava flow—that occurred up until 2010.[24] Likewise, landforms can also be a tourism resource by providing opportunities for tourism. Mountain resorts around the world are extraordinarily popular destinations for both the scenery (figure 6.4) and activities such as hiking, climbing, and winter sports.

Figure 6.4. Set against the dramatic landscape of the Teton Range, Jackson Hole, Wyoming, consistently ranks among the best mountain resorts in North America. *Source:* Velvet Nelson

Box 6.2. Experience: A Physical Geographer's Travels

Travel is an important part of geography. We study the world, but ultimately we have to experience it for ourselves. When we go to other places, we get to compare our knowledge of various patterns and processes with what is before our eyes and under our feet. Christina has been fortunate to participate in a number of field courses during her studies. She describes her experiences below.

As a physical geography student, the physical features of the landscape have played an important role in my travels. In Ainsa, Spain, I was able to see the vivid thrust sheets of the Pyrenees mountain range, and textbook examples of synclines and anticlines. With just a short walk up the hill, we explored and dined in the medieval citadel. With cobble stone roads, historic churches, and a castle, the place truly is a living museum. On the Big Island of Hawaii, I was able to witness island-forming volcanism. In Big Bend National Park in southwestern Texas near the Mexican border, I was able to explore the barren landscape that exposes multiple geologic eras as well as processes. Reading the landscape as a physical geographer can bring about a better understanding of a place, including its current and past environment. Discovering fossils or seeing an active volcano demonstrates time and space at a scale that cannot be obtained in a classroom setting for students. In my opinion, geography brings the physical environment's past and current culture together to create a particular place. Place allows geographers to see the interconnections between the physical and human environment.

In my first semester of graduate school, I took a regional field studies course in which we spent spring break in Big Bend National Park. The assignment was to keep a journal and expand on a specific topic of the Chihuahuan Desert. I chose El Solitario, which is actually located in Big Bend State Park Ranch. El Solitario is an impressive geologic structure. It is a dome that has collapsed and eroded to show one of the longest continuous sequences of geologic time. These rock formations describe the past climate history of this place, thus, telling us a story. The current "page" suggests that the climate is very hot and dry; therefore, this place cannot sustain. The residents of the Big Bend National Park region have a deep connection with the landscape that is expressive of each love, loss, and isolation. Terlingua is known for being a "ghost town" because of its small population and famous cemetery that serves as the resting place for many miners. The main attraction in town that helps to define this place is "The Porch." Locals come to play music while tourists watch in awe of the individuality of the people and situation. It was here I saw a man dressed as a pirate, and he definitely believed he was one!

After a week in the desert, our trip came to an end. We watched the people who called the Chihuahuan Desert home sing and dance, all the while the sun was setting to create the illusion of multiple purple and orange horizons against the mountains.

Throughout my experience, traveling creates an experience imperative to understanding and appreciating the physical landscape. As a graduate student now teaching geology labs, I have added a field trip to a local state park as part of the course. The majority of my students were unaware of the physical processes that formed their local environment. Now, hopefully, some students can create their own personal field experiences with a little more knowledge of their physical environment. For me, a trip into the field is where the learning (and fun) actually begins!

—*Christina*

Landforms can also present a barrier to tourism, primarily by preventing people from reaching the place of destination. In particular, mountain destinations around the world have long had to manage accessibility issues. It can be a difficult and costly process to extend ground transportation lines to a resort or to construct the necessary infrastructure for air transport. In one example, Yeager Airport in the Appalachian Mountains of West Virginia was built on a hilltop in the 1940s. The project cost US$8.3 million and required two million pounds of explosives to move nine million cubic yards of earth to create a large enough area of level land for a runway.[25] At the same time, natural hazards caused by the dynamic processes of landscape formation have the potential to damage the tourism infrastructure and disrupt tourism activities. At a local scale, tourists were evacuated from El Hierro, one of the Canary Islands, due to the threat of a volcanic eruption in mid-2011. Globally, the massive ash cloud from the eruption of Iceland's Eyjafjallajökull volcano in 2010 had a ripple effect, disrupting travel and tourism around the world.

Tourism stakeholders will benefit from the expert knowledge of geomorphologists during the TRA not merely to identify the potential features of interest but also to provide guidance on how these features can be developed for tourism. At the same time, scientific information about landform processes should be used to help the destination create an effective disaster response plan.

Climate Change Impacts

Tourism depends on a high-quality landscape and natural attractions. Indirect climate-induced environmental changes have the potential to undermine this. For example, glaciers contribute to an overall landscape aesthetic and constitute an attraction for millions of tourists around the world. However, warming temperatures have accelerated glacial retreat. This has the potential to change the appearance of mountain scenery and to reshape landforms. For example, retreating glaciers may leave behind moraines or accumulations of rock and soil that were pushed along by the glacier as it moved. In addition, this process could increase the risk of natural hazards. Slopes exposed by glacial retreat may be unstable and prone to landslides, especially with a trigger event such as an earthquake. There may be an initial increase in melt-water, but over time, retreating glaciers provide less water to the rivers downstream. This has further implications for the species, ecosystems, people, and economic activities that depend on the water from those rivers.

BIOGEOGRAPHY AND TOURISM

Biogeography, combining principles from both biology and geography, is the study of living things. Alexander von Humboldt, a leading figure in the development of modern geography, is widely considered to be the founder of biogeography. This topical branch considers the spatial patterns and physical processes of these living things in the collection of ecosystems contained within the earth's biosphere. Biogeographers are interested in the extent of diversity among the earth's species, broadly described by the term biological diversity (biodiversity). Moreover, biogeographers are concerned with explaining the changes in these patterns and processes that have taken place over time and understanding the impact of human activities on the diverse species and their habitats.

Biodiversity as a Resource and a Barrier for Tourism

The biogeography of a place is primarily considered a resource for tourism. For example, the presence of unique animal species and/or plant species becomes a key resource for products such as nature and wildlife tourism. Tourists to southern Africa are interested in viewing the "Big Five" game animals in their natural habitat: the African elephant (figure 6.5), Cape buffalo, leopard, lion, and rhinoceros. Tourists from the temperate zones have long been interested in the tropical rain forest biome.

These characteristics of physical geography can, in some cases, present a barrier to tourism. This is largely based on perception. Consider destinations that lack the "right" kind of vegetation—or the vegetation that tourists expect. Tropical island destinations around the world have been subject to Western perceptions of an island paradise

Figure 6.5. Wildlife is a vital resource for tourism at many African destinations. Tourists look for the "Big Five" game animals, such as this elephant on Lake Kariba, located on the border between Zimbabwe and Zambia. *Source:* Velvet Nelson

(i.e., lush, green environments typically epitomized by palm trees). Of course, not all of these islands exhibit these patterns. The comparatively flat "ABC" islands in the southern Caribbean (i.e., Aruba, Bonaire, and Curaçao) are characterized by desert scrub and cactus vegetation. To some extent, this barrier can be overcome as destinations, and particularly resorts, artificially plant nonnative trees and flowers in an effort to create the perceived desired appearance. However, this is often an unsustainable solution due to limited water resources in such environments.

Tourism stakeholders will benefit from the knowledge generated by biogeographers about the factors that contribute to the success or failure of a particular species so that they may adequately protect these resources. In particular, research in biogeography has examined the potential for tourism to be used as a tool in environmental preservation with the goal of preventing habitat and/or species loss. Much of this work has focused on the ecotourism concept introduced in chapter 3. In addition, biogeographers have been instrumental in studying the effects of tourism on ecosystems. This will be discussed further in chapter 10.

Climate Change Impacts

Loss of biodiversity is one of the key concerns associated with indirect climate-induced environmental change. Rising temperatures and shifting precipitation patterns will affect ecosystems. Plants and animals will have to adapt or migrate, and those that are unable to do so will become extinct. This may present a short-term opportunity for some nature-based destinations, as tourists travel with the intention of seeing endangered species in their native habitat while they still can (see last chance tourism below). If species decline and loss cannot be prevented, the primary reason for tourism in that place may be eliminated. Species of pests may also migrate based on changing conditions. For example, infestations of nonnative insect species can cause further damage to ecosystems. The expansion of dry areas and longer droughts also has the potential to increase the risk of wildfire. This will impact landscape (e.g., forests) and infrastructure, both of which can be detrimental to the tourism industry.

Climate Change and Tourism

In addition to the direct and indirect impacts discussed above, climate change will have broader implications for the sustainability and growth of tourism in the future. As we discussed in chapter 2, changes in one part of the tourism system will have ramifications on others. As destinations experience various impacts, revenues from tourism may become more unreliable, while operating costs are likely to increase. Depending on the type of changes experienced, a destination may have to rely more heavily on cooling systems for a longer period of the year, use snowmaking equipment more often, use irrigation and watering systems, and/or pay higher hazard insurance premiums. If these costs are passed on to tourists, travelers will have to reevaluate their decisions, not only about where they go but also whether they go. Those tourists who are able to continue to travel may be forced to choose different destinations for their desired activities or to participate at different times of the year.

Climate change is predicted to have a negative impact on global economic growth by the middle of this century. As such, we could expect to see a decrease in discretionary income. In addition, reduced access to vital natural resources has the potential to increase political instability and conflict. These factors would contribute to a decrease in effective demand for tourism.[26] This reduced demand will have a negative impact on people in tourism-dependent places. At the same time, these places are also likely to be confronting other challenges associated with climate-induced environmental change. Many areas projected to experience the greatest changes are the most dependent on tourism. Five regions of the world have been identified as tourism vulnerability hotspots: the Caribbean, the Mediterranean, the Indian Ocean Small Island Nations, the Pacific Ocean Small Island Nations, and Australia/New Zealand.[27]

In light of media representations of environmental changes occurring in these places, and others, there have been growing concerns that such places might be fundamentally changed or destroyed altogether. To some extent, this has been viewed as an opportunity in tourism. Playing upon concerns, tourism stakeholders have begun to encourage tourists—at least those who have the means—to see such places before they are "gone." This is sometimes referred to as disappearing tourism, doom tourism, or **last chance tourism**.[28] Last chance tourism provides individuals with the opportunity to see a particular place, geographic feature, or species in its natural habitat while they still can. It also allows them to witness the changes that are taking place and, ultimately the end, firsthand. For some tourists, last chance tourism is a manifestation of their genuine interest in the specific resource and concern for its impending demise. Destinations may choose to promote their vulnerability as a means of generating attention and aid in efforts to protect their vanishing resources. In fact, there are positive examples in which tourism has contributed to the recovery of environments or species. Yet, for other tourists, this is considered to be an expression of egocentrism. Destinations may capitalize on these tourists' desire for exclusivity, and their willingness to pay for the privilege of rarity.

Nearly all of the types of resources discussed in the sections above have become the focus of last chance tourism in various parts of the world. Although tourists have long visited the UNESCO World Heritage Sites identified earlier, there is a new imperative to scuba dive on the Great Barrier Reef while it is still one of the world's most biodiverse ecosystems and to hike in Glacier-Waterton International Peace Park while there are still glaciers. Tourists are interested in skiing historic resorts under natural conditions and seeing endangered wildlife in their natural habitat, whether it is polar bears in Canada or mountain gorillas in Rwanda. They are increasingly visiting the Arctic and Antarctic regions for the experience of sea ice before it melts and possibly even to witness the drama of a calving glacier. They want to have the opportunity to sit on the beaches of small island destinations like Tuvalu or the Maldives before they are submerged. That such tourists might ultimately be contributing to the demise of these destinations will be discussed in chapter 10.

Box 6.3. In-Depth: Climate Change Adaptation in Tourism

Climate change is projected to have widespread effects on peoples and places. Because tourism is dependent on weather and other physical resources, it is considered to be highly sensitive to these effects.* As tourism stakeholders at destinations all over the world will be affected, all need to adapt to changes. **Climate change adaptation** refers to the technological, economic, and sociocultural changes intended to minimize the risks and capitalize on the opportunities created by climate change. Of all tourism stakeholders, tourists generally have the greatest ability to adapt because, at least in theory, they have the ability to choose when and where they travel. Local community members and business operators have the least ability to adapt because they must work within the constraints of the destination's resource base.†

The tourism industry has proven to be relatively adaptive to various circumstances, including natural disasters, disease epidemics, financial crises, and political crises. Thus, the industry has the potential to adapt to the effects of climate change as well. Studies show that awareness of climate change impacts has been highest among stakeholders at the most vulnerable locations, the ones that have already experienced changing conditions. For example, stakeholders at winter sports destinations are not only aware of their risks but have taken a relatively proactive approach to adaptation. Strategies to maintain their viability as a tourist destination have involved both technological (e.g., enhanced snowmaking equipment) and economic changes (e.g., diversifying activities).‡ While research indicates that the majority of tourism stakeholders have had low levels of awareness and concern for the implications of climate change, this is starting to change.§

Stakeholders in some destinations may have the luxury of considering the new opportunities created by changing conditions. As the majority of destinations will face new challenges, it is important to begin the process of identifying their risks and developing a plan for adaptation. The first step is to engage a wide range of stakeholders at all applicable scales and in all relevant industries (e.g., stakeholders in industries that might be affected such as transportation). This not only capitalizes on a diverse set of knowledge and experience but also increases the effectiveness of the adaptation process. The second step is to identify the current and potential risks to tourism resources or infrastructure or changes to the industry. The third step is to assess adaptive capacity. This might include available technologies, resources (e.g., natural, human, financial), institutional support, and previous adaptive experience (e.g., developing and/or implementing a disaster response plan).

In the fourth and fifth steps, stakeholders should identify, evaluate, and select adaptation options. These steps should combine scientific knowledge with local expertise. To determine whether a strategy is appropriate, stakeholders need to consider the following questions: How much does the strategy cost to implement and maintain? How easily can this strategy be implemented? What is the time frame for implementation and outcomes? Do we have the skills needed to implement this strategy or will we need education and training? Will it be effective in solving our problems? Who will benefit? Will there be any negative social or environmental impacts? Are we willing to accept it? The sixth step is to implement the chosen course of action, and the final step is to monitor and evaluate the strategies used.**

---

* Andrew Holden, *Environment and Tourism*, 2nd ed. (London: Routledge, 2008), 215.

† Daniel Scott, "Climate Change Implications for Tourism," in *The Wiley Blackwell Companion to Tourism*, eds. Alan A. Lew, C. Michael Hall, and Allan M. Williams (Malden: Wiley Blackwell, 2014), 474.

‡ Daniel Scott and Christopher Lemieux, "The Vulnerability of Tourism to Climate Change," in *The Routledge Handbook of Tourism and the Environment*, eds. Andrew Holden and David Fennell (London: Routledge, 2013), 247.

§ Scott, "Climate Change Impacts for Tourism," 474.

** Murray C. Simpson, Stefan Gossling, Daniel Scott, C. Michael Hall, and Elizabeth Gladin, *Climate Change Adaptation* and Mitigation in the Tourism Sector: Frameworks, Tools and Practices (Paris: UNEP, University of Oxford, UNWTO and WMO, 2008), 35–45.

There are many challenges to adaptation. While the need for it is widely recognized, local tourism stakeholders have received relatively little support from the scientific community, government organizations, or development agencies.[††] These stakeholders would benefit from technical expertise and financial assistance to help understand their risks and develop appropriate adaptation strategies that will minimize risks without affecting the quality of their tourism resources.[‡‡] When adaptation is prioritized, it is integrated into national strategic planning processes; however, implementation is done at the local scale and must consider local conditions.[§§] This means that a nationwide course of action may not be appropriate for all places, and one place cannot simply replicate what was successful in another. Additionally, adaptation strategies cannot be based on past evidence because conditions are changing. Instead, they are based on scenarios and projections rather than actual impacts. This makes long-term planning difficult, and any strategies implemented now must be flexible to adjust for changes as they occur.[***]

Stakeholders at the most vulnerable destinations are already experiencing impacts and are looking for ways to adapt before it is too late. For example, visit the link below to read about a program to promote climate change adaptation through tourism in the Maldives. Projects to coordinate adaptation efforts, and research on these projects, have been on the rise, but more work needs to be done in both regards to promote more systematic adaptation efforts in the tourism system in the future.

Discussion topic: What arguments would you use to convince a stakeholder to participate in adaptation for future climate change impacts?

Tourism on the web: United Nations Development Programme, "Increasing Climate Change Resilience of Maldives through Adaptation in the Tourism Sector," http://www.mv.undp.org/content/maldives/en/home/operations/projects/environment_and_energy/tourism adaptation-to-climate-change/

[††] Scott, "Climate Change Impacts for Tourism," 474.
[‡‡] Scott and Lemieux, "The Vulnerability of Tourism," 250.
[§§] Simpson, Gossling, Scott, Hall, and Gladin, *Climate Change Adaptation*, 34.
[***] Wolfgang Strasdas, "Ecotourism and the Challenge of Climate Change: Vulnerability, Responsibility, and Mitigation Strategies," in *Sustainable Tourism & the Millennium Development Goals: Effecting Positive Change*, ed. Kelly S. Bricker, Rosemary Black, and Stuart Cottrell (Burlington: Jones & Bartlett Learning, 2013), 212.

Conclusion

As a place-based phenomenon, tourism is shaped by and to some extent dependent on the earth's physical features and processes. These things can be either a resource that allows for tourism to take place or a barrier that prevents it. The factors that determine whether something is a resource or a barrier vary between places, societies, and even periods of time depending on the particular circumstances, perceptions, and perhaps level of technology. The topical branches of physical geography provide the means of examining the earth's physical systems across the atmosphere, hydrosphere, lithosphere, and biosphere. The knowledge generated by meteorology, climatology, hydrology, oceanography, geomorphology, and biogeography can be used to better understand how elements in the physical system affect patterns of tourism. This knowledge will become even more important in the future in light of global environmental change.

Key Terms

- biogeography
- climate change adaptation
- climatology
- geomorphology
- hydrology
- last chance tourism
- meteorology
- tourism resource
- tourism resource audit

Notes

1. Derek Gregory, Ron Johnston, and Geraldine Pratt. *Dictionary of Human Geography*, 5th ed. (Hoboken, NJ: Wiley-Blackwell, 2009), 649.

2. Kerry Godfrey and Jackie Clarke, *The Tourism Development Handbook: A Practical Approach to Planning and Marketing* (London: Cassell, 2000), 66.

3. Godfrey and Clarke, *The Tourism Development Handbook*, 72.

4. Tom McKnight and Darrel Hess, *Physical Geography: A Landscape Appreciation* (Upper Saddle River, NJ: Prentice Hall, 2000), 67.

5. C.R. De Freitas, "Tourism Climatology: Evaluating Environmental Information for Decision Making and Business Planning in the Recreation and Tourism Sector," *International Journal of Biometeorology* 48 (2003): 45.

6. Gómez Martín, Ma. Belén Martín, "Weather, Climate, and Tourism: A Geographical Perspective," *Annals of Tourism Research* 32, no. 3 (2005): 587.

7. Gómez Martín, "Weather, Climate, and Tourism," 576.

8. Visit Scotland, "Climate & Weather in Scotland," accessed July 27, 2016, https://www.visitscotland.com/about/practical-information/weather/

9. Carey Goh, "Exploring Impact of Climate on Tourism Demand." *Annals of Tourism Research* 39, no. 4 (2012): 1869.

10. Jelmer H.G. Jeuring and Karin B.M. Peters, "The Influence of the Weather on Tourist Experiences: Analysing Travel Blog Narratives," *Journal of Vacation Marketing* 19, no. 3 (2013): 214–15.

11. Daniel Scott, "Climate Change Implications for Tourism," in *The Wiley Blackwell Companion to Tourism*, ed. Alan A. Lew, C. Michael Hall, and Allan M. Williams (Malden, MA: Wiley Blackwell, 2014), 469.

12. "Climate Change Could Bring Tourists to UK—Report," *The Guardian*, July 28, 2006, accessed July 27, 2016, https://www.theguardian.com/travel/2006/jul/28/travelnews.uknews.climatechange

13. Daniel Scott and Christopher Lemieux, "The Vulnerability of Tourism to Climate Change," in *The Routledge Handbook of Tourism and the Environment*, ed. Andrew Holden and David Fennell (London: Routledge, 2013), 245–6.

14. Scott, "Climate Change Implications," 469.

15. Tim Davie, *Fundamentals of Hydrology*, 2nd ed. (London: Routledge, 2002), xvii.

16. Steve Kershaw, *Oceanography: An Earth Science Perspective* (Cheltenham, UK: Stanley Thornes, 2000), 5, 17.

17. Robert W. Christopherson, *Geosystems: An Introduction to Physical Geography*, 7th ed. (Upper Saddle River, NJ: Pearson Prentice Hall, 2009), 516.

18. Las Vegas Convention and Visitors Authority, "2015 Las Vegas Year-to-Date Executive Summary," accessed July 27, 2016, http://www.lvcva.com/includes/content/images/media/docs/ES-YTD-2015.pdf

19. Andrew Holden, *Environment and Tourism*, 2nd ed. (London: Routledge, 2008), 222.

20. World Travel & Tourism Council, "The Tohoku Pacific Earthquake and Tsunami," accessed July 27, 2016, http://www.wttc.org/-/media/files/reports/special-and-periodic-reports/japan_report_march_update_v7.ashx

21. Scott, "Climate Change Implications," 469.

22. Holden, *Environment and Tourism*, 222.

23. Peter Burns and Lyn Bibbings, "Climate Change and Tourism," in *The Routledge Handbook of Tourism and the Environment*, ed. Andrew Holden and David Fennell (London: Routledge, 2013), 414–15.

24. "Arenal Volcano Costa Rica," accessed July 28, 2016, http://www.arenal.net

25. Central West Virginia Regional Airport Authority, "Yeager Airport History," accessed July 28, 2016, http://www2.yeagerairport.com/history/

26. Scott, "Climate Change Implications," 469–71.

27. Murray C. Simpson, Stefan Gossling, Daniel Scott, C. Michael Hall, and Elizabeth Gladin, *Climate Change Adaptation and Mitigation in the Tourism Sector: Frameworks, Tools and Practices* (Paris: UNEP, University of Oxford, UNWTO and WMO, 2008), 14.

28. Raynald Harvey Lemelin, Emma Stewart, and Jackie Dawson, "An Introduction to Last Chance Tourism," in *Last Chance Tourism: Adapting Tourism Opportunities in a Changing World*, ed. Raynald Harvey Lemelin, Jackie Dawson, and Emma J. Stewart (London, Routledge, 2012).

CHAPTER 7

The Human Geography of Tourism: Resources and Barriers

When we consider potential vacation destinations, our assessment of the human geography of those places is probably more instinctual than that of the physical geography. If we dislike the cultural characteristics of a particular group of people, we will simply avoid those destinations where we are most likely to encounter them. If we find big, crowded cities overwhelming and stressful to navigate or if we find the rural countryside boring and uneventful, we will not consider such places in our destination search. If a country is going through a bloody civil war, it would be unlikely to enter our thought processes to consider it a vacation spot. From the demand perspective, clearly the human characteristics of a place play an important role in shaping what we want or expect from the destinations we visit. From the supply perspective, these characteristics are also important in determining, first, if tourism will occur in a place, and second, how it will occur.

As we established in the previous chapter, tourism resources are those components of *both* the physical and human environment of a destination that have the potential to facilitate tourism or provide the basis for tourism attractions. While the previous chapter focused on the physical components of the environment, we turn our attention to the human components in this chapter. In particular, we introduce several new topical branches in geography—including cultural, urban, rural, and political geographies—for the purpose of identifying and examining the human resources that provide the basis for tourism as well as the human factors that present a barrier to tourism. These topical branches are clearly interrelated. Many of the resources and barriers discussed through each of the branches below could easily be approached from a different perspective through the framework of another.

Cultural Geography and Tourism

The subject of cultural geography and issue of cultural tourism are both widely discussed and the focus of entire books. Most, quite naturally, begin with a discussion of culture. The concept of culture is considered problematic, hard to define, and open to multiple interpretations. Culture is global and local, historic and contemporary, material and symbolic. It can be considered high (oriented toward a select audience educated to appreciate it), or it may be defined as mass, popular, or low and consumed by a wide audience. It is dynamic and ever changing. Thus, we can think of culture as encompassing the way of life for a group of people, with its roots in the past but evolving with present circumstances. Everything, from their artistic expressions to their daily activities, contributes to this way of life and helps create and re-create the meanings and associations they have, as well as their values and identity.

In geography, culture has long been an important topic as we try to understand the world. Cultural geography is a dynamic topical branch that has often been influential in shaping trends in wider geographic research. Today, it is one of the most widely recognized topical branches; in fact, it is sometimes considered synonymous with human geography as a whole. To some extent, this is a reflection of the wide-ranging approaches to and issues in cultural geography. For example, **cultural geography** may be considered the study of how cultures make sense of space, how they give meanings to places, how they create landscapes, how they spread over space, how their identities form, how they are different from others, or how institutions shape culture.

Culture has also long been an important tourism resource, whether in ancient times or today. Because the concept of culture is so ambiguous, there has been much debate about the definition of cultural tourism and/or cultural tourists. Some favor a narrow, idealized view: "To be a cultural tourist is to attempt, I would suggest, to go beyond idle leisure and to return enriched with knowledge of other places and people. . . . In this way cultural tourism is clearly demarcated as a distinct form of tourism."[1] Others argue that, under a broad definition of culture, almost all of tourism today could be considered cultural tourism. We can, perhaps, consider a point somewhere in between. Certainly the cultural resources for tourism are extraordinarily important for much of tourism, and there may be significant overlap between cultural tourism and other products. However, it may be simply a matter of emphasis.

Given the vast scope of cultural geography, there has been considerable interconnection between it and the geography of tourism. Within the context of this discussion, we can use the framework of cultural geography to help us identify cultural resources for tourism and analyze potential barriers that exist between cultures. While the role of cultural geography is highlighted here, it is important to note that the branch also has a significant part to play in helping us understand the effects of tourism on societies (discussed in Part III) and factors that shape interactions between tourists and places (discussed in Part IV).

CULTURE AS A RESOURCE AND A BARRIER FOR TOURISM

The cultural resources for tourism are virtually limitless. Remnants and symbols of a place's cultural heritage have been some of the most significant resources for tourism throughout history. For example, ancient Roman tourists and modern international tourists alike have been fascinated by the Pyramids of Giza, both for the spectacle of the archaeological site and for its mythology. Cultural heritage resources for tourism can be specific features within a place that hold significance (e.g., Stari Most or Old Bridge in Mostar, Bosnia and Herzegovina, a UNESCO World Heritage Site based

Figure 7.1. The UNESCO World Heritage Site of Ephesus, in modern-day Turkey, has a long history dating back to the Greek and Roman Eras. Prior to recent regional instability, approximately two million tourists visited the site a year. *Source:* Don Albert

on its symbolic importance in connecting and reconnecting the ethnically divided city). Such resources in a place may also reflect different time periods and cultural influences (e.g., Ephesus, Turkey, a UNESCO World Heritage Site that hosted significant Greek and Roman settlements; figure 7.1). Religious sites, which are often tied to cultural heritage, can also be resources for tourism. Cathedrals, mosques, and temples all over the world are recognized within their respective belief systems for their religious importance. However, they are also more widely appreciated for their history and their aesthetic design. Thus, well-known sites like the Barcelona Cathedral, Hagia Sophia, or Potala Palace have become significant tourism attractions visited both by adherents of the particular belief system and other international tourists.

Not all elements of cultural heritage need to have a long history or be significant in the greater scope of world affairs to be considered a potential resource. For example, Abbey Road's place in England's cultural heritage dates back to 1969 with the release of the Beatles' final studio LP recorded at Abbey Road Studios and given the same title. The album cover, showing the band members crossing the street, has become iconic, even among younger generations. Even Harry Potter has become an integral part of England's cultural heritage today, and any place that has a part to play in his story—or which served as the inspiration for a place in the story—becomes a resource. These places, such as Gloucester Cathedral, the Glenfinnan Viaduct, or Leadenhall Market, would have otherwise remained unknown to potential tourists around the world.

Various aspects of the arts, whether based on traditional or modern culture, can serve as a resource as well. While these resources may not be the primary attraction that draws tourists to a place, they constitute a part of the experience and something that tourists may see or do during their visit. Visual arts are often used as a basis for a destination's attraction: famous museums (e.g., the National Gallery of Art in Washington, D.C.), city-level art districts (e.g., the South Main Arts District in Memphis, Tennessee), and open-air sculpture parks (e.g., Laumeier Sculpture Park in St. Louis, Missouri). Likewise, the production of arts and crafts—and these items themselves—can be a resource for tourism. For example, tourists may visit a traditional "factory" in Tunisia to watch the skilled craft workers make carpets and potentially purchase one to take home with them. The same applies to the performing arts. Tourists may try to see a play or musical (for some, any play or musical will do) in a famous theater district, such as New York City's Broadway or London's West End. In other cases, tourists may wish to see the performances that are specific to the place visited. For example, tourists to Chengdu, China, may attend a Sichuan opera, the highlight of which is typically *bian lian*, or face-changing. In this unique and highly protected art form, performers rapidly change a succession of brightly colored masks. Tourists may also seek to observe or participate in cultural festivals (figure 7.2).

Figure 7.2. The Official Tourism Site of the City of New Orleans declares New Orleans to be the Festival Capital of the World.* While Mardi Gras is the most famous, others include the Jazz & Heritage Festival, French Quarter Fest, Wine and Food Experience, and Po-Boy Festival. *Source:* Scott Jeffcote

* New Orleans Tourism Marketing Corporation, "2016 New Orleans Festivals," accessed September 12, 2016, http://www.neworleansonline.com/neworleans/festivals/festivals.html.

The varied characteristics of traditional cultures—such as distinctive appearances, clothing styles, livelihood patterns, housing types, cuisines, and more—can individually or collectively be considered tourism resources. These may be indigenous peoples, minority groups, or other populations that live outside of the wider society such as the Amish in parts of Pennsylvania, New York, and Ohio. At the same time, the characteristics of a modern society can be a resource, as international tourists seek to do things like experience a "typical" Irish pub or ride one of London's double-decker buses. Likewise, aspects of a society's popular fashion, musical culture, culinary styles, and more can constitute tourism resources. Additionally, less tangible elements of a place's culture can either contribute as a resource for tourism or constitute a barrier. For example, language is a basic element of culture. A common language (i.e., the native language of that place or one that is widely spoken) between the sending and receiving countries or regions may be considered a resource for that particular destination. In contrast, the lack of a commonly spoken language can become a barrier. This, of course, is perceptual. For many tourists, the idea of not being able to communicate with people at the destination is a source of anxiety and stress; thus, they will be more likely to choose destinations where they feel confident they know the majority of people will be able to speak the same language.

The same can apply to religious beliefs or cultural value systems. Many societies, particularly those seeking to develop tourism, are open to and tolerant of cultural differences in physical appearance, patterns of dress, or codes of behavior. However, this is not always the case. Tourists are often requested to observe the norms of the society they visit, which may involve changes in the ways they dress (e.g., wearing more conservative clothing) or the ways they act (e.g., refraining from holding hands with one's partner or other public displays of affection). Many tourists are willing to respect these practices so that they may have the experience of that place and culture. However, others may be reluctant to visit a place where they feel they are restricted or are concerned about reports of harsh punishments for those who, perhaps unintentionally, violate one of these social rules.

AUTHENTICITY

Over the past fifty years, there has been much debate about the relationship between authenticity and tourism. In the 1960s, Daniel Boorstin bemoaned the decline of the traveler and the rise of the tourist. Of course, this derisive use of the term "tourist" was not new. He elaborated that the tourist was someone who was satisfied with experiences that were contrived and inauthentic. He called these experiences "pseudo-events."[2] More than a decade later, Dean MacCannell argued that it was life in the modern world that was increasingly superficial, inauthentic, and ultimately inadequate. As a result, he claimed that people feel the need to search for authenticity, in other times or other places, and tourism aids this quest. However, he also recognized that, even though tourists may be motivated by a desire for authentic experiences, they might not always achieve such experiences. He explained this through the concept of staged authenticity.[3]

MacCannell's concept was based on sociologist Erving Goffman's structural division of social settings. Goffman's theory stated that places have front regions and back regions. Front regions are those that are open to and intended for outsiders. This is the part of a place that is carefully constructed to present a certain image to outsiders, and it is where these outsiders interact with the insiders who function as hosts or service providers. Back regions are those that are reserved for insiders. This is the part of the place that facilitates insiders' daily activities; it is where they can be themselves rather than putting on a show or providing a service. Because the back region is generally closed to outsiders, this helps maintain the illusions presented to outsiders in the front. As he applied this model to tourism, MacCannell expanded this dichotomy into a continuum of stages in between front and back regions. While the first three stages are front regions in that they are intended for tourists, they are increasingly designed to give the appearance of a back region and therefore a more authentic experience. The final three stages are back regions in that they were not intended for tourists, but they make varying accommodations for tourists.[4]

The first stage begins with the **front region** as described by Goffman. It refers to places that have been entirely constructed for the purpose of tourism. This includes all-inclusive tourist resorts and theme parks. These sites have little, if any, relation to the character of the larger place in which they are situated. For example, Fort Fun Adventure Park is a Wild West–themed amusement park in Germany. People in such places are either tourists or employees doing a job. In the case of Fort Fun, this ranges from the ticket taker to the blonde-haired girl dressed in a "Native American" dance costume performing in the live Western show. The second stage is used to describe a front region that is decorated in a style reminiscent of a back region at that destination. Although these places are not likely to be mistaken for the real thing, they are not intended to be. Patrons of a waterfront seafood restaurant with the façade and décor of a fishing boat may enjoy the atmosphere but will be under no illusions about where they are.

The third stage is still a front region, but it is designed to simulate a back region. In contrast with the second stage, areas in the third stage may be intended to convince visitors that they are, in fact, visiting a back region. Tourist ranches may exist within the same type of setting as working ranches. This contributes to the appearance of authenticity, but the tourist ranches are romanticized—and most likely sanitized—versions of ranch life that are re-created for visitors to experience. Tourists may not be aware of the difference between the actual and simulated experiences, either due to the provider's attention to detail or simply their own lack of knowledge. Therefore, it may be difficult to distinguish the third stage from the fourth stage. The fourth stage is a back region that has been opened up to outsiders. It was not explicitly constructed for tourism, but it has been altered in some ways to accommodate tourism. Working ranches that also provide experiences for tourists would fall into this category.

The fifth stage is nearer to Goffman's idea of back region, but under various circumstances, visitors are allowed into these places. Because of this, places might be "cleaned up" a bit. In some destinations with a poorly developed tourism infrastructure, there may be a lack of formal restaurants; therefore, tourists may take meals in residents' homes. Although this is typically done on an informal basis, there are some examples of destinations that recognize this practice and have legal regulations for such in-home "restaurants" (e.g., Cuba's *paladares*). The final stage is the true **back region**. These are places that are not intended for, or expected to receive, outsiders. As a result, the nature of the place and insiders' patterns of behavior remain largely unchanged by tourism. For example, the home provides a refuge for individuals or families who live in a destination region and work in the tourism industry. It is a place of their own that need only meet their own expectations and allows them to do what they choose. On occasion, tourists who develop a relationship (of any nature) with local people may be invited into the home. This gives the tourists access to the back region and insight into life in that place.

Tourists have been generalized as and criticized for being people who travel to another place without really experiencing it. In other words, the tourist's experiences are in the first stages of the continuum. The parts of a place he or she experiences are staged for the benefit of tourists and do not reflect the real character of the place; thus, they are considered to be inauthentic. Some tourists, such as the organized mass tourists, are content with these experiences. Their primary motivation for travel may be to escape their normal environment as opposed to experiencing a new place. Yet, other tourists, such as the drifters, will find such experiences unsatisfying. They will continue to search for entry into back regions for more authentic experiences of place.

Authenticity is generally used to refer to the genuineness of something or the accuracy of the reproduction of that thing. It is often equated with traditional culture. However, scholars point out that authenticity is a socially constructed concept. People will have different ideas about what is authentic and what is not, and they will use different criteria to make those judgments. For some, authenticity may refer only to what is original (e.g., a structure from the period in which it was built, an artifact from the period in which it was used). For others, the idea may allow for people from the original ethnic/culture group to produce things in traditional ways (e.g., crafts made by people from that group using typical materials and tools, classic performances by people from that group). Still others may look for expert confirmation or external certification to tell them that something is authentic.[5] Additionally, authenticity is negotiable. Ideas about what is authentic may change over time. People may forget or romanticize the origins of a cultural practice.[6] They may simply accept that culture changes over time and with various influences. Something that was once viewed as inauthentic may eventually come to be seen as an authentic expression of culture.[7]

Some tourist scholars have argued that staged authenticity can play a role in tourism. For example, it can help destination stakeholders control interactions with visitors and protect elements of their culture they do not want to see used for the purposes of tourism. It can also be used to adapt cultural practices for tourists who lack the knowledge to be able to understand them in their original form.[8] For example, in the examination of a Scottish heritage festival held in the United States, a research team determined that the "staging" of the event in a new time and place did not necessarily mean it was superficial and inauthentic.[9] However, an additional issue that has generated much debate in the tourism literature is the effect that tourism has on culture and authenticity. This will be discussed in chapter 9.

Urban Geography, Rural Geography, and Tourism

Urban geography and rural geography are distinct topical branches of geography that study specific geographic areas. Yet, these areas—and the studies of them—are not unrelated. Ideas about and definitions of urban and rural areas are often contingent upon each other. They may be negatively defined (i.e., the definition of one is predicated on *not* being the other) or simply defined in opposition with one another (i.e., the characteristic of one is the opposite for the other). For example, using the 2010 United States Census Bureau urban and rural classifications as an example, the basic criteria for an urban area is at least 2,500 people. Correspondingly, a rural area "encompasses all population, housing, and territory not included within an urban area."[10]

Urban geography and rural geography are topical branches in human geography that have clear ties to population and economic geographies. In addition, these branches deal with themes that are shared by social, cultural, political, and even environmental geographies. **Urban geography** may be defined as the study of the relationships between or patterns within cities and metropolitan areas,[11] while **rural geography** may be defined as the study of contemporary rural landscapes, societies, and economies. As long as the majority of the world's population lived in rural areas and was dependent on environmental resources for their survival, greater attention was given to these issues. Although the earliest cities emerged some six thousand years ago, it has been only recently that the majority of the world's population has lived in cities. Reflecting these changing spatial patterns, urban geography emerged as a topic of inquiry in the mid-twentieth century. Subsequently, less attention was given to rural geography; however, specialists in this area continue to work to understand the changing nature of rural areas (e.g., changing patterns of rural production and consumption or rural poverty).[12]

Both cities and the countryside have long been sites for tourism. It could be argued that urban tourism dates back to the Grand Tour in classical cities such as Paris, Vienna, Venice, and Rome. The pastoral ideal was popularized in art and literature based on beautiful rural settings and romanticized country life. Interest in experiences of the rural gained even more momentum with the Romantic Movement, which was primarily a response to the tremendous rise in both industrialization and urbanization. Today, tourists often consider the potential resources and barriers of cities and/or the countryside in their destination decision-making process.

CITIES AS A RESOURCE AND A BARRIER FOR TOURISM

The urban tourism product is based on a wide range of resources. In fact, one of the strengths of this product is the tremendous extent of urban resources that can be made into various categories of attractions (human—not originally intended for tourism; human—intended for tourism; and special events). Thus, urban tourism overlaps with many other products (e.g., cultural, heritage, VFR, MICE, etc.), and it draws from a number of different tourist markets (e.g., leisure tourists, business tourists, etc.).

Many attractions are derived from the history of cities around the world, from the ancient (e.g., the Acropolis in Athens, Greece, used as a fortress from the second millennium BCE, with the present temples dating back to the fifth century BCE, now a UNESCO World Heritage Site and major tourism attraction) to the more modern (e.g., Robben Island, offshore from Cape Town, South Africa, used as a prison until the mid-1990s with prisoners such as Nelson Mandela, now a UNESCO World Heritage Site and museum). Indeed, these attractions have become so vital that in many cases a city's "old town" or historic district is also the primary tourist district. The spatial concentration of attractions prompts the development of other tourist facilities and services to meet the needs of this readymade market. Likewise, markets or bazaars that were key places for residents to obtain needed products now serve as an attraction for tourists (e.g., Pike Place fish market in Seattle, Washington) and a place to buy souvenirs (e.g., the City Market in Charleston, South Carolina). Ethnic neighborhoods once served a distinct purpose for immigrants; today, they are highlighted in tourism as part of the city's unique and colorful character (e.g., Little Havana in Miami, Florida).

Cities all over the world have recognized the potential to create new attractions for tourism from existing resources.[13] Tourism has been seen as an important vehicle for urban revitalization, such as the redevelopment of a harbor or former industrial site into a fashionable shopping and/or entertainment district (e.g., the in-progress Harbor Island Redvelopment project in San Diego, California). Tourism also stimulates new ways of looking at urban infrastructure. For example, Rio de Janeiro, Brazil, has been at the forefront of slum tourism, an offshoot of the urban tourism product that involves tours of the city's infamous favelas. Tourism can be a factor in cities' decisions to host special events, from local festivals to hallmark events like the Olympic Games. Such events generate tourist visits and help raise the profile of the destination.

Urban destinations depend heavily on their reputations, both nationally and internationally. A widespread reputation is a distinctly positive factor, as it creates a demand for the experience of that specific destination. However, based on reputation alone, cities may suffer from the perception that they have a finite number of resources to offer. Every major city has certain attractions that are well known; tourists to these destinations will be sure to see or experience them. In fact, it may be considered a "crime" to visit a city without seeing them—whether it is the Alamo in San Antonio, Texas, or the Little Mermaid statue in Copenhagen, Denmark. However, these types of destinations are prone to one-time visits, where tourists feel that they have "been there, done that" and are ready to experience attractions in other cities. Destinations have to work to create new attractions, revitalize existing ones, or promote lesser-known and alternative attractions. The corporate blog for London Connection, a company that rents flats in the city, reads: "If you are a frequent visitor you may feel that you've seen it all. Amazing museums, historical monuments, and the pomp and ceremony of the royal guards are all old hat to you. Could there be anything else to see? Rest assured, London will never run out of things to see and do."[14]

The principal barriers to urban tourism are typically perceptual, often based on stereotypes of major cities, although there may be a real basis behind this. Tourists may feel that pollution (e.g., smog or litter) renders urban destinations unattractive. They may be put off by having to experience high-profile destinations with hundreds, if not thousands, of other tourists. They may be concerned by reports of a local population that is hostile to foreign tourists or by high crime rates. While tourists are more likely to be affected by petty crime in tourist areas (e.g., pickpocketing or scams) than violent crimes, it can be a deterrent nonetheless.

COUNTRYSIDE AS A RESOURCE AND A BARRIER FOR TOURISM

Understanding the rural ideal—what it is and why it is appealing—is vital in tourism. Much of this is geographically contingent. In large countries with vast areas of undeveloped land, such as the United States or Canada, interest in rural tourism has more often focused on lands protected as national parks. As such, there may be considerable overlap between natural tourism resources and rural ones. In smaller countries with extensive cultivated landscapes, such as those in Europe, there is a greater appreciation for experiences oriented around rural economic activities.[15] Farm tourism, in particular, has been immensely popular.

Rural tourism serves as an umbrella for a diverse set of more specialized tourism products. Although the extent of resources is, perhaps, less varied than in urban areas, each resource is used in a specific way to meet the demands of a particular group of tourists. For example, both farm tourism and wine tourism are based on agricultural production, but tourists interested in farm stays are different from tourists interested in vineyard tours and wine tastings. Each experience is considered unique; these tourists may participate in the same activities at different farms or estates. Similarly, the tremendous geographic scale of rural areas means that activities are often time consuming, and tourists only have the opportunity to experience a part of the landscape. If they are satisfied with this experience, they may be interested in returning to experience more. In fact, rural tourists often develop an attachment to place based on their experiences that lead them to return to the same places and possibly even purchase or build a second home in their preferred rural settings (e.g., a country house or a lake house).

While the diversity of activities has the potential to appeal to different tourist markets, rural tourism draws almost exclusively from leisure tourists; opportunities to receive business tourists are far more restricted than in urban areas. Rural tourism businesses are more likely to be locally owned and to operate on a small scale. Rural tourism also tends to be highly seasonal in nature. The majority of these activities take place out-of-doors, and they may be dependent on the stages of agricultural production (e.g., planting or harvest). Some of the greatest barriers to rural tourism relate to accessibility. With the exception of vast publicly owned spaces, found in areas such as the western United States, much of the rural landscape is privately owned. Even when some owners are willing to give tourists access to their land, activities may be disrupted or even rendered impossible by owners who do not permit access. Rural areas are also less connected than urban ones. It may be more difficult and more time consuming to reach these places, and it is dependent on transportation. Personal vehicles must be used to reach areas that are not served by public transportation, such as a rail network. This becomes a form of social exclusion, where certain groups, such as the lower socioeconomic segments of the population, are not able to participate.[16]

Box 7.1. Case Study: Vancouver, Canada—A City "Spectacular by Nature"

Some destinations are a clear fit with either urban or rural tourism products. Tourists who visit New York City are clearly interested in experiencing all that the city has to offer, from the lights and busy streets of Times Square to the high-end shopping of Fifth Avenue to the nightlife of East Village. In contrast, tourists who visit Maine will be primarily interested in experiencing the state's rural character, from small towns and villages to pick-your-own orchards to scenic drives. Yet, not all destinations are this clear-cut. In fact, many places around the world thrive on their ability to offer visitors experiences of both urban and rural attractions and opportunities to participate in activities in both urban and rural areas.

Vancouver, Canada's third largest city, is an urban destination that claims attractions rivaling other such destinations. As one of the country's cultural centers, the city has many cultural resources for tourism, such as performing arts and festivals. Vancouver's high level of ethnic diversity gives the city a unique character and provides many of its attractions. Various museums and art galleries highlight elements of First Nations (i.e., Aboriginal) culture. Vibrant ethnic neighborhoods, such as Chinatown and the Punjabi Market, bring in visitors with distinctive shops and restaurants as well as temples (Buddhist and Sikh, respectively) and special events like the Chinese New Year celebration or Vaisakhi, an ancient harvest festival that originated in northern India. Other urban tourism resources can be found at Robson Street, the city's premier shopping district with popular, high-end, and boutique stores, and Granville Street, the main entertainment district with its concentration of bars and nightclubs. Many of these popular urban districts can be explored on foot; however, some visitors use the "Big Bus"—the city's fleet of vintage double-decker and open-top sightseeing buses. The hop-on, hop-off tour allows visitors to maximize their time by providing direct access to more than twenty of the city's best-known attractions.

Yet, despite all of these resources and potential, tourists rarely visit Vancouver solely for the purpose of urban tourism. On the edge of downtown, Stanley Park is one of the city's greatest attractions visited by approximately eight million people a year (figure 7.3). This urban greenspace is over 1,000 acres in size and contains half a million trees.* Other significant parks and gardens in or adjacent to the urban center include the Dr. Sun Yat Sen Classical Chinese Garden, Bloedel Floral Conservatory and Queen Elizabeth Park, VanDusen Botanical Garden, University of British Columbia Botanical Garden and Greenheart Canopy Walkway, Nitobe Memorial Garden, and Pacific Spirit Regional Park.

Some of Vancouver's most frequently visited attractions are not within the city center but within the surrounding area. These attractions, based on a range of physical and human resources, can be reached in as little as fifteen minutes from downtown or within an hour or two. Promoted rural tourism activities consist of visits to farms, farmers markets, and wineries, as well as scenic drives along the Sea-to-Sky Highway and in the Okanagan Valley during the fall colors season. Outdoor activities can be found all across the area, including the popular Capilano Suspension Bridge and Grouse Mountain recreation areas as well as Capilano River Regional Park, Lynn Canyon Park, Lighthouse Park, and many others farther afield. Fort Langley National Historic Site, twenty-four miles from the city, pays homage to the province's heritage at the site of a Hudson's Bay Company trading post, originally constructed in 1827, through reconstructed facilities and demonstrations. And, certainly, the area around

* The Metro Vancouver Convention and Visitors Bureau, "Stanley Park," accessed August 4, 2016, http://www.tourismvancouver.com/activities/stanley-park/

Figure 7.3. Stanley Park is a distinctive attraction for Vancouver, British Columbia. Parks and gardens are an integral part of this urban destination. *Source:* Velvet Nelson

Vancouver is a well-known destination for winter sports, highlighted by the city's selection as the host site for the 2010 Olympic Games.

The destination's promotional literature makes the most of this dual character. The provincial tourism authority Tourism BC claims: "Renowned for its scenic beauty and endless opportunities for outdoor activities, Vancouver is also a cosmopolitan city with all the urban amenities—fine dining, shopping, museums, galleries, music and theatre."[†] This shows potential tourists the city's range of resources, which appeal to a wide tourist market and create an attractive idea of a distinct place. Along the same lines, Tourism Vancouver simply uses the tagline: Spectacular by Nature. Thus, unlike tourists visiting other major urban areas, those visiting Vancouver come with the expectation of experiencing more than just an urban destination.

Discussion topic: Take and argue a position: do you think urban and rural tourism are compatible or incompatible tourism products?

Tourism on the web: Tourism Vancouver, "Vancouver: Spectacular by Nature," at http://www.tourismvancouver.com

---

[†] Tourism BC, "Vancouver Things to Do," accessed August 4, 2016, http://www.hellobc.com/vancouver/things-to-do.aspx

CREATIVITY

Creativity has become an important topic in geography and tourism. One of the leading scholars on the subject, Richard Florida, identified a creative class comprised of individuals working in arts, media, culture, science, and technology that has been reshaping patterns of geography by seeking out urban centers.[17] This creative class has been drawn to cities for many reasons. For example, these individuals generally place a high value on access to a variety of urban amenities, such as restaurants, museums, or music venues. They also seek places that are diverse, with people of different ages, races, ethnic backgrounds, religions, and sexual orientations. Such diversity is not only indicative of tolerance and open-mindedness but also opportunities to get new ideas and have new experiences.[18] These factors are considered creative resources that make a place attractive to residents as well as to visitors.

Intangible creative resources have become important for cities that lack significant tangible resources (e.g., former manufacturing cities). Such resources allow for the development of new types of attractions and help shape place reputations (see box 13.1). This increases places' competitiveness in tourism as well as other industries. Essentially, creativity becomes the basis for places to be considered "cool."[19] Austin, Texas, is a prime example; city attractions include live music, diverse restaurants, and a generally distinctive (sometimes weird by their own admission) cultural environment.

Based on early examples, the emergence of creative communities also appeared to help drive urban regeneration.[20] Cities in North America, Europe, and Australia subsequently began to designate cultural quarters, defined spaces characterized by creativity, to stimulate this process. In many cases, this was seen as a means to an end—increased property values, investment, and tourism—rather than an opportunity to support creative livelihoods. The increased concentration of artists in such quarters led to increased competition and decreased profitability. At the same time, rising property values began to force out not only the local working classes but also the low-income artists. Only the most commercially successful artists were able to stay, which limited the creative synergy that such places generally depend on.[21]

In comparison to cities, small towns have generally been perceived to be homogenous and boring. However, in recent years, creative individuals have begun to reconsider small towns, which also reshapes patterns of geography. For many, especially visual and performing artists who could no longer afford to live in urban arts districts, this was initially for practical reasons. Yet, these individuals have been able to find small town places where people were open-minded and interested in embracing the creative industries. They have found different types of amenities in these areas and new sources of inspiration. Visual artists have adopted new materials from the environment; performing artists have drawn upon the issues of the rural working class.[22]

Arts towns attract urban residents on day-trips and weekend getaways, and they serve as attractions for tourists to a larger region. For example, Manitou Springs, Colorado, is easily accessible to Denver-area residents, but it is also a potential point of interest for tourists in the region for other attractions such as the Pike's Peak Cog Railway. Not all small towns or rural communities will be accepting of such change, but there are many examples of small towns that have benefitted from this type of arts-driven economic development and revitalization.

Political Geography and Tourism

Traditionally, political geography has focused on the study of the spatial structure of states and their struggle for territory and resources. Political geography has existed as an academic study for more than a hundred years, but the changes it has experienced over the years reflect changes in both the field of geography and the world.[23] Today, globalization has become one of the greatest processes of change in the world and requires new ways of looking at the policies of and connections between places. Thus, we might consider **political geography** to be the study of the ways states relate to each other in a globalized world. Clearly this topical branch overlaps with others in human geography, namely, economic and social geography, but also urban geography.

Studies in tourism have considered the political factors in tourism; however, few studies have focused on the geography of political factors in tourism. Examinations of tourism from the perspective of political geography have primarily considered the relationship between national identity and tourism.[24] Yet, there are various ways we can examine the policies that shape tourism through the framework of geography.

POLITICAL FACTORS AS A RESOURCE AND A BARRIER FOR TOURISM

So far, we have primarily considered tourism resources as those components of a destination's (physical or human) environment that have the potential to provide the basis for tourism attractions, but we can also consider those factors that have the potential to facilitate tourism. For the most part, this applies to political tourism resources. A government's policies at the national, regional, or local scale can shape tourism development. The public sector may determine which areas will be targeted for development. For example, in the 1970s, the Mexican government selected the state of Quintana Roo to be the site of the country's first master-planned resort: Cancún.[25] The public sector can support tourism development by investing in the construction or upgrade of basic infrastructure (e.g., transportation facilities, electricity, water supply, sewage, etc.) and protection of the appropriate resources (e.g., natural, cultural, heritage, etc.). It may also offer tax breaks or subsidies to encourage private sector investment in the tourism industry. At the same time, a government can create barriers to tourism development through bureaucratic red tape. Good international relations between countries and open entry policies, such as eliminating visa requirements for some or all inbound international tourists, can improve accessibility and facilitate the movement that is a fundamental component of tourism. In particular, open borders can be advantageous for tourism. The Schengen Area is a "borderless" region encompassing twenty-six European countries, primarily European Union members. Tourists entering the Schengen Area must pass through border control, but under normal circumstances any international travel within the area can be done without undergoing these procedures (box 7.2). This greatly facilitates ease of travel, both for international tourists who are citizens of any of the member countries and for other tourists traveling between destinations in the area.

Box 7.2. Experience: Border Controls, Travel Alerts, and Refugees in Germany

With an ongoing crisis in which more than 250,000 people have lost their lives and another 11 million have been displaced, few Americans would consider traveling to Syria. Seemingly far away from the turmoil of that country, Europe is a different story. Some 12 million Americans travel to Europe every year. Yet, Europe isn't so very far from Syria. In 2015, more than one million migrants and refugees crossed into Europe, with many fleeing the violence in Syria. Germany became the primary destination. By mid-September, some estimates suggest that the country was receiving up to 10,000 people per day. This volume of arrivals put a strain on infrastructure and services, heightened tensions, and prompted heated debate over policy. Tourists to a country generally perceived to be safe, stable, and relatively easy to navigate faced new potential challenges. Kay was one of these tourists. She was fortunate that the situation created only minor disruptions in her travel plans, but she also made sure she had access to the most up-to-date information about issues that might affect her trip.

I love to travel, so I jumped at the opportunity to visit friends in Germany. I scheduled the trip for late September/early October. In the time leading up to it, I watched the unprecedented situation in Europe unfold. I didn't think I needed to cancel my trip, but I decided to register with the US State Department's Smart Traveler Enrollment Program (STEP). I had learned about the program from a friend who was on a tour of Egypt in 2011, the time in which the Arab Uprisings were occurring. The company she was traveling with had asked everyone in the group to register, as it helps the US Embassy to contact travelers, and provide assistance as needed, in the event of an emergency. While the company did a good job of keeping them away from any potential hotspots, she also got notifications from STEP about what was happening in the country. Since I was traveling independently, I thought it would be wise to make sure I got the same type of information so that I could change my plans at the last minute if necessary. I signed up to have cell service while I was there, and with the Smart Traveler app on my phone, I could get any alerts and warnings while I was on the go.

To visit my friends, I stayed in Winnweiler, a small town in western Germany. But I knew I wanted to see more of the country. I could have used their car to explore, but in my previous trips to Europe, I have always gotten along just fine traveling by train. So I got a rail pass and traveled across the country to Berlin and then south to Nuremburg. I had planned to continue south to visit the famous "fairy tale" castle, Neuschwanstein, before crossing into Austria. Under normal circumstances, this would have been easy. Germany and Austria are part of the Schengen Area, which permits free movement across the internal borders of the European Union. However, a few weeks earlier, Germany made the decision to temporarily reinstate border controls along this part of its boundary to try to manage the thousands of refugees who had been arriving each day.

Before I got anywhere close to the border, my phone started buzzing with travel alerts. These alerts warned me that, if I continued on into Austria as I had planned, I would likely experience long delays at the new checkpoint to get back into Germany. It was already becoming apparent that the trains were more crowded in that part of the country. Even on those trains that were supposed to have reserved seating, people would be standing or sitting in the aisles. I was disappointed that I wasn't going to be able to do all that I had wanted, but I decided it was going to be too difficult to try to keep going. So I turned back westward and visited places like Heidelberg, where refugees were occupying an abandoned US Army base, Kaiserslautern, Ramstein-Miesenbach, and others.

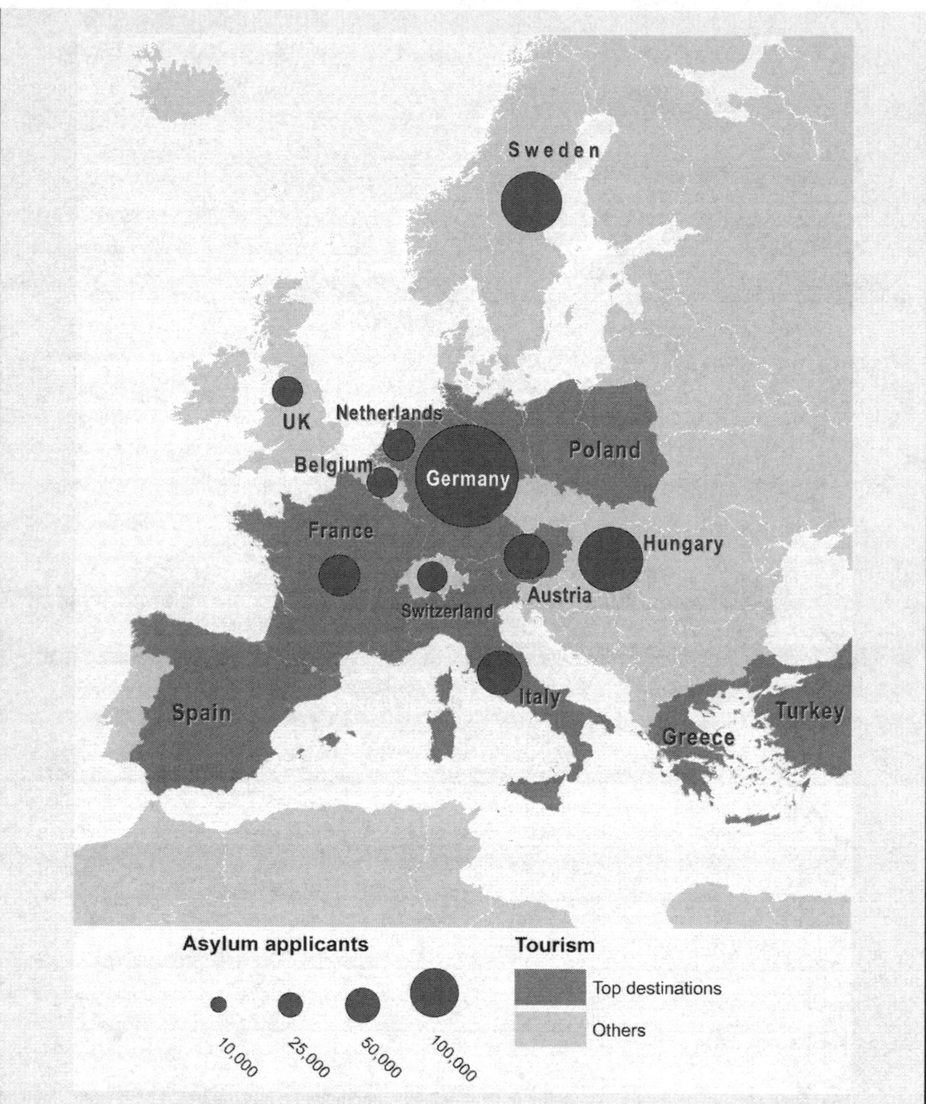

Map 7.1. This map shows us the top ten European destinations for international tourist arrivals and the top ten European countries for asylum seekers. *Source:* Gang Gong

I saw both sides of the Germans' response to the refugees. In Berlin, there were some protests. I didn't see any of the extreme demonstrations, mainly people gathering to voice their concerns about the number of refugees who were arriving. I would say that I saw more examples of people treating refugees with courtesy. With the language barrier, there wasn't much communication, but there were little things like making sure a pregnant woman was able to sit in a seat on the train. In fact, I think the Germans showed more hostility to those foreign tourists who weren't respecting the local norms, like speaking softly and not disturbing others on the train.

Traveling in Germany at the height of the refugee migration was certainly an extraordinary experience. I was glad that I had ready access to information about what was happening in the places I wanted to go so that I wouldn't get caught up in something unexpectedly. I did have to change my plans because of the situation, but ultimately it was still a great trip.

—*Kay*

In some cases, poor relations between two countries will simply restrict travel from one to the other but not necessarily have an adverse impact on the industry as a whole. For example, American policies have restricted travel to Cuba (to varying degrees depending on the administration) since the Cuban Revolution; thus, the island has been deprived of this potentially large tourist market with a good relative location. Yet, Cuba's tourism industry has attracted approximately three million visitors a year from Canada, Mexico, and various other countries in South America and Europe. In March 2016, the Obama administration announced that it would be lifting the long-standing ban on American tourists to Cuba, and major American air carriers were subsequently approved to begin scheduled flights to destinations like Havana (figure 7.4) and Varadero. Due to the pent-up demand for the experience of Cuba by Americans, this policy decision is expected to generate a significant increase in tourist visits. Similarly, sanctions on Iran were lifted in January 2016. With greater access, and expectations of new investments from Western hospitality corporations, the Iranian government is projecting twenty million tourist arrivals by 2025. This represents an increase of 300 percent over present arrivals.[26]

Figure 7.4. A long-standing tourism ban prevented Americans from traveling to Cuba under most circumstances. In 2016, the U.S. government began the process of easing restrictions. As many Americans have a vivid mental image of the destination based on things like music and cars, tourism to Cuba from the United States is expected to increase in the future. ***Source:*** **Velvet Nelson**

Of course, the "forbidden" nature of tourism under such political restrictions still has the potential to generate tourist visits, even with the threat of fines and/or imprisonment. However, the tourism industry will be seriously affected in a country that has been widely sanctioned by the international community for its policies and actions. At the same time, tourists from those sanctioned countries may also experience problems if they try to travel abroad.

In light of various issues at a destination, such as politically motivated violence or riots, countries may issue temporary travel alerts or warnings. For example, as of June 2016, the U.S. State Department issued a travel warning for Ukraine in which travelers were advised to defer travel to Crimea and the separatist-controlled areas of Donetsk and Luhansk. In addition to over nine thousand casualties from clashes between government and separatist forces, there have been reports that American travelers have been threatened, detained, and kidnapped in these areas.[27] In the following month, the Bahamas' Ministry of Foreign Affairs and Immigration issued a travel advisory for their citizens traveling to United States. In the aftermath of two fatal shootings of African American men within a week, the advisory recommended that Bahamian tourists, especially young men, exercise caution in any interactions with the police.[28] Perceptions of other internal issues, such as crime or corruption, can also present a barrier to tourism. Destination stakeholders can try to address this barrier through the creation of a visible and accessible tourist police force (box 7.3).

Box 7.3. In-Depth: Tourist Police

Safety and security are fundamental to tourism, but it is not uncommon for tourists to become victims of crimes. As strangers in a place, tourists lack an understanding of local circumstances. Thus, they are less able to prepare for various situations and less able to respond should those situations arise.[*] In the twentieth century, destination stakeholders were slow to address issues of crimes against visitors. While acknowledging the impact of crime on tourism, industry stakeholders had little opportunity to take preventative measures.[†] At the same time, local police forces often lacked the capacity to effectively respond to situations pertaining to tourists. In 1991, Thailand was one of the first countries to establish a tourist police department.[‡] The number of designated tourist police departments, divisions, units, and stations has increased significantly in recent years. This is particularly the case in developing countries with significant tourism industries, from Argentina to Egypt, Kenya, India, and Nepal.[§]

Tourist police forces generally work toward the safety of tourists and their belongings. This involves maintaining a presence in tourist areas in an effort to prevent crime; however, officers must be able to support tourists who have become victims of crime. Tourist police

[*] Seung-Il Moon, Sean Il-Kwon, In-Ho Choi, and Ju-Seok Park, "Research on the Introduction of Tourist Police System," *Proceedings, the 2nd International Conference on Information Science and Technology* 23 (2013): 354.

[†] Abraham Pizam, Peter E. Tarlow, and Jonathan Bloom, "Making Tourists Feel Safe: Whose Responsibility Is It?" *Journal of Travel Research* 36, no. 1 (1997): 24.

[‡] Pimmada Wichasin and Nuntiya Doungphummes, "A Comparative Study of International Tourists' Safety Needs and Thai Tourist Polices' Perception Towards International Tourists' Safety Needs," *International Journal of Social, Behavioral, Educational, Economic, Business and Industrial Engineering* 6, no. 7 (2012): 1938.

[§] Kristin Lozanski, "Desire for Danger, Aversion to Harm: Violence in Travel to 'Other' Places," in *Tourism and Violence*, ed. Hazel Andrews (London: Routledge, 2016): 37.

may also be used to minimize instances of exploitation (e.g., scams targeted at tourists) or harassment (e.g., aggressive hawkers in tourist areas) in an effort to increase visitor comfort and satisfaction. Tourist police may participate in the investigation of crimes against tourists or international crimes that may affect the tourism industry. In recent years, violence against tourists has been met with swift prosecutions and harsh sentences. In a few cases, such as Sri Lanka, tourist police activities also protect local people from issues that can arise with tourism (e.g., the sex trade).**

Interactions with police at a destination—whether it involves asking for directions or reporting a crime—can have a significant impact on tourists' impressions of a place. It is recommended that tourist police officers be knowledgeable about the destination, sociable and patient, and often multilingual.†† Because tourist police are intended to ease tourists' anxiety, destination stakeholders may want them to be highly visible. For example, they may wear distinctive uniforms and have a presence in high-traffic areas such as transportation terminals or concentrated tourist districts. However, some scholars note that this presence could have the opposite effect; tourists' fear may increase because of speculation about why a conspicuous police presence is necessary.‡‡

It is important to educate tourists, prior to or upon their arrival, about the role of tourist police (see, for example, the link below to learn more about Nepal's tourist police). Still, tourists may be unsure who to trust in an unfamiliar environment. For instance, tourists may have preconceived ideas about police and corruption in other countries and therefore reluctant to get involved with them.

Recently, China and Italy formed an unprecedented police cooperation agreement that could add another dimension to this discussion. Accounting for approximately three million tourists a year, China has become a significant source of visitors to Italy. Due to cultural and linguistic differences, these tourists may experience more confusion or anxiety than other European or North American tourists, and they may have greater difficulty accessing help if it is needed. The Italian government recognized the need to improve services for Chinese tourists. As part of the new agreement, four Chinese police officers traveled to Italy in May 2016 to patrol tourist areas in Rome and Milan. These officers did not have arrest powers; they were primarily there to provide assistance as needed and to reassure Chinese tourists with their presence. While this experiment was small-scale and short-term, it has been suggested that it could lead to a more sustained Chinese police presence at major Italian destinations.§§

Discussion topic: Do you think dedicated tourist police forces are necessary? Explain.

Tourism on the web: Government of Nepal, "Tourist Police," at http://www.tourism.gov.np/en/category/tourism/tourist_police

** Chootima Longjit and Douglas G. Pearce, "Managing a Mature Coastal Destination: Pattaya, Thailand," *Journal of Destination Marketing & Management* 2 (2013): 170; Lozanski, "Desire for Danger," 37–9.

†† Peter E. Tarlow, "Tourism Police Help Create Destination Image," *Tourism-Review*, August 25, 2014, accessed August 1, 2016, http://www.tourism-review.com/travel-tourism-magazine-tourism-police-create-the-image-of-the-destination-article2450

‡‡ Reiner Jaackson, "Beyond the Tourist Bubble? Cruiseship Passengers in Port," *Annals of Tourism Research* 31, no. 1 (2004): 54.

§§ Jim Yardley, "For Chinese Police Officers, Light Duty on Tourist Patrol in Italy," *The New York Times*, May 12, 2016, accessed August 1, 2016, http://www.nytimes.com/2016/05/13/world/europe/chinese-police-rome-italy.html

Entry regulations and border controls can also effectively serve as a barrier to tourism. A country may discourage travel to their destinations by imposing strict regulations on travel, such as requiring all tourists to check in with the police upon entry or traveling with a guide at all times. The same is true if the country has difficult and confusing procedures, long wait times, and/or high fees for applying for entry visas. A country's reputation for long lines at border checkpoints and rigorous customs and security inspections may also serve as a deterrent.

CONFLICT AND TERRORISM

Logic dictates that a stable political environment is an important precondition for tourism; thus, an unstable political environment is a barrier to tourism. Tourists' concerns for political safety pertain to issues such as workers' strikes, closure of transportation routes or terminals, and political unrest.[29] Most tourists will avoid any destination that is unstable or that they perceive to be unstable—at least until the situation changes. At best, tourists at destinations experiencing political instability may find their vacation disrupted; at worst, they may find themselves caught up in the middle of a conflict. Early in 2011, over three thousand British tourists were forced to cut short their vacations in Tunisia, a popular Mediterranean destination.[30] This was due to unexpected protests and riots that marked the beginning of the Arab Uprisings. Shortly thereafter, many governments around the world issued warnings for their citizens to avoid all nonessential travel to the country.

There are numerous examples in which conflict has had a negative impact on tourism. However, the editors of a recent volume on tourism and war argue that the relationship between the two is not quite as simple as it seems.[31] Certainly it is expected that tourism in a place will decline during times of conflict, but global tourism continues, typically unabated, during that conflict. If that place previously received tourists, other destinations might benefit from its conflict. Tourists who might have chosen to visit that place will consider alternative destinations. In some cases, the growth of these destinations may be short-lived, as tourists return to the original place once the conflict is over. In other cases, destinations benefit from increased exposure during that time as a result of new tourist visits and the social media representations of their experiences.

As we have already seen in chapter 3, sites associated with past conflicts or incidents of terrorism can become dark tourism attractions due to curiosity or commemoration. In recent years, there has also been a growing niche tourism market for places with active conflicts. Highly specialized tour operators create experiences in which tourists can see what is happening in these places. This is described as a hybrid product with elements of both adventure tourism and dark tourism. Tourists are motivated by the risk involved as well as the desire for unusual and "authentic" experiences. This is typically a high-end tour product due to the logistical planning for such an excursion and security precautions (e.g., armored vehicles, security guards). There is considerable debate about this type of conflict tourism. Some argue that it is an educational experience in which participants gain a deeper understanding of the issues involved in the conflict. Others cite the ethical issues associated with the voyeurism in watching the devastation of other peoples and places.[32]

While such opportunities exist, conflict and terrorism most often have a negative impact on tourism. In fact, terrorist organizations may specifically target the tourism industry. In areas with tourists, terrorists are less likely to stand out as outsiders who may arouse suspicion. Tourists and the tourism infrastructure (e.g., attractions, hotels, shopping venues, cafés and restaurants, etc.) are viewed as "soft targets." They are relatively vulnerable, and attacks against them have the potential to generate maximum terror and chaos. Targeting tourists (especially foreign) and the tourism industry ensures a high level of publicity. Attacks on the industry has an immediate, and potentially long-term, economic impact.[33]

Some places are more resilient to terrorism than others.[34] Places that face recurring issues with violence will struggle to retain a positive reputation and tourist confidence. Turkey's tourism industry had experienced significant growth. In 2014, the country received 39.8 million international visitors, making them the sixth largest international tourist destination in the world.[35] Between June 2015 and June 2016, there were fourteen major attacks linked to Kurdish militants and ISIL that resulted in the deaths of more than 280 people. Three of these attacks occurred in tourist areas of Istanbul, and one took place at the Istanbul airport.[36] By June 2016, some resorts were posting as low as 10 percent occupancy rates and projections were estimating a drop in tourism by at least 40 percent by the end of the year.[37] Then, on July 15, 2016, a coup d'état was attempted against the government that resulted in the deaths of more than three hundred people. The long-term effects from this pervasive violence remain to be seen. While France, the world's leading international tourist destination, has been somewhat more resilient, it too has suffered from repeated attacks at destinations such as Paris (2015) and Nice (2016).

Tourism can play a role in conflict and terrorism. There may be a preexisting dispute between peoples or places that is exacerbated by tourism. Both parties may compete for the potential benefits of tourism, or one may react negatively to the perceived costs associated with tourism. For example, tourism geographer Dallen Timothy discusses the case of an eleventh-century Hindu temple complex and UNESCO World Heritage Site situated within a contested section of the border between Thailand and Cambodia. While there are other issues in this conflict, the economic implications of tourism at such an attraction cannot be discounted.[38] Finally, tourism may factor into conditions for conflict. When tourism development hurts local people, economically or socially, it may spark retaliation and violent confrontation. Likewise, when tourists' behaviors are incompatible with local culture and traditions, it may prompt conservative members of society to take extreme actions.[39]

Conclusion

Just as the physical features and processes of place shape tourism, so will the varied human features and processes of that place. There are countless factors that will act as resources for or barriers to tourism; these chapters have barely skimmed the surface. Nonetheless, we can begin to see how we can use a geographic framework to help us identify and explore these issues. In part III, we will build on this foundation by examining the geographic effects of tourism. The nature of these effects for any destination will undoubtedly be influenced by the characteristics of both the physical geography and the human geography of that place. Thus, we will continue to draw upon these topical branches that we have already discussed, even as we turn our focus to a few new ones.

Key Terms

- back region
- cultural geography
- front region
- political geography
- rural geography
- urban geography

Notes

1. Kevin Meethan, *Tourism in Global Society: Place, Culture, Consumption* (Houndmills, UK: Palgrave, 2001), 128.
2. Daniel J. Boorstin, *The Image: A Guide to Pseudo-Events in America* (New York: Vintage Books, 1961; 50th Anniversary Edition, 2012), 84–5.
3. Dean MacCannell, "Staged Authenticity: Arrangements of Social Space in Tourist Settings," *American Journal of Sociology* 79, no. 3 (1973): 597.
4. MacCannell, "Staged Authenticity"; Dean MacCannell, *The Tourist: A New Theory of the Leisure Class* (New York: Schocken Books, 1976; reprinted with foreword by Lucy R. Lippard; Berkeley: University of California Press, 1999).
5. Chris Halewood and Kevin Hannam, "Viking Heritage Tourism: Authenticity and Commodification," *Annals of Tourism Research* 28, no. 3 (2001): 568.
6. Deepak Chhabra, Robert Healy, and Erin Sills, "Staged Authenticity and Heritage Tourism," *Annals of Tourism Research* 30, no. 3 (2003): 706.
7. Erik Cohen, "Authenticity and Commoditization in Tourism," *Annals of Tourism Research* 15 (1988): 374.
8. Stephen Williams, *Tourism Geography* (London: Routledge, 1998), 161.
9. Chhabra, Healy, and Sills, "Staged Authenticity," 715.
10. United States Census Bureau, "2010 Census Urban and Rural Classification and Urban Area Criteria," accessed August 4, 2016, http://www.census.gov/geo/www/ua/2010urbanruralclass.html
11. David H. Kaplan, Steven R. Holloway, and James O. Wheeler, *Urban Geography*, 3rd ed. (Hoboken, NJ: Wiley, 2014), 7.
12. Michael Woods, *Rural Geography: Processes Responses and Experiences in Rural Restructuring* (Los Angeles: Sage, 2005), 32.
13. T.C. Chang and Shirlena Huang, "Urban Tourism: Between the Global and the Local," in *A Companion to Tourism*, ed. Alan A. Lew, C. Michael Hall, and Allan M. Williams, (Malden, MA: Blackwell, 2004), 227.
14. London Connection, "Think You've Seen It All...London Will Surprise You!" accessed August 4, 2016, https://londonconnection.com/think-youve-seen-it-all-london-will-surprise-you/
15. Gareth Shaw and Allan M. Williams, *Critical Issues in Tourism: A Geographical Perspective*, 2nd ed. (Malden, MA: Blackwell, 2002), 279.

16. Richard Sharpley, "Tourism and the Countryside," in *A Companion to Tourism*, ed. Alan A. Lew, C. Michael Hall, and Allan M. Williams (Malden, MA: Blackwell, 2004), 380.

17. Richard Florida, *The Rise of the Creative Class, Revisited* (New York: Basic Books, 2012), 11.

18. Florida, *The Rise of the Creative Class, Revisited*, 293–4.

19. Greg Richards, "Creativity and Tourism: The State of the Art," *Annals of Tourism Research* 38, no. 4 (2011): 1230.

20. Florida, *The Rise of the Creative Class*, Revisited, xiii.

21. Gordon Waitt and Chris Gibson, "Tourism and Creative Economies," in *The Wiley Blackwell Companion to Tourism*, ed. Alan A. Lew, C. Michael Hall, and Allan M. Williams (Malden, MA: Wiley Blackwell, 2014), 233–5.

22. Charlotte Higgins, "Art in the Countryside: Why More and More UK Creatives Are Leaving the City," *The Guardian*, August 26, 2013, accessed August 4, 2016, https://www.theguardian.com/artanddesign/2013/aug/26/art-countryside-uk-creatives

23. Mark Blacksell, *Political Geography* (London: Routledge, 2006), 3.

24. See, for example, Duncan Light, " 'Facing the Future': Tourism and Identity-Building in Post-Socialist Romania," *Political Geography* 20 (2001), 1053–1074.

25. Rebecca Maria Torres and Janet D. Momsen, "Gringolandia: The Construction of a New Tourist Space in Mexico," *Annals of the Association of American Geographers* 95, no. 2 (2005): 315.

26. Art Patnaude and Nicolas Parasie, "Next Big Travel Destination: Iran?" *The Wall Street Journal*, May 17, 2016, accessed August 2, 2016, http://www.wsj.com/articles/next-big-travel-destination-iran-1463490999

27. United States Department of State, "Alerts and Warnings," accessed August 2, 2016, https://travel.state.gov/content/passports/en/alertswarnings.html

28. Azadeh Ansari, "Bahamas Tells Its Citizens Traveling to U.S.: Be careful," *CNN*, July 10, 2016, accessed August 2, 2016, http://www.cnn.com/2016/07/09/travel/bahamas-us-travel-advisory/

29. Pimmada Wichasin and Nuntiya Doungphummes, "A Comparative Study of International Tourists' Safety Needs and Thai Tourist Polices' Perception Towards International Tourists' Safety Needs," *International Journal of Social, Behavioral, Educational, Economic, Business and Industrial Engineering* 6, no. 7 (2012): 1940.

30. "3,300 British Tourists Evacuated from Tunisia as Riot Chaos Deepens After President Flees," *Daily Mail*, January 14, 2011, accessed August 2, 2016, http://www.dailymail.co.uk/news/article-1347112/Tunisia-riots-3-300-British-tourists-evacuated-travel-companies-23-die.html

31. Richard Butler and Wantanee Suntikul, "Tourism and War: An Ill Wind?" in *Tourism and War*, ed. Richard Butler and Wantanee Suntikul (London: Routledge, 2013), 3.

32. Gada Mahrouse, "War-Zone Tourism: Thinking Beyond Voyeurism and Danger," *ACME: An International Journal for Critical Geographies* 15 no. 2 (2016): 331–6.

33. Sevil F. Sönmez, "Tourism, Terrorism, and Political Instability," *Annals of Tourism Research* 25, no. 2 (1998): 424–5.

34. Shrabani Saha and Ghialy Yap, "The Moderation Effects of Political Instability and Terrorism on Tourism Development: A Cross-Country Panel Analysis," *Journal of Travel Research* 53 no. 4 (2014): 509.

35. United Nations World Tourism Organization, *Tourism Highlights 2015 Edition* 8, no. 1 (2015), accessed June 8, 2016, http://www.e-unwto.org/doi/pdf/10.18111/9789284416899

36. "Airport Attack in Istanbul Is the Latest in a Year of Terror in Turkey," *The New York Times*, June 30, 2016, accessed August 2, 2016, http://www.nytimes.com/interactive/2016/06/28/world/middleeast/turkey-terror-attacks-bombings.html?_r=0

37. Mark Lowen, "Turkey Tourism: An Industry in Crisis," *BBC News*, June 17, 2016, accessed August 2, 2016, http://www.bbc.com/news/world-europe-36549880

38. Dallen J. Timothy, "Tourism, War, and Political Instability: Territorial and Religious Perspectives," in *Tourism and War*, ed. Richard Butler and Wantanee Suntikul (London: Routledge, 2013): 17.

39. Sönmez, "Tourism, Terrorism, and Political Instability," 426.

Part III

THE GEOGRAPHY OF TOURISM EFFECTS

While tourism began as an activity, it quickly became an industry. Peoples and places all over the world—now more than ever—looked to tourism as a means of development. The promise of economic benefits from tourism, namely, job creation and income generation, has been extremely alluring. Likewise, with the evolution of the environmental movement, the potential for environmental preservation has also been a strong motivator. Today, these arguments for tourism can be heard in places all over the world, whether on a small island with chronic high unemployment rates or in a remote wilderness area under pressure from the extractive industries. Tourism is held up as the panacea for all sorts of problems. Indeed, properly planned and developed, tourism can have a positive impact on both the peoples and the places involved. However, this is not always the case. The benefits of tourism are not always evenly distributed, and it can have unforeseen consequences. Both the potential costs and benefits for a particular place must be carefully considered and weighed to understand the net result of tourism. Ultimately, this knowledge should be used to determine the most appropriate strategies to maximize the benefits of tourism at the destination and to minimize the costs.

This section examines the geography of tourism effects. In particular, chapter 8 discusses the economic geography of tourism. Chapter 9 considers the social geography of tourism, and chapter 10 explores the environmental geography of tourism. Each of these chapters uses the tools and concepts of the respective topical branches to help us understand both the benefits and the costs of tourism on the human and physical resources of the destination. In addition, they will address some of the factors that play a role in determining what the outcome of tourism will be for a particular place. Finally, chapter 11 brings all of these topics together as it looks at the relationship between tourism and sustainable development.

CHAPTER 8

The Economic Geography of Tourism

Tourism is big business. The UNWTO estimates that the international tourism industry—without considering the value of domestic tourism in countries around the world—generated US$1.245 trillion in 2014.[1] With figures like this, it is not surprising that the economic impact of tourism is considered so important. However, neither the economic benefits nor the economic costs of tourism are evenly distributed between countries, communities, or even segments of the population. Consequently, who benefits from tourism and who is hurt by it are issues that need to be carefully considered.

The economic geography of tourism gives us the means to examine the economic effects of tourism at the individual, local, and national scales. **Economic geography** is a topical branch in human geography that is related to the field of economics and intersects with other branches such as social, political, and urban geographies. Broadly, economic geography is the study of the spatial patterns, human–environment interactions, and place-based effects of economic activities. Economic geography has a long-standing focus on issues of production. Traditionally, production has been used to describe production in the primary (e.g., agriculture) or secondary (e.g., manufacturing) economic sectors. Yet, with the tremendous rise of the tertiary sector (i.e., services) in the modern world, the study of economic geography has adapted to reflect this change.[2]

Given this emphasis on understanding the patterns that have developed with the service sector, it would stand to reason that there should be a close relationship between economic geography and the geography of tourism. In today's world, tourism is undoubtedly one of the most significant economic activities and arguably the most significant service sector industry. As such, tourism geographers have naturally drawn from the theories and concepts of economic geography. Yet, the exchange has not always been mutual. Despite calls for greater connections between economic geography and the other topical branches of geography[3] and, specifically tourism geography,[4] economic geographers have given little attention to either tourism geography research or tourism as a topic of inquiry.[5] Tourism geographer Dimitri Ioannides suggests several potential barriers to greater interaction between the two branches, including the inability in many cases to distinguish between tourism and other related services, the conglomeration of industries and services that make up the production of the tourist experience, and the greater emphasis that has been placed on demand-side perspectives in tourism studies.[6]

Economic geography has a vital part to play in our understanding of the geography of tourism. The potential economic benefits of tourism are extraordinarily important in the development of tourism destinations around the world. Yet, the promise of such benefits should not be adhered to blindly; they must be weighed against the potential costs to determine whether net benefits will, in fact, be received. This chapter utilizes the tools and concepts of economic geography to consider the potential for tourism to contribute to economic development at a destination, as well as the failure of tourism to live up to this potential or have other negative consequences for the destination. Additionally, it also discusses the factors that influence the outcome of these effects.

Economic Benefits of Tourism

Particularly since the post–World War II era, tourism has been seen as an attractive option for **economic development**. Economic development is typically described as a process. It encompasses the various changes that create conditions for improvements in productivity and income and therefore the well-being of the population. Essentially, economic development has the potential to bring many changes to the economic geography of a place. For many less developed countries based on predominantly low-income primary sector activities, tourism has provided new opportunities for economic diversification. For example, countries that were not considered to have a cost-effective location for industrial development might now be identified as having attractive locations and/or resources for tourism. This allows the development of tertiary activities, which may be accompanied by an increase in income. As such, the benefits of tourism have primarily focused on job creation and the interrelated factors of income, investment, and associated economic development.

TOURISM EMPLOYMENT

One of the principal benefits of tourism is job creation. This is particularly important for countries that have traditionally experienced problems with high unemployment rates, as well as rural and peripheral regions of countries where jobs are limited. For example, the Caribbean has a history of chronic unemployment and high rates of labor-based emigration. Thus, the creation of new jobs for tourism has been a distinct advantage for many islands.

There is considerable potential for direct employment in the tourism industry, which is considered to have a relatively high demand for labor. For example, the Venetian Resort Hotel Casino in Las Vegas, Nevada, combined with the Palazzo Resort Hotel Casino, is the largest hotel/resort complex in the United States and employs over eight thousand people.[7] This one complex alone maintains a larger workforce than many traditional manufacturing facilities, which have been experiencing declines in labor demand as a result of increased mechanization. By way of comparison, Hyundai Motor Manufacturing Alabama employs three thousand people.[8]

Direct employment in the tourism industry varies widely. Some people may be employed, often in the private sector, to facilitate destination planning, development, or promotion. Others provide services to tourists by working at local information offices or serving as guides. Hotels and resorts employ countless people; depending on the scale of the hotel and the services provided, these employees may function as valet parking attendants, bellhops, check-in clerks, concierges, housekeepers, groundskeepers, maintenance crews, security forces, bartenders, servers, kitchen staff, salespeople in in-house retailers, spa therapists, casino dealers, entertainers, and even the gondoliers in the case of the Venetian. Tourist attractions also employ a range of staff to maintain facilities and to facilitate the tourism experience.

In addition, there is considerable potential for indirect employment generated by the tourism industry. In some cases, these jobs support tourism development but are not directly involved in serving tourists. This includes jobs in the construction industry that are required to build both the general infrastructure that will allow tourism (e.g., airports or highways) and the specific tourism infrastructure (e.g., hotels). Likewise, this can include manufacturing jobs that produce the goods that are sold to tourists. In other cases, these jobs are created in related service industries that both support and benefit from tourism but do not solely cater to the tourist market. This includes jobs in transport services, general retail businesses, local restaurants, or others (figure 8.1).

The tremendous diversity of services provided in the context of tourism constitutes an added benefit: it allows jobs to be created in a variety of capacities and at different skill or education levels. This opens up tourism employment to a wider range of people, rather than a subset of the population. For example, a higher proportion of tourism-related jobs go to women compared to jobs in other modern industries.

Figure 8.1. Atlantis Paradise Island Resort in the Bahamas reportedly accounts for an estimated 8,000 direct jobs and contributes to another 10,000–12,000 jobs indirectly.* *Source:* Scott Jeffcote

* Taneka Thompson, Uncertainty Over the Future of Atlantis," *The Nassau Guardian*, January 23, 2012, accessed June 24, 2016, http://www.thenassauguardian.com/index.php?option=com_content&id=22206:too-big-to-fail&Itemid=37.

Particularly in less developed countries, less skilled work is accessible to women, who may not have had the opportunity to obtain a formal education. In addition, the domestic nature of many of the services provided in tourism may be seen as an acceptable form of employment in parts of the world where women have not traditionally had a place outside of the home and in the formal economic sectors.

INCOME, INVESTMENT, AND ECONOMIC DEVELOPMENT

For most places, the potential financial benefits of tourism are one of the key factors driving tourism development. However, these benefits come in different forms with different effects. Tourism has the potential to bring investment to a place or region. A place may possess the resources for tourism, but varying degrees of development will be required before tourism can take place. This is typically infrastructural development to allow people to reach the destination, to stay there (if appropriate), and to appreciate attractions. The public sector is likely to invest in the basic infrastructure, such as transport systems and utilities, and some attractions, such as local/national parks, monuments, or museums. The private sector is likely to invest in specific tourist infrastructure, such as accommodations, as well as attractions. For many destinations, particularly those in poorer regions seeking to use tourism as a strategy to improve the economy and income levels, the local private sector may not have the capital to invest in tourism development. As such, external—often foreign—investment may be a crucial catalyst for growth and starting the development process in that place, when it would not have been possible otherwise.

Tourism also has the potential to bring currency to a destination. Thus, tourism geographers are concerned not only with the movement of people from one place to another but also with money. In the case of domestic tourism, this signifies a spatial redistribution of currency within the country. This is particularly significant when destinations are developed in poorer, peripheral regions of a country. Tourism is considered an important means of allowing some of the wealth that is typically concentrated in the country's primary urban area to be channeled into these destinations, thereby decreasing regional inequalities. In the case of international tourism, this indicates an influx of currency that contributes to the country's gross domestic product (GDP) and has the potential to improve its trade balance (i.e., increase its surplus or decrease its deficit). Similarly, this signifies a means of redistributing some of the wealth from the more developed countries of the world to the less developed ones.

The **travel account** is defined as the difference between the income that the destination country receives from tourism and the expenditures of that country's citizens when they travel abroad. The trend toward tourism in less developed countries such as Honduras, Mozambique, or Cambodia has allowed these destinations to develop a positive travel account. In other words, such destinations receive international tourists and therefore derive income from these tourists, but because international tourism is beyond the financial means of much of their population, they send relatively few tourists to other countries. In contrast, more developed countries such as Germany and Japan have traditionally had a negative travel account. These countries are significant tourist-generating areas and send substantial numbers of tourists abroad every year who spend money in other countries. As smaller numbers of tourists visit these destinations, the income derived from them is less than the amount being spent abroad.[9]

The **direct economic effect** of tourism refers to the initial introduction of currency into the local economy by tourists themselves. This is in the form of **tourist dollars**, or the money that tourists bring with them and spend at the destination on lodging, food, excursions, souvenirs, and more. The **indirect economic effect** of tourism refers to the second round of spending that is a direct result of the tourist dollars. The recipients of tourist dollars use this money to pay expenses, employees, taxes, and so on, as well as to reinvest in the tourism business. This involves buying goods and services demanded by tourists or new equipment that will allow them to better serve the tourist market. This round of spending is primarily intended to improve or expand the tourism sector in a way that will encourage future visitation and spending (i.e., greater direct effects) at the destination. As long as the additional spending takes place within the local economy (e.g., hiring local workers, buying locally produced goods), it can create additional economic benefits for the destination. Finally, the **induced economic effect** constitutes an additional round of spending. For example, recipients of tourist dollars pay taxes, licenses, or fees on their business; then the government may use this money to subsidize local development projects. Likewise, recipients may use tourist dollars to pay their employees, who then purchase the goods and services that they need for their own consumption.

This process of spending and re-spending may be quantified by the **multiplier effect**. This is typically expressed as a ratio. For example, the ratio of 1 to 1.25 means that for every tourist dollar that is spent directly on tourism, an additional 25 cents is created indirectly in the local economy. The multiplier effect may be used to estimate the economic benefits of tourism because it provides an indication of how the income from tourism is distributed throughout the economy. The greater the ratio, the more likely it is that money is staying within that economy. Consequently, the ratio will generally be higher at the national scale rather than at the local scale, where some money will necessarily be spent outside of the community. Businesses will be required to pay federal taxes, which may not be reinvested in that particular place. Not all goods can be obtained locally, and not all employees live at the destination and spend their earnings locally. The multiplier effect is often criticized because it can be extremely difficult to calculate depending on a wide range of factors. As such, it should be considered with caution, as a general guide to describe the potential for additional economic benefits to the destination as opposed to a precise measure.[10]

The development of tourism has the potential to contribute to further economic development and diversification. The tourism industry can take advantage of existing local industries or encourage the development of new ones to support tourism and provide the goods or services demanded by tourists. These are called **linkages** and are typically part of the indirect economic effects. For instance, tourism has the potential for strong linkages to local agricultural industries where foods typical to a particular region are produced and prepared locally for tourist consumption at markets, restaurants, or resorts. These linkages may be used to create a sense of place-based distinctiveness for the destination, where tourists will have the opportunity to experience things that

Figure 8.2. The Cadushy Distillery on the Caribbean island of Bonaire uses the locally grown Kadushi cactus to make its distinctive liqueur and tea. Visitors tour the distillery and purchase products to be taken home. *Source:* Velvet Nelson

they would not be exposed to at home. In the case of agricultural linkages, this may be a type of food or beverage that is based on local ingredients not readily accessible or commonly used at the tourists' place of origin (figure 8.2). Tourists feel like they tried something new and authentic, and they may purchase the product to enjoy after they return home. At the same time, linkages may be used to appeal to a particular brand of "responsible" tourists, typically those socially conscious consumers interested in things like organic and fair trade products.

Box 8.1. Experience: Wine and Tourism in Chile

Tourism has the potential to generate support for other local economic sectors. The growth of food and beverage tourism in particular is creating new opportunities. Tourists interested in this product want to learn more about the food and/or drink of a place. Tours of places where products are grown or processed, and associated opportunities to sample these products, add value to the destination experience. Income from such tours is used to supplement income generated from the primary industry as well as create connections with potential customers. Having already tried the product, customers, like Barret and his girlfriend in the case below, may choose to purchase it after they return home, if it is available through regular distribution channels, or to special order it via the Internet. This loyalty to the product may be a result of its perceived quality, distinctiveness, or simply the associations one has with it from the original experience.

I cannot say that I ever planned on going to Chile, but when my girlfriend had to go to Santiago for her job, I decided that I would meet her there. After she finished her work, we would spend a few days in the city, but we both wanted to get out and explore some more of the country. I didn't know much about Chile before this trip, but I knew about Chilean wine. I manage a restaurant, and we have a selection of South American wines, including a Chilean Malbec. It was my idea to visit the wine regions in the Central Valley. I did some research and found that the company that produces this wine had several vineyards in the region. I chose one of them to visit; it was the largest of their vineyards and located in the area where we wanted to go, south of Santiago in the less-visited Colchagua Valley.

The smog had been getting worse every day we spent in Santiago, and when we drove out of the city, visibility was low. We couldn't even see the mountains that surround the city. As the morning went on and we got farther from the city, the haze began to burn off. By the time we reached the site, it had turned out to be a beautiful day with clean air and clear skies. We first toured the vineyard; they took us up the vine-covered hillside for a spectacular view of the valley's scenery. Then we toured the winery and had a tasting. They produce mainly red wines at this winery, and we had the opportunity to try the Cabernet Sauvignon, Syrah, and Carmenère, which is a variety I'd never heard of before. We also tried a Sauvignon Blanc from another one of the company's vineyards. Although our guide for the tour had limited English, she was plainly knowledgeable about viticulture. She did her best to answer some good questions from our group about the valley's soils, climate, and varieties of grapes, as well as the processes of hand-picking the grapes and producing the wines.

We were somewhat surprised to learn that taxes make the wines very expensive to purchase locally, so the majority is exported. Because the white wines have a much more limited distribution in the United States, we purchased a few of these bottles at the winery and brought them home with us. In particular, we wanted to try their award-winning Chardonnay, and we didn't think we'd be able to get it here. The reds are pretty widely distributed, though, like the Malbec we have at the restaurant, and some of the varieties can also be found at the grocery stores in town.

Wine is clearly an important part of the economy, but at the same time, I think it really adds to what Chile has to offer in terms of tourism. The two industries are, in fact, a perfect complement to each other. For me, our tour of the vineyard and winery was one of the highlights of our trip. We wanted to see some of the countryside beyond the city, and the landscape scenery of the vineyard set in the valley was exceptionally beautiful. As food and wine are my passion, I am always interested in learning more about and trying things that are unique to the places I visit. It gave us a distinctly local experience as well as something local to bring home with us. In addition, I can tell my customers more about the Chilean wine we sell at the restaurant because I've actually been to the winery, and my girlfriend has been buying these wines at the store since we've been back. While this is partially because she is familiar with them, it also has more meaning for us because it reminds us of our trip.

—*Barret*

Although some existing economic activities might have the potential for linkages as the tourism industry becomes well established, at least initially the local economy may not be sufficiently or appropriately developed to support the tourism industry. Existing activities may require adaptations to meet the specific demands of tourism. The local agricultural industry may make a transition from a single crop (e.g., corn) to more diversified agriculture to supply the range of products in demand by the tourism industry (e.g., lettuce, tomatoes, jalapeño peppers). Likewise, the local fishing industry may need to expand in order to supply the quantity of products demanded by the tourism industry. Additionally, activities may need to be wholly developed; until then, products must be imported. However, to maximize the economic benefits of tourism, local linkages should be created as soon as it is feasible to minimize the dependence on external suppliers.

Box 8.2. In-Depth: Pro-Poor Tourism

In the development of modern tourism, those who are already economically better off in a community have been more likely to receive the benefits of tourism, while the poorest segments of the population have been more likely to experience the costs of tourism. Since the tourism industry, particularly those segments driven by the private sector, is a profit-oriented venture, tourism development has done little to consider the relationship between tourism and poverty or to view poverty elimination as a goal.[*] Moreover, the poor are not a homogenous group; there is a tremendous range of factors that may prevent people from participating in tourism. The poor may not have access to the education and/or training—including language skills—that would allow them to obtain tourism jobs or to understand and work within the legal framework to start a tourism business. They may not have access to the capital or credit they need to start or expand a tourism or related business. Similarly, they may not have access to or control over the land and other resources that could be used for the purposes of tourism. They may be only small-scale producers of goods or services, and tourism businesses may seek to deal with a single, large supplier of the products they need.[†] Additionally, they may be excluded from the tourism planning and development process.

Yet, there is an undeniable relationship between tourism and poverty. In particular, the regions of the world that have experienced some of the highest tourism growth rates are also the regions that have the highest concentrations of poor people. Particularly since the beginning of the twenty-first century, new studies have been undertaken to better understand this relationship. The Pro-Poor Tourism Partnership was formed as a collaborative research initiative to develop the concept of **pro-poor tourism** (PPT). PPT is defined as tourism that results in increased net benefits for poor people and ensures that tourism growth contributes to poverty reduction.[‡] PPT is not a tourism "product" like those discussed in chapter 3; rather, it is an approach to tourism. It seeks to expand the economic benefits of tourism to the poor by addressing those barriers that prevent them from participating in tourism.

Proponents of PPT argue that the tourism industry is better suited to contribute to poverty reduction than many other economic activities because it is a diverse, labor-intensive industry that often utilizes freely available resources (e.g., natural environments or elements of culture). Tourism brings the consumer to the producer, which opens up opportunities for poor people who may not have the ability to take their goods or services to a place of consumption. In addition, tourism has the potential to complement and supplement existing economic activities (e.g., agriculture or fishing), thereby increasing income-earning potential.[§]

[*] Caroline Ashley, Charlotte Boyd, and Harold Goodwin, "Pro-Poor Tourism: Putting Poverty at the Heart of the Tourism Agenda," *Natural Resource Perspectives* 51 (2000): 1.

[†] Rebecca Torres and Janet Henshall Momsem, "Challenges and Potential for Linking Tourism and Agriculture to Achieve Pro-Poor Tourism Objectives," *Progress in Development Studies* 4, no. 4 (2004): 300.

[‡] Caroline Ashley, Dilys Roe, and Harold Goodwin, "Pro-Poor Strategies: Making Tourism Work for the Poor," *Pro-Poor Tourism Report* 1 (2001): viii.

[§] Ashley, Boyd, and Goodwin, "Pro-Poor Tourism," 1–2.

Critics of PPT argue that few studies have demonstrated the effectiveness of the concept in practice and question whether it can be successful without private sector support and broader changes within the tourism industry.** With a focus on accruing net economic benefits, other social and environmental impacts may be overlooked.†† In addition, wider tourism impacts are not considered. For example, PPT encourages travel from more developed countries to less developed ones. This typically takes the form of long-distance international air transport. The greenhouse gas emissions incurred in this travel contributes to climate change, and impoverished people in the less developed countries are highly vulnerable to the effects of climate change.‡‡

Although there may appear to be little incentive for a tourism venture to take PPT measures, especially profit-oriented private sector ventures, tourism is nonetheless dependent on a supportive and stable local community. The issues associated with high rates of poverty have the potential to erode this foundation. As we are encouraged to become more socially conscious and to consider the effects of our actions in our daily lives, more tourists may be willing to support products incorporating PPT objectives. In general, small-scale, socially responsible niche tourism products (e.g., community-based ecotourism) have been the ones to incorporate PPT; however, it is argued that the full potential of the PPT approach will not be realized until its strategies are also expanded to mass tourism.

** Richard Butler, Ross Curran, and Kevin D. O'Gorman, "Pro-Poor Tourism in a First World Urban Setting: Case Study of Glasgow Govan," *International Journal of Tourism Research*, 15 (2013): 443.

†† Christian M. Rogerson, "Informal Sector Business Tourism and Pro-Poor Tourism: Africa's Migrant Entrepreneurs," *Mediterranean Journal of Social Sciences* 5, no. 16 (2014): 153–4.

‡‡ Jordi Gascón, "Pro-Poor Tourism as a Strategy to Fight Rural Poverty: A Critique," *Journal of Agrarian Change* 15, no. 4 (2015): 500.

Economic Costs of Tourism

Places all across the globe would like to get a share of the trillion-dollar international and domestic tourism industry. Consequently, much emphasis is placed on the economic benefits of tourism. Yet, the *potential* for these benefits must be carefully considered to understand what the *actual* effect will be on a destination (i.e., who will receive the benefits and what will be the consequences). As such, we need to take a closer look at the nature of the jobs created in the tourism industry at a destination, the extent of economic effects, and the changes that occur in the local economy.

TOURISM EMPLOYMENT

The jobs created in the construction industry, the tourism industry, and the various supporting industries can attract immigrants from other regions of a country or other countries. This is particularly the case when tourism is developed in peripheral areas where there may not be a large enough local population to fill the new demand for labor. However, this labor-induced migration may quickly outstrip available jobs and result in a labor surplus. In fact, areas with a high dependence on tourism may have higher levels of unemployment than surrounding regions. This is due to the number of people who move to the destination based on the promise of employment or with the hope of higher-paid employment than would be available to them in their home community.

In addition, tourism is both seasonal and cyclical, which has the potential to affect employment patterns. Even popular tropical destinations, such as the islands of the Caribbean, have a distinct tourist season. Higher rates of precipitation during the wet season, as well as the increased risk of hurricanes, are considered barriers to tourism (chapter 6). At the same time, the majority of tourists come from the Northern Hemisphere during the winter months for the inversion of cold to warm weather (chapter 2). Consequently, at these destinations, there is a high demand for labor during peak months, and tourism businesses may need to hire additional workers. Yet, during the low season, some of these workers may have to find temporary employment in other sectors or face unemployment.

Tourism employment is entirely based on the success of the industry. When a destination is experiencing growth, new jobs will be created in both tourism and related industries. For example, when a destination is developing or expanding, a tremendous amount of local and/or tourism infrastructure may be built, creating a boom in the construction industry. However, when the industry as a whole, or a specific destination, is experiencing a decline, jobs will quickly disappear.

One of the most common criticisms of the tourism industry is based on the type of jobs created. Many jobs are unskilled and low-wage, which may have few benefits for the local population. Tourism businesses may seek to minimize costs by providing services to tourists on an on-demand basis. Therefore, instead of a staff of full-time employees, businesses may rely on independent contractors who are paid an hourly wage or per job and who receive no benefits like health care or pension plans. For example, rather than maintaining a salaried staff of tour guides who are available to give regularly scheduled tours whether or not any tourists have signed up, a company is likely to hire guides for specific tours once they have filled. In addition, many jobs at tourism destinations are in the informal sector of the economy, like street vending (figure 8.3).

Many countries do not have minimum wage laws, do not enforce these laws, or generally have a low wage standard. Like other industries, multinational tourism companies take advantage of this. Tourism is an extremely competitive industry; there are many destinations around the world competing for investment from the same companies and visits by the same tourists. If the cost of wages at a destination reduces companies' profit margins, they will choose to locate elsewhere. If it raises the cost of vacationing at that destination too much, tourists will choose to visit someplace less expensive. As such, to ensure that the destination continues to receive the income—particularly foreign income—from tourism, local or national governments may be reluctant to pressure companies for higher wages.

Highly paid managerial positions represent a smaller proportion of overall tourism jobs, and these positions may not go to local people. Particularly when tourism is developed in rural or peripheral regions of a country and/or less developed countries, the population may not have the necessary education, training, or experience to be able to hold these types of jobs. Additionally, multinational companies may prefer to import workers who are already familiar with company policies or who may be more likely to understand the needs of the targeted tourist market. Unless a concerted effort is made

Figure 8.3. Street vendors can be found in destinations around the world. This Quechua woman is selling textiles in Cusco, Peru. *Source:* Velvet Nelson

to develop the local resource base to be able to successfully hold higher-level tourism positions, a destination can become dependent on foreign expertise.

Finally, the growth of jobs in the tourism industry may have negative consequences in other areas of the economy. Jobs in the tourism industry may have higher wages than other local employment opportunities. They may be perceived to be less physically demanding than manual labor jobs or more stable than activities like agriculture that are highly dependent on unpredictable environmental conditions. As people choose tourism employment over others, this can contribute to labor shortages and declines in other parts of the local economy. Thus, tourism will not necessarily result in an increase in employment opportunities or produce a net gain in the local or national economy.

INCOME, INVESTMENT, AND ECONOMIC DEVELOPMENT

While tourism has the potential to support other economic activities, the reality is that tourism often grows at the expense of other activities around the destination.

For example, land traditionally used for agriculture may be sold, voluntarily or under coercion, for commercial tourism development. Depending on the terms of the sale, this may present an opportunity cost for the farmer, as he or she exchanges the value of the land for the income that he or she would have derived from the annual produce of that land in the years to come. Furthermore, the potential for creating economic linkages between agriculture and tourism at the destination is diminished or lost altogether.

This can also lead to an overdependence on tourism as the principal source of income in a local or national economy. Rather than promoting economic diversification, this trend toward tourism growth at the expense of other economic activities can result in greater concentration in a single economic sector. As a result, the economy is more vulnerable to fluctuations in the industry, and tourism is a sensitive industry that may be adversely affected by any number of unforeseen factors that the destination has little or no control over.

Although foreign investment may initially provide an important source of income that allows for destination development, an overdependence on multinational companies can have a long-term impact on the nature of economic effects at the destination. The income these companies generate is more likely to be transferred out of the region or country instead of contributing to the multiplier effect. This is referred to as **leakages**. These leakages are the part of tourism income that does not get reinvested in the local economy. Some leakage occurs with each round of spending after tourist dollars are introduced into the economy, until no further re-spending can take place. The direct effect will be mitigated if the tourism businesses are externally owned, because profits will be transferred out of the region or country. The indirect effects will be reduced if tourism businesses seek external suppliers for their products. Likewise, the induced effects will be eliminated if local people spend their earnings outside of the region or country or on imported goods.[11]

The classic example of leakage is seen in the all-inclusive package mass tourism resorts. Multinational companies based in the more developed countries of the world own these large-scale chain resorts. When tourists travel from these same countries and spend their tourist dollars at the resort, those dollars may, in fact, be returning to the country of origin. Take, for example, the context of food. In some cases, despite the higher transportation costs of importing food, the company may be able to obtain food items at a lower cost from a subsidized mass producer at home as opposed to the higher per-unit cost of a small-scale local producer. In other cases, local producers simply may not be able to sell their products to the resorts because they do not have the appropriate contacts, contracts, or the ability to provide buyers with receipts (figure 8.4).[12] Whether it is true or not, the perception exists that mass tourists do not care where their food came from or that they will have concerns about the quality and integrity of local foods. They may suspect that this food was produced and/or processed under unhygienic conditions or that it will not be up to the same standards they are accustomed to at home. These types of resorts may also assume that mass tourists prefer the types of foods eaten at home and will not try new types of foods for fear of getting sick.[13]

The result is that, although countries like Honduras, Mozambique, or Cambodia may have a positive travel account, they may not receive the full economic benefits of

Figure 8.4. When tourism providers are unable or unwilling to work with small-scale agricultural producers, like this vegetable farmer in Guizhou, China, opportunities for creating local linkages are lost. ***Source:*** **Velvet Nelson**

tourism. These less developed countries have little capacity to develop tourism on their own; therefore, they are likely to rely on multinational companies to do it for them. Even though they receive international tourists, there are typically high levels of leakages that reduce the multiplier effect. As a result, little money is retained or reinvested at the destination, and the goal of redistributing wealth between parts of the world remains unrealized.

Moreover, rather than redistributing a country's wealth and economic opportunities, tourism may contribute to a further spatial concentration. This may be in an area that already has a stronger development base, such as a major metropolitan area, or it may be the development of a tourist zone. This can create or contribute to regional inequalities, as that area develops, modernizes, and experiences the economic benefits of tourism while other areas remain marginalized.

Likewise, tourism will not necessarily result in a redistribution of wealth among the population. Much of the money that is not leaked out of a region or country tends to stay in the hands of the upper- and middle-income groups. Despite concepts like pro-poor tourism (see box 8.2) and the potential for tourism to improve the economic well-being of traditionally marginalized populations such as indigenous peoples, few of the economic benefits of tourism actually accrue to the poorest segments of a population. For the most part, tourism development does not incorporate poverty elimination objectives. This is because tourism is largely driven by the private sector and companies whose primary objective is profit.

In fact, tourism has the potential to further harm the local population, especially the poorest segments of that population. As a destination develops, it may experience increasing land costs and costs of living. Land is often assessed for tax purposes on its market value rather than its use. As land at an emerging destination is sold to tourism developers at a relatively high cost, the value of all of the land in the area may be reassessed, resulting in an increase in property taxes. This may make it increasingly difficult for landowners to be able to afford to live on and/or use their land as they had in the past. For example, a farmer may have a piece of land that has been in his family for generations. He would like to maintain this land for both the heritage it represents to him and its use as a working farm. However, his income from the farm may no longer be enough to support his family and pay the increased taxes to keep the land.

The value of goods and services may be adjusted based on the amount of money tourists will pay for such things as opposed to the local population. For example, property owners at a destination may have the opportunity to rent houses or apartments/flats to tourists at a high per-week price. As such, local residents might not be able to afford to rent living space within the destination area, and they will be forced to move elsewhere. Similarly, stores may cater to tourists who are able and/or willing to pay more for certain products. Again, residents might not be able to afford to buy these products, at least from these stores. Consequently, they may have to travel outside of the destination to obtain the products they need, and if this is not an option for them based on transportation constraints, they will have to find alternatives or go without.

Finally, a destination may experience a range of hidden economic costs associated with tourism. For example, there may be a financial cost associated with managing the social or environmental effects of tourism that will be discussed in the following chapters.

Factors in Economic Effects

In some cases, we will be able to clearly see the effects of tourism on the economy. This is most likely in communities or countries with relatively undeveloped or undiversified economies. For example, small island developing states (SIDS), whether they are located in the Caribbean, the Mediterranean, the Indian Ocean, or the Pacific, face similar constraints to economic development (e.g., limited resource bases, lack of economies of scale, reduced competitiveness of products due to high transportation costs, few opportunities for private sector investment, etc.).[14] The development of tourism in the SIDS can result in clearly identifiable economic impacts, such as an influx of foreign investment, a decrease in the unemployment rate, and an increase in the per capita gross domestic product. In other cases, the impact of tourism on a country's overall economy can be more difficult to determine. The development of tourism may reflect a redistribution of resources, such as an investment or employment in tourism instead of other economic activities.

The specific economic effects of tourism at a destination, and the extent of these effects, will likely vary widely. The often interrelated factors that may determine these effects can include the nature of the local economy and the level of development at the destination, as seen in the example above. Similarly, the type of tourism, ownership in the tourism industry, and tourist spending patterns at the destination can have an impact on these effects.

For example, mass tourism destinations are more likely to be characterized by large-scale multinational companies that may use foreign staff and rely on imported supplies. As such, leakages are going to be high. In contrast, niche tourism destinations are more likely to be characterized by small-scale, locally owned tourism businesses with significant local linkages and therefore a higher multiplier effect. However, many destinations have a combination of both, typically a higher proportion of small local businesses (e.g., hotels or restaurants), but the large companies will have the greatest capacity (e.g., the most beds or tables).

Organized mass tourists typically contribute the fewest tourist dollars to the local economy. One of the most prominent examples can be seen in the cruise industry. The large quantities of cruise tourists notoriously contribute very little to the local economy at the visited ports. Tourists may not venture farther than the markets at cruise terminals, which are often characterized by cheap, mass-produced, imported souvenirs like T-shirts. They may be reluctant to spend additional money on destination excursions when the majority of activities that take place onboard are included in the purchase price. Some never even leave the ship when it is in port. In contrast, drifters and explorers, who spend longer at a single destination and rely less on the explicit tourism infrastructure, have greater opportunities to support local businesses.

Knowledge and Education

For the majority of tourists (actual and potential) around the world, money is one of the most important concerns. We may have a demand for tourism, but it is a nonessential expense. We have to determine how much of our disposable income we can devote to travel and what experiences we can afford. While this may be the only consideration for many tourists, there is a subset of socially conscious individuals who would be interested in supporting the local economy at the places they visit. However consumers may be faced with a choice: If they cannot afford a niche destination (one with a higher cost due to the lack of economies of scale but strong local linkages), do they accept a cheaper vacation at a mass destination (one with a lower cost due to economies of scale but higher levels of leakages) or stay home?

Even if tourists are interested in and willing to pay more for destinations that are economically beneficial to local people, they may not be knowledgeable enough to make an informed decision. In contrast to fair trade commodities, there is little in the way of widely known or recognized certification of tourism products. There are a few examples, such as Fair Trade Tourism, which is considered to be the first program to certify that tourism enterprises have fair wages and working conditions and an equitable distribution of benefits.[15] However, it often falls to individual tourism businesses to advertise their own policies on their website or other promotional literature, such as local ownership, fair wages, support of local farmers and craftspeople, and so on.

Tourism businesses must also weigh the higher cost of policies that support the local economy against the cost of not doing so. These are profit-oriented businesses, and paying higher wages or higher costs for locally produced supplies can reduce their profit margin. Alternatively, if these expenses are built into the price of the tourism product, the venture may become less competitive and lose business to other companies or destinations offering similar products. Yet, there can be value in promoting ethical business practices. Moreover, as seen in the discussion of PPT, the local population has the potential to undermine tourism if they feel they are not receiving enough benefits from the industry. Consequently, it is important to understand the nature and distribution of the economic effects at a destination.

Box 8.3. Case Study: Pricing Policies and Willingness to Pay in Zimbabwe

Pricing policies for natural or cultural heritage attractions have generated considerable public and academic debate. Where such things are valued as a public good, a place may choose to make attractions freely available to all. However, since these sites must be wholly supported, typically by the government, this is often impractical. Visitor fees may remain nominal if the site is subsidized, but in many cases, fees are needed to contribute to the costs associated with operating and maintaining the site.[*]

Differential pricing policies, or **price discrimination**, is where different categories of visitors are charged different fees for the same activities or services. For example, sites may differentiate based on age (e.g., children or senior citizens) or income; in international tourism, the primary distinction is made between citizens and non-citizens. Supporters of these policies argue that it opens up opportunities for local people to participate in tourism when they might not be able to otherwise. Using a social equity perspective, it allows local people to appreciate their own natural or cultural heritage instead of reserving it for the exclusive use of foreigners. Additionally, supporters maintain that foreign tourists who are able to visit the country should be able to afford the higher rates and be willing to pay for the privilege of the, perhaps once-in-a-lifetime, experience.[†] Such fees can be used to manage site visitation and promote low-volume tourism. Fewer tourists can generate the same amount of income to maintain the attraction and support the local economy while keeping social and environmental impacts to a minimum.[‡]

There are many issues for destinations and attractions to consider when determining pricing policies. Given that higher fees for foreign tourists may reduce total visitor numbers, this amount must be sufficient to offset lower visitation rates. This low demand may discourage the development of tourism infrastructure or services and have little contribution to the

[*] Jan G. Laarman and Hans M. Gregersen, "Pricing Policy in Nature-Based Tourism," *Tourism Management* 17, no. 4 (1996): 248–50.

[†] Petra Andersson, Sara Crone, Jesper Stage, and Jorn Stage, "Potential Monopoly Rents from International Wildlife Tourism: An Example from Uganda's Gorilla Tourism, *Eastern Africa Social Science Research Review* 21, no. 1 (2005): 10–11.

[‡] Jin Young Chung, Gerard T. Kyle, James F. Petrick, and James D. Absher, "Fairness of Prices, User Fee Policy and Willingness to Pay Among Visitors to a National Forest," *Tourism Management* 32 (2011): 1038.

local economy. Critics caution that international tourism is unstable and should not be relied on for consistent site support.[§] Moreover, such fees must be understood by tourists and remain within their willingness to pay (WTP).

WTP does not equate to ability to pay, and income may not be the most significant factor in determining the maximum amount tourists would pay for a particular experience. Perceived price fairness also plays a role in WTP, and fees that are considered unfair have the potential to generate hostility and resistance.[**] For example, when Costa Rica significantly increased nonresident park fees in the mid 1990s, newspapers featured interviews with angry tourists who claimed they were being exploited. Within a year, nonresident visitation to parks in the country fell by 47 percent.[††]

While price discrimination is not unique to Zimbabwe, the country provides an interesting case for examining its use. Admission at Zimbabwe's most popular tourism attraction—Victoria Falls—is US$30 for foreign visitors, US$20 for visitors from South African Development Community member countries, and US$7 for Zimbabwean residents. Even this relatively small admission fee can be beyond the means of much of the population where the per capita gross domestic product in purchasing power parity is estimated at US$2,100, or about US$5.75 per day.[‡‡] At the same time, foreign tourists may come to resent being charged substantially higher fees on top of all of the other money spent on expected expenses like transportation, accommodation, food, activities, souvenirs, and more. This resentment can compound, considering that they have already found themselves subject to visa fees, airport taxes, and other costs not directly associated with tourism services. Faced with these costs, they may choose alternative destinations. This is particularly applicable in the example of Victoria Falls. Shared with Zambia, the admission fee on that side of the border is US$20 for foreign tourists.

The situation gets more complicated when foreign and domestic tourists travel together. Many upper- and middle-class Zimbabweans (the ones who are most likely to be able to participate in tourism) are educated abroad. They make friends and contacts with foreign nationals whom they may encourage to visit them when they return to their homeland (VFR tourism). Price discrimination practices can make the hosts uncomfortable, as they have to ask their guests to pay more money to experience the same sites or receive the same services.

Moreover, when tourism facilities or services are shared between residents and their guests, the locals may find themselves charged the foreign rate simply because they are traveling with a foreigner. Personal vehicle entry at a national park, for instance, is double for nonresidents of Zimbabwe (US$10 versus US$5 for residents). Accommodations can run as much as US$50 more per night for foreign tourists (e.g., the per-night rate at a bed-and-breakfast located around Victoria Falls is US$120 for foreigners versus US$70 for residents). These concerns lead to the practice of borrowing Zimbabwean identification cards for guests to ensure that everyone receives the local rate. This puts visitors who may be able to pay the higher fees, and understand why they are being asked to do so, in an uncomfortable position. If they insist on paying the foreign rate, they may be condemning their hosts to do the same.

[§] G. Mmopelwa, D.L. Kgathi, and L. Molefhe, "Tourists' Perceptions and Their Willingness to Pay for Park Fees: A Case Study of Self-Drive Tourists and Clients for Mobile Tour Operators in Moremi Game Reserve, Botswana," *Tourism Management* 28 (2007): 1045.

[**] Young Chung, Kyle, Petrick, and Absher, "Fairness of Prices," 1040.

[††] Laarman and Gregersen, "Pricing Policy," 253.

[‡‡] Central Intelligence Agency, "The World Factbook: Zimbabwe," accessed June 23, 2016, https://www.cia.gov/library/publications/the-world-factbook/geos/zi.html

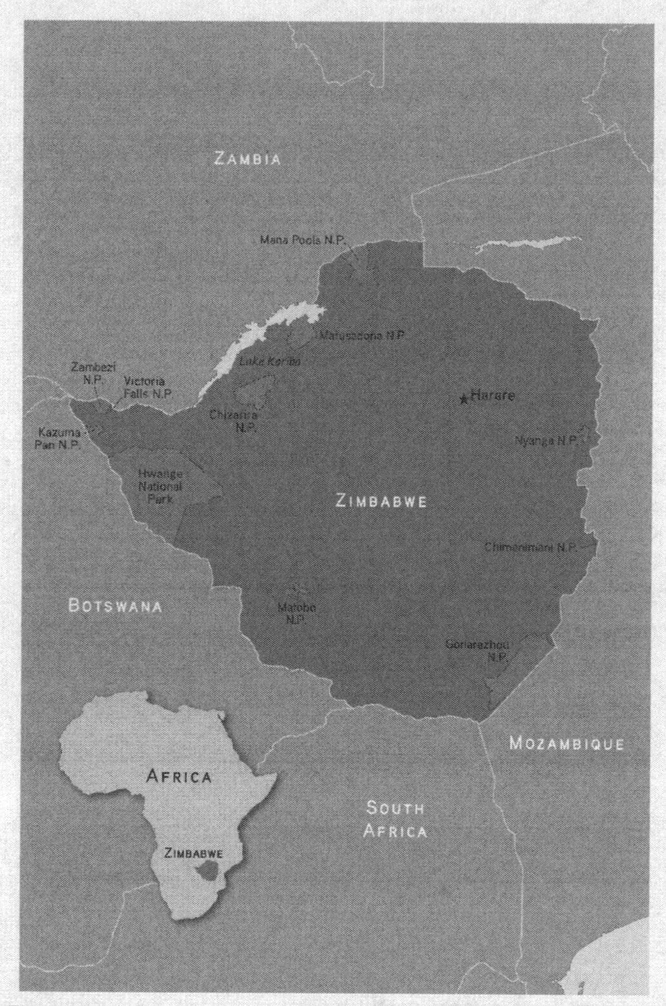

Map 8.1. Zimbabwe. This southern African destination uses a differentiated price structure at attractions such as the national parks. *Source:* XNR Productions

Price discrimination may seem like an obvious strategy to maximize the economic benefits of tourism and, to some extent, social benefits. However, such policies must be carefully considered to try to minimize unintended consequences. In particular, Travelers 2.0 are increasingly accustomed to expressing their opinion online. Approximately 4 percent of English-language reviews on TripAdvisor for Mosi-oa-Tunya/Victoria Falls National Park discuss price. Although a visitor from South Africa did not have to pay the highest fee, he still felt it was too much. A visitor from the United Kingdom who titled his review "daylight robbery at thirty US dollars" complained about the price in comparison to parks in neighboring countries. A visitor from the Netherlands was unhappy that he was going to be charged a different rate because he was from Europe and decided not to see the falls. These reviews reach a potentially limitless audience and can influence future tourists' perceptions about fairness and WTP.

Discussion topic: What factors would affect your WTP to visit Victoria Falls? Explain.

Tourism on the web: Zimbabwe Tourism Authority, "Welcome to a World of Wonders," at http://www.zimbabwetourism.net/

Conclusion

Tourism has emerged as one of the world's most significant economic activities; consequently, a foundation in economic geography is a vital component in the geography of tourism. It can be easy to focus only on the positive economic effects of tourism. In fact, it is often the case that those who are better off to begin with are the ones who are most likely to benefit from tourism. As they have the greatest voice and power in decision making, the positive outcomes may be the only ones considered. Yet, it may be that a substantial proportion of the population will never see the benefits of tourism and may not want to see it take place. Ultimately, both the public and the private sectors should undertake efforts to see that more segments of the population are able to benefit from and therefore support the industry.

Key Terms

- direct economic effect
- economic development
- economic geography
- indirect economic effect
- induced economic effect
- leakages
- linkages
- multiplier effect
- price discrimination
- pro-poor tourism
- tourist dollars
- travel account

Notes

1. United Nations World Tourism Organization, *Tourism Highlights 2015 Edition* 8, no. 1 (2015), accessed June 8, 2016, http://www.e-unwto.org/doi/pdf/10.18111/9789284416899
2. Susan Hanson. "Thinking Back, Thinking Ahead: Some Questions for Economic Geographers," in *Economic Geography: Past, Present, and Future*, ed. Sharmistha Bagchi-Sen and Helen Lawton Smith (London: Routledge, 2006), 25–33.
3. Sharmistha Bagchi-Sen and Helen Lawton Smith, "Introduction: The Past, Present, and Future of Economic Geography," in *Economic Geography: Past, Present, and Future*, ed. Sharmistha Bagchi-Sen and Helen Lawton Smith (London: Routledge, 2006), 1.
4. Dimitri Ioannides, "Strengthening the Ties between Tourism and Economic Geography: A Theoretical Agenda," *Professional Geographer* 47 (1995): 50.
5. Dimitri Ioannides and Keith G. Debbage, "Introduction: Exploring the Economic Geography and Tourism Nexus," in *The Economic Geography of the Tourist Industry*, ed. Dimitri Ioannides and Keith G. Debbage (London: Routledge, 1998), 5.
6. Dimitri Ioannides, "The Economic Geography of the Tourist Industry: Ten Years of Progress in Research and an Agenda for the Future," *Tourism Geographies* 8 (2006): 76–8.
7. Public Relations, e-mail message to author, July 12, 2016.
8. Hyundai Motor Manufacturing Alabama, "About HMMA," accessed July 12, 2016, http://www.hmmausa.com/our-company/about-hmma/
9. Stephen Williams. *Tourism Geography* (London: Routledge, 1998), 87–8.

10. Williams, *Tourism Geography*, 90–1.
11. Williams, *Tourism Geography*, 86.
12. Rebecca Torres, "Linkages between Tourism and Agriculture in Mexico," *Annals of Tourism Research* 30, no. 3 (2003): 555.
13. Rebecca Torres, "Toward a Better Understanding of Tourism and Agriculture Linkages," in the Yucatan: Tourist Food Consumption and Preferences," *Tourism Geographies* 4, no. 3 (2002): 293.
14. United Nations Office of the High Representative for the Least Developed Countries, Landlocked Developing Countries and the Small Island Developing States, "Small Island Developing States," accessed July 14, 2016, http://unohrlls.org/about-sids/
15. Fair Trade Tourism, "About Us," accessed July 14, 2016, http://www.fairtrade.travel/corporate

CHAPTER 9

The Social Geography of Tourism

For those of us who do not live in (or at least in the vicinity of) a tourism destination, we probably give very little thought to what tourism means to the people whose lives it directly affects on a daily basis. In the previous chapter, we began to examine some of the effects of tourism as an industry: We saw how tourism can impact peoples' lives directly, through personal income from work in tourism or related fields, or indirectly, through economic development at the destination resulting from tourism. However, tourism affects peoples and societies in far more ways than this.

The social geography of tourism gives us the means to explore more of the effects tourism has on people and their lives. **Social geography** is a topical branch in human geography that encompasses a range of perspectives on the relationships between society and space. Although society is a commonly used term, it is used in so many different ways that it can be difficult to conceptualize. Broadly, society refers to the ties or connections people have with others. More specifically, we can consider the ties between people either occupying a geographic space or connected by networks across space (e.g., common values, ways of life, political systems, or perceived identities). Thus, social geography might consider space as a setting for social interactions or the ways in which spaces are shaped by these interactions. In particular, the focus in studies of social geography is on issues of inequality.[1]

Social geography is the topical branch most closely related to the discipline of sociology and clearly overlaps with other branches of human geography, including economic geography, political geography, and especially cultural geography. To some extent, social geography has had a weaker relationship with the geography of tourism than some of these other branches. As we saw in the previous chapter, the positive economic effects have been the biggest factors driving tourism; therefore, a strong case may be made for the relationship between economic geography and tourism geography. Likewise, the negative environmental effects have been the most visible effects of tourism; thus, human–environment interactions have long been important in the geography of tourism, as we will see in the next chapter. In contrast, the social effects of tourism have received less attention in tourism geography studies. Historically, tourism took place within societies or between relatively similar societies (e.g., North American tourists in Western Europe), which generated few readily apparent social effects. In the modern world, it can be difficult to distinguish what effects might occur in a society as a direct result of tourism from changes that might occur as a result of the wider processes of globalization.

Nonetheless, social geography has an important part to play in our understanding of the geography of tourism. Tourism presents an opportunity for social interaction; tourism brings together groups of people who have little historic contact and/or little in common. Tourism can also be a catalyst for social change; the development of tourism activities may reshape long-standing cultural patterns and ways of life. As such, this chapter continues our discussion of the geographic effects of tourism by utilizing the tools and concepts of social geography to consider the potential benefits of tourism for tourists and local peoples. It also considers the potential costs of tourism, particularly with regard to people in the community in which tourism is taking place. Finally, it discusses the factors that might determine the type and extent of these effects, as well as possible measures to maximize the positive effects while minimizing the negative.

Social Benefits of Tourism

Much attention is given to the negative implications of the social interactions that take place in the context of tourism and the social changes that result from tourism. Yet, tourism can have positive outcomes as well. At its most idealistic, tourism has been conceptualized as a means of promoting global understanding and international peace.[2] On a large scale, this goal is considered naive and improbable. However, it should not be considered unrealistic that the experience of other places and their peoples can have a beneficial impact on tourists at a personal level. This, of course, is contingent upon both tourists and locals approaching the experience with civility and a willingness to learn about the other.

Tourism may also be argued to have the potential for social development at a destination. Although development is frequently discussed in economic terms, the concept can be extended to nonmaterial indicators as well. As such, tourism can generate positive social changes at a destination, such as new opportunities for segments of the local community or new developments that could improve the quality of life for local people. At the same time, tourism development may serve as a catalyst for movements to protect against wider (negative) social changes by supporting traditional ways of life and reinforcing social identities.

PERSONAL DEVELOPMENT

Tourism can provide the opportunity for interaction between peoples and places that might not otherwise occur. If this is undertaken with an openness and a sensitivity to other peoples and ways of life, tourism can be a beneficial experience for both tourists and local people at the destination visited. Tourists have the opportunity to not only meet other people but to also experience life in other societies firsthand. This allows tourists to gain a better understanding of, and perhaps even empathy for, that society.

Consequently, tourism can—even if only subconsciously—promote a greater geographic literacy of the world. Conversely, these experiences will also give tourists a new perspective on life in their own society and possibly even generate changes in their daily lives. These may be small changes, such as a desire to eat different kinds of foods, but it can also lead to major changes, such as moving to a place that he or she has visited, or marrying someone from that place. In fact, some tourists—particularly explorers and drifters—participate in tourism because of the potential for such transformative experiences. Although such potential effects on tourists are generally recognized, there has been little research on these effects.[3]

Residents at a destination also have the opportunity to interact with other groups of people. This is particularly applicable in the case of countries or communities that have a positive travel account (i.e., they have high rates of inbound tourism and low rates of outbound tourism). Thus, although they may not have the opportunity to travel to and experience other societies, these people still have the opportunity to meet and learn about the tourists who visit their destination. Meaningful engagement with tourists can promote greater cross-cultural understanding beyond stereotyped images.

Social media has the potential to extend this interaction. When tourists "friend" or "follow" people they meet in other places on social media sites, they maintain the connection with those people and continue to learn about each other in ways that would not have been possible at the destination. For example, when connections are made with residents of the destination, those individuals have the opportunity to see what tourists' daily lives are like through the posts they make after they return home. When connections are made with other tourists, both individuals can see what life is like in places they may not yet have had the chance to visit. Additionally, everyone involved gets to see how people in other places react to local or global events, which may help them to consider issues from other perspectives and gain a broader worldview.

In addition, tourism can promote a greater understanding of various issues and events. For example, tourists to a place like Jerusalem will have the opportunity to see significant historic and cultural attractions. However, for many tourists, an equally important part of their trip will be to gain a better understanding of this complex place. They may visit educational and cultural centers to learn about different groups of people and to hear about their experiences (figure 9.1). These tourists will return home with new perspectives on an issue that is often in the news.

SOCIAL DEVELOPMENT

Arguments for tourism have often cited the potential for new opportunities within a community. The economic benefits of tourism, such as jobs and income, can lead to social benefits as well. New types of jobs associated with tourism and the income from these jobs can provide local people with greater freedom to choose not only where they live but also how they live. This, of course, depends on one's perspective. For example, rural communities around the world have been experiencing declining opportunities for employment and livelihood. This creates a push factor, in other words, pushing people out of the region to other areas—typically urban ones—that might have greater

Figure 9.1. This American tourist had the opportunity to talk to Palestinians in the Dheisheh Refugee Camp in the West Bank. The camp was created in 1949 after the Arab Israeli War as a temporary refuge for around 3,000 people. Today, estimates suggest there are as many as 15,000 people living in the camp. *Source:* Larry Hardman

potential for access to more and/or better jobs. In some cases, people, especially those who are young, are attracted by the promise of a new and different life in these areas. For these people, the development of tourism might provide the ability to leave a traditional or stagnated community, with its real or perceived restrictions, for a more modern or vibrant community with access to jobs as well as other opportunities. Yet, in other cases, people prefer to maintain their life with family and friends in their home community. The extension of tourism into rural areas might provide them with this opportunity.

Access to jobs and income can create new opportunities for traditionally marginalized populations. For example, in many parts of the world, women have had few opportunities for employment outside of the home and/or in the formal sector of the economy (i.e., their domain has been in the private sphere). To some extent, the jobs created by tourism have not been incompatible with traditional roles for women, while giving them the opportunity for employment in the public sphere, increased social interaction, increased income to support themselves and/or their families, and possibly even financial independence from their family or spouse. As women gain a greater value in society and greater freedom, this can lay the foundation for additional social changes with the goal of reducing gender inequalities.

We have already seen that income from tourism can be reinvested in the destination. While this investment may be directed at improvements in the destination that will create and maintain an attractive environment for tourism, the local community can also benefit from these improvements. Improved public transportation systems that will facilitate the movement of tourists at the destination can also be used to increase the motilities of local people. Increased police protection and measures to reduce incidents of both petty and violent crimes may be done as a means of trying to promote a safe and stable environment for tourism investment and tourist visitation, but local people can benefit from increased security as well. The creation or beautification of open-access public spaces to ensure an attractive environment for tourism can contribute to increased community pride, satisfaction, and general quality of life among local people.

In addition, there is potential for a destination to use the income from tourism to invest in local and/or domestic social development. Again, these improvements may not be entirely unrelated to the tourism industry, but local people can be the beneficiaries nonetheless. For example, a destination might choose to invest in the human resource base by building or expanding schools, education, and training programs. This increasingly well-educated population will have more opportunities to improve their lives and contribute to the local/national economy, including in the tourism industry.

PRESERVATION

Perhaps the most frequently cited argument for tourism is preservation—in this case, the preservation of local ways of life, traditions, and identities. Tourism is a distinctly place-based activity. The trend away from the standardization of mass tourism has emphasized the need for destinations to be able to offer the unique combination of physical and human attributes that constitute a place. Accordingly, a place seeking to develop this type of tourism will have an impetus to maintain these attributes. In such a case, the place may resist the processes of globalization that contribute to a sameness between different parts of the world—a sense of placelessness (see chapter 12). This place might choose to limit the development of chain restaurants or big box stores and use tourism to support local businesses, like independent restaurants or mom-and-pop shops. Likewise, this need for a sense of distinctiveness at destinations can help reinforce or rejuvenate social identities that might otherwise be lost.

Tourism can support traditional ways of life or simply be more compatible with them than other forms of economic development. This might be in the form of the backward linkages discussed in the previous chapter with local farmers or fishermen who are able to continue their livelihood, with the tourism industry serving as a ready market for their produce. In the case of traditional societies, livelihood and culture may be linked. Therefore, if these societies can continue to practice long-standing patterns of livelihood, then they have greater ability to maintain their cultural heritage.

Tourism is cited as having the potential to maintain or even revitalize aspects of traditional culture, such as artistic performances or crafts. One of the criticisms of globalization is the loss of such traditions. However, certain categories of tourists seek unique experiences of places and souvenirs of these experiences. As such, tourism can provide the motivation for local people to continue to practice rituals, songs, dances, or theatrical performances, when these things might otherwise be abandoned in favor of the patterns of modern global culture. Similarly, the production of local arts and crafts for tourists' consumption has the potential to keep traditions and skills alive

Figure 9.2. When tourists demand high-quality products made at the destination, it allows skilled craftspeople to continue their work and carry on cultural traditions. This artist is working in a handicraft center in the old Tibetan quarter of Lhasa. *Source:* Velvet Nelson

when local craftspeople might otherwise find higher-wage industrial employment and local consumers have increased access to cheap, mass-produced merchandise. Even in light of the production of cheaper copies of traditional crafts, there is still likely to be a demand for high-quality, "authentic" items produced by local people using local materials (figure 9.2).

Social Costs of Tourism

Depending on how tourism is developed and managed at a given destination, there may be missed opportunities to achieve these potential social benefits. In the worst-case scenario, tourism may have social costs for peoples and places. Social interactions through tourism can result in culture clash and misperceptions about the other group.

Tourism development can contribute to a decrease in the quality of life for a local community by marginalizing certain sectors of the population, introducing social problems in communities that had few previously, and limiting local peoples' access to certain sites. Finally, tourism can also contribute to irreversible changes in traditional societies and cultural patterns or, ultimately, their destruction.

MISSED OPPORTUNITIES

For many people, tourism is associated with relaxation and pleasure. To meet this demand, tourism stakeholders may seek to provide entertaining and enjoyable experiences rather than ones that prompt tourists to challenge their opinions or to think about issues that might be difficult or uncomfortable. This represents a missed opportunity for tourism to act as a catalyst for personal or societal development. For example, the history of slavery in the United States continues to be a sensitive topic for many people. Tourism at plantation homes in the American South could be used to initiate a more open and honest discussion of slavery and its implications on modern race relations in the country. However, until the past few years, slavery was a topic that was avoided by the majority of plantation museums in the region (see box 9.1).

Box 9.1. Case Study: Plantation Tourism in the American South

> "As geographers, we are interested in understanding the process and politics of narrating the history of slavery through the South's museums, historical markers, and other places of memory."*

Heritage sites, such as museums, have the potential to engage the public with issues of the past that guide our understanding of the present. Despite the significance of slavery in American history, the issue has long gone unmentioned in the plantation museums that have become a distinct part of the Southern heritage tourism landscape (map 9.1). Slavery remains an uncomfortable subject for many people who do not want to be reminded of the inhumanity and trauma.† Heritage tourism stakeholders have relied on romanticized narratives of the plantation past to produce experiences that will best appeal to potential tourists.‡

Critical research on the topic began in the early 2000s. David Butler conducted a study of tourist brochures to understand what themes stakeholders were using to promote plantation sites to potential markets. He found that the brochures gave considerably more attention to the owners of the plantation, the architecture of the "Big House," the grounds, the crops, and the furnishings than to slavery. He described this as a process of "whitewashing."§

* Derek H. Alderman and Rachel M. Campbell, "Symbolic Excavation and the Artifact Politics of Remembering Slavery in the American South: Observations from Walterboro, South Carolina," *Southeastern Geographer* 48, no. 3 (2008): 339.

† Derek H. Alderman, "Surrogation and the Politics of Remembering Slavery in Savannah, Georgia (USA)," *Journal of Historical Geography* 36, no. 1 (2010): 91.

‡ David L. Butler, Perry L. Carter, and Owen J. Dwyer, "Imagining Plantations: Slavery, Dominant Narratives, and the Foreign Born," *Southeastern Geographer* 48, no. 3 (2008): 299.

§ David L. Butler, "Whitewashing Plantations," *International Journal of Hospitality & Tourism Administration* 2, no. 3–4 (2001): 166–7.

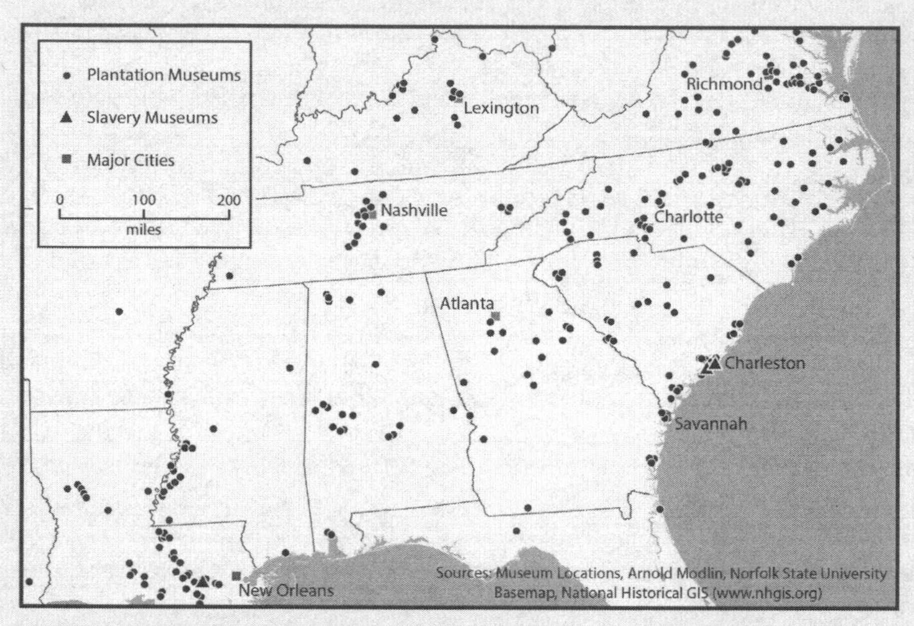

Map 9.1. Plantation museums in the American South. Plantation museums are a part of the heritage tourism landscape in the American South, while slavery museums are a relatively new addition. From "The Transformation of Racialized American Southern Heritage Landscape," NSF Award #1359780, Research Team: Derek Alderman, Candace Bright, David Butler, Perry Carter, Stephen P. Hanna, Arnold Modlin, and Amy Potter. *Source:* Stephen P. Hanna

Similarly, Jennifer Eichstedt and Stephen Small conducted a study of plantation tours. They found that the majority of plantations practiced "symbolic annihilation"; in other words, these sites ignored the institution and experience of slavery. This eliminates the significance of slavery and the enslaved to the region and its history. It implies that the topic is not important enough—to either stakeholders or visitors—to be acknowledged.[**]

Of the plantations that broached the topic, Eichstedt and Small described three additional categories. In the first of these categories, "trivialization and deflection," sites acknowledged slavery and/or the enslaved but primarily relied on romanticized narratives to control the discussion. While admitting that slavery was a part of the landscape, these narratives tap into visitors' ideas of the Southern plantation from popular culture (e.g., *Gone with the Wind*) and shield them from potentially unpleasant conversations. In the next category, "segregation and marginalization," sites offered representations of slavery and/or the enslaved but in separate spaces (e.g., tours of "slave cabins" in addition to those at the "Big House") or separate experiences (e.g., special interest tours with a different script than the standard tour). This also minimizes the subject as not important enough to be included in the typical visitor experience. Only a minority of visitors, generally those already interested in the topic, will undertake the extra effort required to visit additional sites or to plan a special tour in advance. In the final category, "relative incorporation," sites made an effort to address slavery but often failed to fully engage visitors with the issue.[††] Derek Alderman and Rachel Campbell argue that "incorporation" implies that slave histories are simply "added" to the existing narratives. Having been marginalized for so long, it is not that easy. They propose a "symbolic

** Jennifer L. Eichstedt and Stephen Small, *Representations of Slavery: Race and Ideology in Southern Plantation Museums* (Washington, DC: Smithsonian Institution Press, 2002): 10.

†† Eichstedt and Small, *Representations of Slavery*, 10–11.

excavation" in which the identities of the enslaved and their contributions to the plantation and the economy need to be uncovered and possibly even reconstructed.[‡‡]

In his original study, Butler asked the question, "Does it really matter?" The majority of tourists on vacation are looking for enjoyable experiences, and plantations are trying to provide those experiences, resulting in the processes described above. However, in answer to his own question, Butler reminds us that these heritage museums represent the past to visitors.[§§] Ignoring slavery at these sites leaves this past unresolved; ignoring the enslaved disregards them as important historical actors. Thus, the issue is very much connected to current understandings of race and racism.[***]

Although change can be slow, there are signs of progress since those first studies of the millennium. For example, Oak Alley Plantation is a popular site in the River Road area of Louisiana that presented a romanticized image of the antebellum South to visitors. In 2013, the site opened the "Slavery at Oak Alley" exhibit with newly constructed replica slave cabins, period or replica furniture and artifacts, and a wall bearing the names of the enslaved on the plantation. Stephen Hanna notes that visitors' questions and criticisms about the absence of slavery on the plantation—both during tours and in online reviews on sites like TripAdvisor—played an important role in the decision to create this exhibit.[†††] Matthew Cook discusses sites that employ "counter-narrative" strategies. These sites not only address slavery more directly but also work to emotionally engage visitors with the issue. One such site is Whitney Plantation, also in the River Road area. Opened in 2014, this plantation museum is dedicated to slavery.[‡‡‡]

Plantations have an opportunity to critically address slavery and work toward a better understanding of how it has shaped the region and the nation. Geographers, like those mentioned here among others, have taken a leading role in the research to understand and raise awareness about these issues.

Discussion topic: What factors should plantation museums consider in deciding how to represent slavery and the enslaved?

Tourism on the web: Oak Alley Plantation, "Antebellum Mansion, Historic Grounds, Restaurant & Inn," at http://www.oakalleyplantation.com; Whitney Plantation, "Home," http://www.whitneyplantation.com

---

‡‡ Alderman and Campbell, "Symbolic Excavation," 342.

§§ Butler, "Whitewashing Plantations," 173.

*** Christine N. Buzinde and Carla Almeida Santos, "Interpreting Slavery Tourism," *Annals of Tourism Research* 36, no. 3 (2009): 441.

††† Stephen P. Hanna, "Placing the Enslaved at Oak Alley Plantation: Narratives, Spatial Contexts, and the Limits of Surrogation," *Journal of Heritage Tourism* 11, no. 3 (2016): 224.

‡‡‡ Matthew R. Cook, "Counter-Narratives of Slavery in the Deep South: The Politics of Empathy Along and Beyond River Road," *Journal of Heritage Tourism* 11, no. 3 (2016): 304.

CULTURE CLASH

Although tourism can have a positive effect on tourists and locals if both approach the experience with an openness and a respect for others, it can also have a negative effect if one or both groups are closed-minded and/or hostile to the other. Tourists and local people alike have been guilty of this. Tourists may travel to a place in which they are not familiar with the language or customs. Whether this is merely a source of annoyance to the local community or a serious offense, tourism may come to be viewed as a detriment to the community rather than an asset. Local people may not treat outsiders with civility. The way tourists are received by local people can have a significant impact on how they view the destination and represent the destination to other potential visitors. As such, a place can develop a poor reputation, which can affect future tourism.

When different social groups come together—especially groups with significant differences in languages, ethnicities, religions, or lifestyles—the potential exists for misunderstandings and culture clash. For tourists, the interaction with these different groups can contribute to culture shock. This may be as minor as a feeling of uneasiness in the unfamiliar setting of the destination or as significant as a sense of complete disorientation. Culture shock is most likely to occur among tourists who have had little previous opportunity to interact with other social groups or those who are ill prepared for the differences they encounter. The greater the culture shock, the more likely it is that the tourist will be dissatisfied with the experience.

To some extent, local people may also experience culture shock; however, culture shock generally occurs with the initial encounter of another social group. Locals are likely to have more sustained contact with the tourist culture. Thus, local people are more likely to experience significant effects from tourism. For example, the **demonstration effect** is a term used to suggest that local people will experience changes in attitudes, values, or patterns of behavior as a result of observing tourists. Although it may be argued that the demonstration effect can be positive, it is far more likely to be negative.[4] In particular, the image that is presented by tourists on vacation is often substantially different from their patterns of behavior and consumption at home. This can be seen in the tourist inversions discussed in chapter 2, where tourists may dress more casually, behave more freely, and spend more money on food, alcohol, or luxury items than they would at home. As such, local people may develop considerable misconceptions about life in other parts of the world and, given the demonstration effect, may strive to emulate the worst parts of tourist culture.

The young people in a community are generally most susceptible to the demonstration effect, as they are quick to adopt outside values, dress codes, or lifestyles. This has the potential to cause significant problems within the local community. These new patterns may conflict with traditional views held by older generations, which can contribute to the creation of new social divisions between young people and elders. Yet, even as people aspire to emulate a particular lifestyle with attendant material possessions (e.g., technology), these things may remain inaccessible to them. Such items may be deemed incompatible with traditional lifestyles by other community members or by tourism stakeholders interested in promoting that lifestyle. Certain products may be unavailable through local distribution channels or simply too expensive for local income levels. Consequently, seeing this lifestyle on a regular basis without being able to obtain it can lead to frustration and resentment.

LOCAL DECLINE

While tourism can create new job opportunities when it is developed in a community, tourism developed outside a community can create a push factor for migration. This contributes to problems of brain drain, where these people are no longer contributing their human capital to their home community. Certain subsets of the population may be more likely to migrate. Depending on the location, this may be a gendered migration, where cultural norms (e.g., men are responsible for earning wages to support the family) or social opportunities (e.g., certain types of jobs, like housekeeping, target female workers) may make it easier for one gender to migrate for work than the other. This can contribute to a breakdown of family systems. Young people may be attracted to destination regions for the opportunities of new or higher-paying jobs, greater wealth, access to material possessions, or the ability to change their lifestyles. As a result, the remaining aging community may experience problems with stagnation and decline, further perpetuating the problem of emigration.

Tourism may introduce new preferences and patterns of behavior or consumption that can contribute to a decline in quality of life. For example, a tourism destination may experience an increase in fast food restaurants and restaurants catering to Western tastes. As local preferences change accordingly, people may eat more of these high-calorie, processed foods as opposed to fresh, healthy local options. This can create new problems like obesity, type 2 diabetes, and heart disease. Likewise, tourism can create the conditions that allow social problems that may or may not already exist—such as alcoholism, the sale of illegal drugs, prostitution, and/or gambling—to flourish. The arrival of affluent tourists may provide the inducement for local people to get involved in the provision of one of these activities. Conversely, wage labor from tourism may give local people increased disposable income to participate in such activities.

While some destinations may take measures to decrease crime, tourism can also contribute to an increase in crime. Differences in wealth between tourists and the local community may lead to a rise in robbery and muggings, and tourists are often seen as easy targets for these types of crimes in destinations around the world. The introduction of or increase in the drug trade, prostitution, or gambling may be associated with the rise of organized crime at a destination. Moreover, tourists indulging in the excesses of a destination have the potential to be either the victims or the perpetrators of physical and/or sexual assault.

Although tourism development can lead to improvements at a destination, local people may not always benefit from these improvements. The construction of tourist facilities or the designation of parks or reserves can lead to the displacement or segregation of local people. For example, in less developed countries, groups of people, often ethnic minorities, may have lived in a certain area for generations without having modern, legal land ownership. As this land becomes targeted for tourism development, the lack of legal ownership allows the government to sanction their removal. Similarly, tourism developments may limit local peoples' access to certain sites or facilities, such as beaches or parks. In some cases, this may be a result of the transition from public to private lands where the new owners prohibit local people. In other cases, this may be a function of a newly introduced fee structure that local people are unable to afford. The destination might see an increase in new goods available at local stores, but these goods may replace the ones traditionally used and may be beyond the means of local people.

CULTURAL EROSION

If efforts are not consciously made to preserve culture and ways of life, they can be eroded or even destroyed by tourism. The development of standardized mass tourism, particularly by multinational companies with globally recognized brands, can overwhelm and fundamentally alter the unique character of a place. A rural area or village may become more urbanized, and local businesses may be replaced. Traditional patterns of livelihood may be diminished or supplanted as people turn to tourism as easier (i.e., less physically demanding) or more lucrative employment.

Elements of local culture may be lost as people adopt elements of the tourist culture. For example, it is typically unrealistic to expect that tourists will learn a significant amount of the language spoken at their intended destination. It is, of course, also unrealistic to expect local people to learn all of the native languages spoken by the tourists they receive. As a result, there is a distinct need for a **lingua franca**, or a language that is used for the purpose of communication between people speaking different languages. As local people learn and speak this lingua franca to communicate with tourists in various service capacities, less emphasis may be placed on using local languages.

The very elements of culture that are intended to be preserved through tourism may also be destroyed by it, as they are changed to meet tourists' demands. Local cuisine may be made more palatable to tourists' tastes. Local objects that once had use and meaning within a society may be turned into something to be produced and sold as souvenirs (box 9.2). The mass production of these objects—possibly outsourced to factories with cheap labor—may result in a decline in the need for local skilled craftspeople. Rituals, songs, dances, or theater shows may be held for profit rather than for their original spiritual or social functions. Such events may be reformulated to make them easy to be performed for and understood by tourists. In some cases, they may be created specifically to fit tourists' preconceived ideas. Eventually, those within the society, including the craftspeople and performers, may lose their understanding of the original significance and/or meaning.

Finally, even those practices that continue to be undertaken by local people may be disrupted by the presence of tourists and tourism activities. For example, churches, temples, mosques, and other sacred sites may be presented to or perceived by tourists as attractions to be seen rather than places for practitioners to worship. These tourists may fail to respect religious ceremonies taking place or disturb those who came to pray (figure 9.3). Although these sites may not be logistically off-limits to local people, as in the case of site privatization, they become effectively unavailable to people who wish to use them for their original purpose.

Figure 9.3. Tourist information and guidebooks often highlight churches, temples, or mosques as tourism attractions and encourage visitation for historic and/or cultural reasons. As such, these sites may be perceived as open for tourists' pleasure as opposed to a place with a specific function for local people. *Source:* Tom Nelson

Box 9.2. In-Depth: Museumization and Commodification of Culture in Tourism

In our discussion of the potential effects of tourism, we are generally looking at extreme positions—the best- and worst-case scenarios. Reality is seldom that clear-cut. Tourism scholars delve more deeply into the issues associated with sociocultural impacts of tourism. Some ask questions about the nature of cultural change. Cultures are dynamic and evolve based on a wide variety of factors. For example, a community comprised of a particular ethnic group that feels its cultural heritage is eroding as a result of pressures to assimilate or the processes of globalization may seek to find ways to prevent it from being lost entirely. Tourism presents an opportunity for the community to ensure that traditions are maintained. In such cases, community members have the ability to identify the aspects of culture they wish to recognize and the ways in which these aspects will be presented to tourists.

However, even attempts to preserve culture can have negative consequences. In realizing the economic value of heritage tourism, the community may become museumized, with its culture essentially frozen at a point in time to match tourists' expectations. Particularly with ethnic or indigenous tourism, tourists may hold a certain image of "traditional" culture (e.g., patterns of dress or lack of technology). These tourists may be looking for authentic experiences, but they may not have complete or accurate information about the culture to be able to make an effective judgment about its authenticity. Likewise, external tourism stakeholders seeking to create a unique experience for tourists may manipulate cultural aspects to emphasize differences with modern or mainstream culture. Aspects of culture that do not fit this image, such as those that are recent developments or influenced by other groups, may be suppressed. This museumization prevents cultures from evolving naturally and can result in dissatisfaction as local people may come to resent being viewed as "backward."[*]

This leads to additional questions about the use of culture in tourism. The term **commodification** (sometimes also referred to as commoditization) is used to describe the way in which something of intrinsic or cultural value is transformed into a product with commercial value that can be packaged and sold for consumption (i.e., a commodity).[†] Although almost all aspects of a place—physical and human—may be commodified for the purposes of tourism, attention has particularly been focused on the commodification of culture. Commodification of culture can take place for many different reasons, but, as one scholar notes, "The most important characteristic is its *purposeful production* for tourism consumption."[‡] In other words, a tourism company or a local community may deliberately make changes or adaptations to aspects of culture based on the potential economic benefits that can be derived from tourism.

Some scholars argue that tourism inevitably results in cultural commodification. This is seen as particularly applicable in cases of mass tourism, in which the emphasis is on mass production and consumption of both tangible goods and experiences.[§] As the objects, performances, and so on are commodified for tourists' consumption, they are irrevocably changed. Many scholars have viewed these changes as a loss of authenticity. Yet, others argue that this need not be the case. A society can use—and perhaps adapt—an element of their culture in tourism, to receive the economic benefits of the industry. At the same time, they will be able to keep the most important, closely guarded elements of culture for themselves, with their original meanings or for their intended purposes. Indeed, there are some cases in which communities have entirely created cultural works (e.g., items to be sold as souvenirs) or practices

---

[*] Philip Feifan Xie, *Authenticating Ethnic Tourism* (Bristol: Channel View Publications, 2011), 109–11, 239.

[†] Erik Cohen, "Authenticity and Commoditization in Tourism," *Annals of Tourism Research* 15 (1988): 380.

[‡] Milena Ivanovic, *Cultural Tourism* (Cape Town: Juta, 2008), 121 (emphasis added).

[§] Robert Shepherd, "Commodification, Culture, and Tourism," *Tourist Studies* 2, no. 2 (2002): 185.

(e.g., festivals) for the purpose of tourism that have no traditional basis in the culture. These works or practices are empty of meaning, but they may be presented to tourists as authentic.**

Tourism's sociocultural impacts are complex and continue to prompt theoretical debate among scholars as well as practical case-based research. As we have already discussed, much of these effects are contingent upon how tourism is developed and managed in a place, who is involved in the process, and what are their objectives.

Discussion topic: Do you think culture should be turned into a saleable commodity? Do you think it is ethical to present something as authentic to tourists when it isn't?

** Nicola MacLeod, "Cultural Tourism: Aspects of Authenticity and Commodification," in *Cultural Tourism in a Changing World: Politics, Participation, and (Re)presentation*, ed. Melanie K. Smith and Mike Robinson (Clevedon: Channel View Publications, 2006), 177–8.

Factors in Social Effects

Similar to the economic effects, there are some instances in which it is easy to see the direct effects of tourism on a community. This is particularly the case when tourism activities are developed in relatively isolated and undeveloped communities. For example, the development of tourism in remote Amazonian indigenous communities will bring a host of changes to their society. This might include the construction of new infrastructure to support tourism (e.g., roads or modern bathroom facilities) and the importation of new products for tourists' consumption (e.g., bottled water), which might create further problems like package waste. It might require minor changes, such as in traditional patterns of dress (e.g., women covering their breasts), but it might also bring devastating consequences, such as diseases. Yet, in many cases, it may be difficult to separate what effects are a direct result of tourism and what would have occurred as a result of large societal changes. This is particularly the case when tourism activities occur in areas that are already well connected to modern, global culture. These destinations may be experiencing an erosion of local culture and unique social identities, but tourism's contribution may be indistinguishable from the effects of multinational corporations and the global media.

The specific social effects of tourism at a destination, and the extent of these effects, will vary widely. The often interrelated factors that determine these effects can include the type of tourists a destination receives, the number of tourists, the capability of the destination to handle these tourists, the spatial distance between tourists and local communities at the destination, and the type of interaction that takes place between tourists and local people. In addition, other factors might include the extent of similarities between tourists and locals, the origins of tourists, and the duration of exposure to other cultures.

The general typology of tourists discussed in chapter 2 (drifters, explorers, individual mass tourists, and organized mass tourists) may give us an indication of the number of each type of tourists a destination receives, the character of the destination, and/or the type of interaction that will take place between tourists and locals. Drifters and explorers typically arrive at a destination in relatively small numbers, whereas the categories of mass tourists account for large numbers of visitors at a destination. Thus, it might be anticipated that larger numbers of mass tourists will have more effects on the local community at the destination than smaller numbers of independent tourists. In some cases, this assumption might be accurate. For example, local residents in a small, emerging destination may be willing to welcome the drifters and explorers who seek to immerse themselves in the community and thereby have minimal negative impacts. In contrast, once that destination is "discovered" and increasingly visited by mass tourists who are less conscious of their impact on the destination, the community may begin to experience more negative social effects.

However, the number of tourists alone does not provide a complete picture. The character of the destination will affect its ability to handle the tourists it receives. In the example above, the emerging destination may have little infrastructure in place to accommodate even the slightest temporary increases in population associated with its tourism growth (e.g., overcrowding on local roads and public transport or at local restaurants and establishments). Tourists will be much more conspicuous, and local people may have little experience in dealing with outsiders. Consequently, the potential for incidences of culture clash is increased. In contrast, a well-developed tourism destination or a destination in a large urban area may be able to receive large quantities of mass tourists with little effect. The infrastructure is already in place to accommodate the demands of tourists without adversely affecting the needs of local people. Tourists may blend into existing population densities of the area, or they may simply be considered to have a normal presence in the community. Likewise, local people may be accustomed to dealing with tourists on a regular basis.

In addition, some mass tourism destinations were developed specifically to spatially isolate tourists from the local community. In this case, large quantities of tourists may visit the destination, but they will be concentrated within designated areas. This is particularly applicable in the case of enclave resorts (see chapter 11) and self-contained hotel complexes, such as those characterizing many popular 3 and 5S destinations like the Dominican Republic, which were constructed separate from existing communities. The only local people who have interactions with tourists are those who are employed in the resort community. As such, the effects of tourism are largely spatially contained, and local people may be able to live their lives as they choose and experience relatively few negative consequences from tourism. However, the potential for positive social exchange between tourists and locals will also likely be lost.

The type of tourist will also affect the type of interaction between tourists and locals. These interactions typically fall into one of three broad categories. The first category is the most formal and clearly demarcates the difference between tourists as consumers and local people as service providers. In this case, interaction takes place as tourists purchase goods and/or services from local people from street stands, at shops, in restaurants, or within the hotel/resort complex. In the second category, the distinction between insider and outsider becomes more blurred as both tourists and local people visit and use the same facilities, including beaches, parks, restaurants, or other entertainment venues. This spatial proximity increases the opportunity for contact between tourists and locals but does not necessarily indicate that meaningful interaction will occur. Finally, in the third category, tourists—and in some instances locals—seek interaction for the purpose of talking to, getting to know, and exchanging ideas with the other. This might take place in a structured experience, for example, when tourists take a guided tour not only to experience a place but also to gain knowledge about that place from the perspective of a local guide. This can also be something far more intimate and personal, such as when a local person invites a tourist to his or her home for a meal.

As we have already seen, organized mass tourists are primarily motivated by relaxation and self-indulgence and less interested in experiencing the place visited. They are more likely to stay at large multinational resorts. Local people typically have very little presence at these resorts, with the exception of those who are employed by the resort, due to financial barriers as well as physical ones (e.g., walls and gates). With a range of amenities available to them, these tourists have little need or desire to leave the resort. As such, their opportunities for contact with local people are extremely limited and most likely fall under the first category. Individual mass tourists and explorers may have greater interaction in the second category as they seek new places to experience outside of the resort/tourist areas. In addition, the nature of the destinations visited by explorers and drifters lends itself much more to this type of interaction. Because these destinations are less developed with tourism infrastructure, tourists will necessarily share spaces and facilities with local people. This automatically creates opportunities for interaction. Moreover, these tourists tend to be more interested in the experience of place, including experiences with people in the local community. Thus, they are most likely to seek out the type of interaction described in the third category.

The similarity or difference that exists between both the cultures and the levels of development for tourists and local people may also play a role in the type and extent of social effects from tourism. When tourists visit places where there are few major differences in cultural characteristics and levels of socioeconomic development, the potential for tourism to have distinct social effects is lessened. Tourists who have a similar appearance and patterns of dress and speak the same language as people at the destination are less likely to stand out as outsiders. Historically, this was often the case, as international tourism developed among the societies of Western Europe and North America. As people traveled within these regions, there was little evidence of social effects that would not have otherwise occurred within these societies. Yet, once these tourists began visiting new destinations in the less developed parts of the world, the social effects of tourism became far more apparent. In particular, the greatest social effects are likely to be seen in destinations where the local community is relatively small, isolated, and less developed both socially and economically.[5]

To some extent, the social effects discussed above are based on the assumption that the tourists who visit a particular destination are coming from similar regions of origin and therefore have a common culture. This is the case for some destinations. Based on the particular cultural influence, the destination begins to adopt and reflect that particular culture, as in the example of British tourists to Magaluf on Palma de Mallorca, Spain, discussed in chapter 3. However, destinations frequently receive tourists from different regions who all bring their own distinct cultural characteristics. This diversity of cultural patterns may weaken the influence of any one group of people on the destination, or it may simply reshape the destination in different ways.

Finally, one of the key differences between the impacts on tourists and locals is the duration of exposure to other peoples, cultures, and ways of life. The concept of **acculturation** is used to describe the process of exchange that takes place when two groups of people come into contact over time. Yet, this is rarely an equal exchange. One group is likely to have more of an impact on the other, and the second group will experience the greatest changes. Tourist–local interactions present an interesting case in acculturation. Although the potential exists for tourists to be influenced by what they experience at the destination, they are less likely to be affected and experience any real changes to their daily lives because each individual tourist experiences only short-term exposure to the destination culture. In contrast, local people experience sustained (at least for a portion of the year) exposure to tourists and their patterns. As a result, the local community is more likely to adopt these patterns and to experience more significant cultural changes.

Box 9.3. Experience: Life around Tourism

Given the various costs and benefits discussed in this chapter, residents of destinations often have justifiably conflicted feelings about tourism. Even if they do not directly benefit from the industry, they most likely receive some indirect benefits from the income and investment tourism brings to the area. At the same time, they will have to make accommodations for tourism in their community, from minor inconveniences to major life changes. While experiences will vary from place to place, and even person to person, Nancy's account below gives some idea of what her life is like in a tourism-dependent area.

Although I'm not from Alabama originally, that's where I live now. My town is a small one and definitely not on any tourist map, but the county it is in is very much shaped by tourism, with destinations such as Gulf Shores and Fairhope. Gulf Shores is the region's primary destination with its location on the Gulf of Mexico. I work for a local real estate company that manages approximately 1,600 properties in the area, mostly houses and condos that are used for vacation rentals. It's a pretty laid-back, warm-weather destination, where the beach is the main attraction. It is far less commercialized than other popular southern coastal resorts, like Myrtle Beach, South Carolina. There isn't much in the way of big hotels. In fact, there are only a handful of hotels on the beach at all. Most of the restaurants are independent and locally owned as well. Tourists who want to eat at one of the well-known chain restaurants have to drive 10 or 12 miles north of Gulf Shores to Foley.

Fairhope is a very different type of destination in the county, located just off Mobile Bay. It's an attractive small town that has become a center for the arts. A local committee is dedicated to making works of art available to the public by placing them around town along a walking trail. The downtown business district is made up of art galleries, high-end boutique shops, and nice restaurants. Some tourists will visit for the day just for these things, but the town also attracts visitors for the monthly art walks and the many different art shows, fairs, and festivals that they host over the course of the year.

You quickly learn the ins and outs of living and working around here. Summer is the big season, when we get a steady supply of families coming for a week. During this time, I know it's a good idea to leave for work early in the morning and to take back roads. I definitely avoid State Route 59 on Saturdays from late morning on; most tourists come Saturday to Saturday so traffic is always heavy. If I want to go out to eat for lunch, I need to make a reservation. Otherwise, it's going to take longer than my lunch hour with the increased wait times to get a table and to be served. In the winter, Gulf Shores sees a smaller number of snowbirds who come down from the north and stay for maybe a month, maybe three or four. Because of these tourists, the place doesn't "shut down" like some other summer destinations. Business is obviously slower, and some stores and restaurants change their hours of operation. A few places close their doors for a while: Jake's Steakhouse and Grill puts out a sign that reads "Gone fishing, eat at Bubba's (Seafood House)," which is next door and has the same owner.

Some of these things associated with the tourism industry can be an inconvenience when it comes to living in and around this area. But, in the end, I like life here. I appreciate the type of destinations we have; they are places I would like to visit if I were a tourist. I'm able to enjoy the same amenities as the tourists who visit for a week or a winter, whether it is the beach, the nature trails, the art galleries, the shopping (even if it's just window shopping when I cannot afford the upscale boutiques), the fresh seafood, and more. I would actually prefer to live in Gulf Shores. Since much of my time is spent there with work and other daily activities, I would love to be closer and cut down on my commute time. There's not much difference in the cost of living across the county. In fact, the primary disadvantage to living in Gulf Shores is not related to tourism at all: The cost of home ownership is considerably higher due to its coastal location and risk of hurricanes.

I had a friend who grew up in Gulf Shores once point out to me that they—the locals—had done something amazing with the tourism industry. They somehow managed to convince tourists to come in large numbers during the most uncomfortable (hot and humid) time of the year. This means that the locals are able to make a living off of the tourists during the season when they didn't really want to do anything outside anyway. Then, once the summer ends, the masses go home, and the locals are free to enjoy the beach, the town, and all of the other tourism resources during the best parts of the year!

—*Nancy*

Knowledge and Education

In comparison to the economic and environmental effects of tourism, there is relatively little knowledge about the social effects. Because the private sector is typically most concerned with economic effects, it has traditionally done little in the way of assessing the potential social effects of tourism development. If a private sector developer does undertake any form of assessment, it is most likely mandated by the public sector at the local, regional, or national scale.[6] The public sector may have a greater stake in ensuring the social well-being of its population; however, it too has often neglected to consider the social effects.

Although these social effects may seem to be distinct from the economic ones, they are interrelated. In particular, a successful tourism destination, which is often judged on economic criteria, depends on the support of the local community. These are the people who will have to deal with the consequences of tourism. If the local community is concerned about the negative social effects of tourism, they will not support its development. Moreover, if the local community experiences these negative effects, they may actively undermine or sabotage the tourism industry. While it is extremely difficult to predict what will happen as tourism develops at a particular destination, there is nonetheless a clear need for both the public sector and the private sector to investigate and understand what consequences might emerge from tourism in that place.

As there are traditionally few efforts to assess potential social effects, these effects are generally poorly incorporated into the planning process. Yet, both the public and the private sector can use a better knowledge of the social geography of the community

Figure 9.4. This sign is posted in Hulhumale, Maldives, where the predominantly Muslim community seeks to keep visitor behavior within accepted norms. Tourists who wish to dress and behave more freely will choose resort islands. *Source:* Nick Wise

under consideration to maximize the social benefits of tourism and to minimize the costs. This can contribute to general public policy decisions that protect local people and their rights, such as landownership and access to public lands, resources, or sites. It might involve destination policy decisions that seek to manage both the numbers of tourists and the circumstances in which interaction between visitors and locals takes place. Likewise, the destination might seek to establish policies that will control tourists' behavior (e.g., dress codes, codes of conduct, etc.) to fit within the cultural norms of the resident population (figure 9.4).

Education can go a long way in preventing the negative outcomes of tourist–local interactions at a destination. One of the most common complaints levied against tourists is ignorance of the place, its people, and their customs, which contributes to the process of culture clash. At the same time, this ignorance can be one of the key contributors to culture shock. Tourists are almost always encouraged to learn about a place before they visit. This helps ensure that the tourists are able to make an informed decision that their chosen destination will meet their expectations. Moreover, it helps the tourists understand what is expected of them so that they do not generate undue hostility toward themselves or cause offense to people in the local community. In addition, they should be willing to learn about the place through their experience of it. Although the situation is a bit more complicated for local people, education about tourists can help reduce misperceptions that also have the potential to contribute to culture clash.

Conclusion

Although the tourism literature has increasingly recognized the sociocultural impacts of tourism, there has been relatively little interface between social geography and the geography of tourism. While there may be some exceptions of tourism in extremely remote wilderness areas with little to no population, tourism will impact the local community in a multitude of minor and major ways. As is often the case, the worst examples of tourism—ignorant tourists, hostile locals, poorly planned developments—and their negative consequences typically get the most attention. Yet, these consequences are not necessarily unavoidable. Concerted efforts can be made by both tourists and tourism stakeholders to ensure that the negative social effects of tourism are minimized.

Key Terms

- acculturation
- commodification
- demonstration effect
- lingua franca
- social geography

Notes

1. Rachel Pain, Michael Barke, Duncan Fuller, Jamie Gough, Robert MacFarlane, and Graham Mowl, *Introducing Social Geographies* (London: Arnold, 2001), 1.
2. Stephen Pratt and Anyu Liu, "Does Tourism Really Lead to Peace? A Global View," *International Journal of Tourism Research* 18 (2016): 83.
3. Susan Horner and John Swarbrooke, *International Cases in Tourism Management* (Burlington, MA: Elsevier Butterworth-Heinemann, 2004), 22.
4. Stephen Williams, *Tourism Geography* (London: Routledge, 1998), 152–3.
5. Williams, *Tourism Geography*, 156.
6. Michael C. Hall and Alan Lew, *Understanding and Managing Tourism Impacts: An Integrated Approach* (New York: Routledge, 2009), 58.

CHAPTER 10

The Environmental Geography of Tourism

Tourism frequently gets linked to much-discussed environmental issues in the mass media. It is cited as an economic alternative to logging in the Amazon rain forest. It is used to argue against drilling for oil in the Arctic National Wildlife Refuge. It is considered to be the best chance for protecting rare and endangered wildlife species in sub-Saharan Africa. This connection between tourism and environmental issues brings together some of the topics that we have already discussed. In particular, it recognizes that the physical resources of a place can constitute very powerful attractions for tourism. It also recognizes that tourism is a viable economic activity that can be as profitable as or, in fact, more profitable in the long term than other, less environmentally sustainable economic activities.

The environmental geography of tourism allows us to explore this connection. Like tourism geography, **environmental geography** is a topical branch of geography that can be difficult to place within the field. Some scholars consider environmental geography to provide the geographic perspective on environmental science and therefore approach the topic as a "hard" science. Yet, this approach neglects a crucial component of environmental geography: people. Environmental geography is distinguished from other branches of physical geography in the recognition of and focus on the earth as the human environment. In other words, it considers the ways in which the environment affects people and people affect the environment. As such, environmental geography lies at the intersection of human and physical geography. Human-environment interaction is one of the long-standing traditions in geography and one of the key themes identified in chapter 1. Environmental geography provides the means of exploring this theme. While some geographers may approach the topic from a physical geography background (e.g., the science of human-induced climate change), others will do so from a human geography background (e.g., the human response to climate change).

Environmental geography has an important part to play in the geography of tourism. Natural attractions based on physical resources have long provided the basis for different tourism products in destinations around the globe. In the modern world, where millions of people live in highly developed urban areas, tourism provides a distinct opportunity for people to interact with the environment. To some extent, the relationship between tourism and the environment may be described as symbiotic: because tourism benefits from being located in high-quality environments, those same environments ought to benefit from measures of protection aimed at maintaining their value as tourism attractions. However, the incredible growth of the tourism industry has made it difficult to sustain this symbiosis. While tourism does indeed have the potential for enhancement and protection of the environment, it has also, in many cases, become a major source of environmental problems that have threatened to destroy those resources on which tourism depends.

This chapter continues our discussion of the geographic effects of tourism. Specifically, this chapter utilizes the tools and concepts from environmental geography to consider the possibility for tourism to positively contribute to the maintenance of high-quality environments, as well as the potential negative environmental consequences of tourism. It also discusses the factors that shape the nature of these effects and the need for education to maximize the positive effects while minimizing the negative effects.

Environmental Benefits of Tourism

The actions of tourists in a place are unlikely to result in any direct benefits for the environment. Essentially, when we undertake any type of activity—including tourism—in an environment, we cannot help but impact it in some way. Instead, it is tourism planning and development that is often seen as holding the potential to improve the environmental quality of the destination, maintain environmental standards, and/or preserve the environmental resources of that destination.

IMPROVEMENT

Tourism can provide a distinct impetus for cleaning up the environment of a place. Obviously the environment must be safe enough to allow tourist visits. Thus, the destination must ensure an appropriate level of environmental quality. When an existing destination has experienced damage or contamination as a result of, say, a natural disaster or an industrial accident, the affected sites must be cleaned up and/or restored before tourists can return. For example, in the aftermath of the 2004 Indian Ocean tsunami, affected destinations had to manage the disposal of debris and other solid wastes and to purify water sources that were contaminated by damaged septic tanks and sewage treatment infrastructure. Yet, sometimes even recovery efforts are not enough. In 1984, a toxic chemical leak at a Union Carbide pesticide plant in Bhopal, India, poisoned an estimated half a million people in what has been regarded as one of the worst industrial disasters in the world. Bhopal once had a reputation as a tourism destination based on its history, culture, and natural landscape. The perception of the city as "poisoned" persisted long after cleanup efforts.

Box 10.1. Case Study: A Proposed Oil Refinery on the "Nature Island" of Dominica

Dominica is a volcanic island in the eastern Caribbean, characterized by rugged mountains and lush tropical rain forest vegetation. Despite these restrictions in physical geography, the island was developed for export-oriented agriculture as early as the late eighteenth century and remained dependent on these industries for nearly two hundred years. However, agriculture fell into decline in the second half of the twentieth century with political independence and the end of preferential access to the United Kingdom market for bananas. As with the other islands in the region at that time, Dominica looked to tourism as a potential economic alternative. Unlike many of these islands, though, Dominica lacked the white sand beaches upon which mass 3S tourism is based. Thus, the national tourism organization sought to utilize the island's stunning natural landscape to develop small-scale nature-based tourism products. As such, they created a new identity for themselves as the "nature island," which was used to promote these tourism products.

Yet, Dominica's development policies have not always been consistent with this identity and vision. For example, in an effort to increase visitation rates, the government made significant investment in its cruise terminal facilities to allow more—and bigger—ships. However, mass cruise tourism is a less sustainable form of tourism, which brings large quantities of tourists to a destination for a short time. These tourists overwhelm the environment and may provide little economic benefit for the destination.

Then, the government of Venezuela proposed the construction of an oil refinery on Dominica. The government of Dominica forged a temporary agreement with Venezuela, contingent upon additional information, including an environmental impact assessment. While some local stakeholders argued for the economic benefits the refinery would bring to the island, others argued that any economic benefits would be negated by the loss of tourism. Organizations such as the island's Waitukubuli Ecological Foundation and the Dominica Hotel and Tourism Association perceived the proposed project as a dual threat: the threat of actual environmental degradation that would destroy the quality and aesthetic appearance of the nature island and the threat that the *idea* of an oil refinery would have on the concept of Dominica as the "nature island."

Among project opponents, it was taken for granted that an oil refinery would pollute the island's environment—particularly the air but also potentially land and water as well. More than that, however, opponents felt that an oil refinery was inconsistent, incongruous, and quite simply incompatible with Dominica's nature island identity and the nature-based tourism products that had been built on that identity. It was argued that even the possibility of constructing a refinery would irreparably damage Dominica's international reputation and that the tourists who had been drawn to the island's unique environment would no longer visit. This argument was supported by online reactions on both news and travel sites where international contributors—some who had previously visited and others with no prior experience with the island—claimed that they would, indeed, go somewhere else if the refinery was built. It proved to be a powerful argument.*

Although the refinery was never built on Dominica, the case highlights the importance of environmental quality—or at least perceptions thereof. No environmental

---

* Velvet Nelson, "'R.I.P. Nature Island': The Threat of a Proposed Oil Refinery on Dominica's Identity," *Social & Cultural Geography* 11, no. 8 (2010): 912–15.

Figure 10.1. Dominica is characterized by rugged terrain, steep cliffs, and black volcanic sand beaches. Although the island lacks the resources for mass 3S tourism development that characterize other Caribbean islands, it clearly possesses a different set of tourism resources that are particularly well suited to nature-based tourism products. *Source:* Velvet Nelson

impact assessment was conducted to determine projected effects of the project. As the world has been exposed to media images of smokestacks and oil slicks, it is not too hard to imagine opponents—locals and potential tourists—conjuring up their own ideas of what could happen on this island. Yet, this was not the first proposed oil refinery on a Caribbean island. Refineries can be found on other islands throughout the region, such as Aruba and Jamaica, that have thriving tourism industries. This leads us to our discussion questions.

Discussion topic: Do you think that the negative response to a refinery is unique to Dominica? Do you think tourism and industry are incompatible?

Tourism on the web: Discover Dominica Authority, "Discover Dominica, the Nature Island," at http://www.dominica.dm

A potential destination must restore the environmental quality of a brownfield site (i.e., land previously used for industrial purposes that may have been contaminated by low levels of toxins or pollutants) before tourism can be developed. Mines, factories, warehouses, and other industrial facilities may be abandoned after their operations have been shut down. While these places are often considered a form of visual or aesthetic pollution, there may also be a correlation between such derelict facilities and physical pollution or contamination. The land itself may have been damaged by industrial uses; hazardous chemicals may have leached into the soil or water sources; or the decaying infrastructure, such as buried or rusted pipes, may continue to contribute to environmental degradation. As long as the quality of the environment has not been irreparably damaged, these abandoned facilities may be reclaimed and redeveloped for tourism and recreation in a number of forms.

In some cases, the infrastructure may be preserved, essentially in its original state with some modifications to accommodate visitation, as a tourism attraction to highlight the heritage of the industry in that place. For example, mining in Cornwall and West Devon (United Kingdom) occurred from 1700 to the onset of World War I. This economic activity not only had a profound impact on the region and its culture, but its innovations influenced industrialized mining operations throughout the world. In recognition of this significance, the regional landscape has been granted UNESCO World Heritage Status. The site extends across ten different areas and features a range of educational and recreational experiences.[1]

In other cases, some of the infrastructure may be maintained but adapted for new purposes. Baltimore's Inner Harbor revitalization project (Maryland) is cited as a prime example of transforming derelict waterfront warehouses into an attractive shopping and entertainment district that, to some extent, maintains the industrial character of the area. Visit Baltimore, the official destination development and marketing organization for Greater Baltimore, states that the Inner Harbor "has many stories to tell, from its heyday in the 18th century as the nation's leading shipbuilding site, to a major site for oyster canning, steel working, railroad building, immigration port and military supply center…Today, the Inner Harbor is a major tourist destination and port of call for cruise ships."[2]

Finally, landscapes may also be cleared, leveled, contoured, and replanted to reestablish native flora, recreate habitat for native fauna, and develop an appropriate landscape for recreation, such as multiuse paths. Although landscape reclamation can be a costly process, laws increasingly require it in many parts of the world for environmental protection. In addition, grants and other resources may be made available for local communities to convert these areas for tourism as an alternative means of development once nonrenewable resources have been exploited by other economic activities. The Wilds provides an example of a tourist attraction on nearly ten thousand acres of reclaimed mine land in the state of Ohio. A nonprofit organization initially received the land as a donation from the Central Ohio Coal Company in 1986 and has since redeveloped it as a conservation center and open-range habitat for rare and endangered species from all over the world. Visitors to The Wilds can take a "safari"—an interactive wildlife tour—stay at the lodge, or participate in outdoor recreation activities like mountain biking or fly fishing.[3]

The environment must also be attractive enough to encourage and sustain tourist visits. This may be as simple as cleaning up litter on a beach (figure 10.1) or along a nature trail. However, it may also be as involved as improving wastewater treatment systems to prevent untreated discharge from reaching the ocean. This is fundamental in improving the quality of water at beaches that might be used for recreation. Otherwise, these areas would, at best, be perceived by tourists as dirty; at worst, visitors would be at risk for contracting waterborne diseases like gastroenteritis, hepatitis A, dysentery, and typhoid. In either case, tourists would be dissatisfied with their experience of the destination. As they express this dissatisfaction during the post-trip stage—either personally to family and friends or, in the age of Travel 2.0, publicly on blogs and travel-rating sites—the destination will obtain a negative reputation that can be extremely difficult to overcome.

Figure 10.2. Places seeking to become tourist destinations must ensure that their environment is attractive enough to encourage and sustain tourist visits. Belize's Southeast Coast is promoted as a destination for beachside activities, but this type of natural and human debris can present a significant deterrent. *Source:* Tom Nelson

MAINTENANCE

Tourism can provide the means of maintaining the environmental quality of a place. Tourism is often accompanied by infrastructure development. For example, the development of hotels and resorts, as well as the corresponding influx of tourists that temporarily increases the size of the population at a destination, can overwhelm environmental quality systems, such as wastewater treatment facilities. As a result, the developer may be required by the applicable government agency to either construct or contribute to the construction of these facilities. This may be a new facility that did not previously exist at the destination or an expanded facility better equipped to handle increased usage. This may also be an improved facility to ensure that the quality provided meets the standards of foreign tourists. Although tourism is the explicit reason for these changes, local residents may benefit from them as well.

Tourism revenues may also be reinvested in an environment. As tourism activities will likely have an impact on the environment in which they take place, a portion of the income from these activities may be allocated for measures to minimize impacts or repair damage from tourism. For example, nature trails need to be adequately planned and subsequently maintained to limit the extent of erosion, especially in areas expected to receive large quantities of visitors (figure 10.2). Such practices include stabilizing slopes, using natural vegetation to form buffers, and maintaining erosion control measures. These measures need not be expensive or high-tech: in the case of Grenada, the Caribbean island uses nutmeg shells—one of their primary agricultural products—as an organic means of mulching paths that are prone to get muddy and slippery with high traffic.

PRESERVATION

Tourism can also provide a clear rationale for preserving the environmental resources of a place. Environmental preservation has been one of the most significant arguments for tourism development. Many places around the world would be lost to industrial, commercial, or residential development if they were not set aside for the purpose of tourism. Tourism constitutes a viable economic alternative to these other, often more damaging forms of development. As a result, the land can be made economically productive while it is kept, more or less, in its original state. Trees are a resource that can be exploited by removing them from the land and selling them to paper and pulp mills, furniture manufacturers, the construction industry, and so on. However, the forest as a whole may be seen as a resource to be enjoyed by hikers, birdwatchers, and other nature enthusiasts. If tourism and recreation in a place is thought to be as valuable—or perhaps even more valuable in the long term—then the argument for preservation has greater weight, and the landscape can be maintained as a whole.

In some cases, private tourism stakeholders will recognize the potential for protecting the natural features of a place, and they will invest in nature tourism with the intention of ultimately generating a profit. The private sector has an important part to play in tourism. Particularly in less developed countries where local and/

Figure 10.3. Boardwalks on nature trails can serve many functions, such as making the trail more accessible, allowing for drainage, protecting vegetation, and controlling foot traffic. *Source:* Velvet Nelson

or national governments may have few resources to devote to preservation efforts, private individuals and companies may be better able to achieve these goals. For example, the Makasutu Culture Forest in the West African country of The Gambia is a project that began in 1992, when two individuals initially purchased four acres of land to build a small backpackers' lodge. When the surrounding forest became the target of deforestation, they realized the need for preservation of a much wider area. Today, the forest is a thousand-acre private reserve and the site of nature-based tourism activities.[4]

In other cases, the public sector takes a leading role in the landscape preservation and resource protection that is needed to create the foundation for tourism. Local and/or national governments may invest in or subsidize preservation, and at the same time, income generated from operator licenses or visitor fees can help finance site maintenance, resource protection from developers or poachers, and additional preservation. There are many categories of **protected areas**. The Convention on Biological

Diversity's definition of a protected area, accepted by 187 countries, is "a geographically defined area which is designated or regulated and managed to achieve specific conservation objectives."[5] The six overarching categories represent different levels of protection and allow for different types of activities. Category I includes strict nature reserves and wilderness areas. These areas are typically considered to be the most ecologically fragile, and therefore activities within them are the most restricted. The strict nature reserves are generally limited to scientific study (e.g., Snares Island Nature Reserve, New Zealand); wilderness areas preserve largely uninhabited and unmodified lands but may be managed to allow some visitation (e.g., Red Rock–Secret Mountain Wilderness, Arizona). Category II consists of national parks, which are perhaps the most commonly recognized protected areas. These are typically areas with unique natural and/or scenic qualities that are intended to preserve natural heritage and may be managed for scientific, educational, and/or recreational use (e.g., Galápagos National Park, Ecuador).

Box 10.2. Experience: Twenty Years in the National Parks

In the U.S. National Parks System, there are 412 areas, including everything from national parks to historical parks, battlefields, monuments, recreation areas, seashores, scenic trails, and more. The different categories of protected areas within and between units provide diverse opportunities for tourism for the more than 275 million visitors a year. Some of these visitors will make a point to see major attractions, such as the Old Faithful Geyser in Yellowstone National Park, while others, like Kim, want to fully explore these places that have been recognized as important to our national heritage. She has made this her travel priority, and she has had some amazing experiences over the years.

I have had the privilege of visiting 290 units in the National Park System. The "Passport to Your National Parks" program was started in 1986, where you can purchase a "passport" book that includes information about all of the areas in the park system and a space for your book to get stamped at each park you visit. It's a fun way of recording where you've been. I began visiting parks shortly after the program was started, and now I'm on my fifth book.

I think it's neat that we, as a country, have understood that our heritage is important and worth preserving. I'm glad that we have protected our many unique and diverse environments and that I have had the opportunity to experience them. There are so many simply wonderful parks, including places that few people know exist. I enjoy visiting all categories of parks, but I really love the big, wide-open backcountry parks in the West. These are the ones I will visit year after year.

Yellowstone National Park is one of my favorites. I've been there twelve times in the past twenty years. I usually go for about a week and either camp in the park or stay in a lodge. I get up early to see the sunrise, and then I spend my days hiking on some of the backcountry trails. The landscape is so unique, with great opportunities to view wildlife, especially if you're willing to get off the main road. Some people think that they can see the whole park from the loop road, and they want to know when the animals will "come out." In all of my years visiting and exploring the trails—including some old trails that are no longer maintained or published—I know I haven't seen the entire park. But I've seen lots of animals, including mountain goats, bighorn sheep, deer, antelope, elk, moose, buffalo, and bears. In fact, one of the most amazing (and a little scary) moments was when we witnessed a herd of buffalo charging because they were being chased by a bear!

I've seen a lot of changes over the years. Some changes are cyclical, and you can definitely tell when there's money being spent on the parks and when there isn't. I've seen the infrastructure get run down and facilities closed, but then a few years later, things will be open and repaired again. Other changes reflect the development of new—and often better—policies. My mother has some pictures of Yellowstone from the 1960s where bears are being fed out of car windows. They had been allowed to feed out of the park's trash dumps since the nineteenth century, and this was considered a tourist attraction. It wasn't until the 1970s that Yellowstone implemented new policies that kept people from feeding the bears and required visitors to properly store and dispose of food in the park. Today they are very careful, and everything is recycled that can be. This both cuts down on waste in the park and generates a small amount of income.

Some trails that could once be driven are now closed to cars. I think there are more areas that are open for easy access, but more of the backcountry areas have been closed off to the general public. You have to have a permit to go into the backcountry, which makes sense. It keeps the casual person from going into areas that they really shouldn't be in, and it makes sure the park rangers know where people are in case anything happens. I also like the fact they get updates from rangers in the backcountry, so when I check in before a hike, I get some additional information about what's going on in that area. Of course, some of these trails are closed during parts of the year as well, based on the condition of the trail or on the wildlife in that area. For example, I know that the Pelican Creek Trail is closed to hikers during the bears' mating season to prevent any interference or disruption of mating patterns. I think they've also done a good job at keeping these trails in good condition. On one of my last trips, we actually met one of the backcountry rangers on a trail and hiked with him for a while. He told us that he now spends a couple of months each year in the backcountry to do maintenance.

I appreciate the ability to go into the backcountry. I am most interested in getting "off the beaten path," hiking, and experiencing the quiet of nature. That's just me and my personality; I have to find my own way. However, there are places for everyone in the park; you just have to know where to go. At the big, easily accessible camping sites, you'll find lots of people and the big, luxury RVs. At the more primitive, remote sites, you'll find just a handful of hard-core campers. At the major entry points and scenic spots, you'll find tour buses unloading people from all over the world, but get a mile off the main road on a hiking trail, and you might be the only one out there.

—*Kim*

Category III is designated for natural monuments, which are intended to conserve specific features that have unique natural or cultural value (e.g., Devil's Tower Nation Monument, Wyoming). Category IV includes habitat and species management areas protected to prevent loss of biodiversity directly or indirectly due to the loss of habitat (e.g., Haleji Lake Wildlife Sanctuary, Pakistan). Category V comprises protected landscapes and/or seascapes and is intended to maintain the quality of human-environment interactions in that landscape that often take place in the form of tourism and recreation (e.g., Logarska Dolina Landscape Park, Slovenia). Finally, Category VI includes managed resource protection areas that have the least amount of restrictions. These areas need to be managed to allow for multiple uses that might combine sustainable resource harvesting with recreation activities (e.g., Tamshiyacu-Tahuayo Communal Reserve, Peru).

Each of these potential benefits—improvement, maintenance, and preservation—is just that: potential. Concerted efforts must be undertaken to recognize the value of the physical resources for tourism and to ensure the existence of a high-quality environment for tourism. Although this may be considered a necessary prerequisite for tourism, these efforts require knowledge, planning, and financial resources that destinations do not always have. Thus, it is important to carefully consider what efforts are being undertaken at a destination and how they weigh against the negative effects—or environmental costs—of tourism.

Environmental Costs of Tourism

The majority of research on the environmental effects of tourism has focused on costs, as the interactions between tourists and the environments of the places they visit are far more likely to have negative consequences. This includes resource consumption, pollution, and possibly even landscape destruction.

RESOURCE CONSUMPTION

Tourism—particularly large-scale mass tourism—can place a heavy demand on local resources. These demands are likely to be in competition with other local economic activities and/or residential uses. In many cases, tourism as an economic activity is given priority, which means that these resources will be unavailable for other uses. In the worst-case scenario, the high demand for resources from tourism activities depletes that resource, not only to the detriment of future tourism but to the detriment of all activities undertaken in that environment. This can include land, construction resources, water, fuel, and/or power supplies, among others.

Land is a resource that is needed for tourism infrastructure, which can be extensive, including airports, roads, accommodation facilities, entertainment venues, and more. While this extent of infrastructure may not necessarily be a prerequisite for tourism, it accompanies tourism development and facilitates tourism activities. As a result, local people may be displaced and other economic activities supplanted in prime tourism development land. Competition for land between tourism and local uses is typically most intense in small island destinations, such as those in the Caribbean or the South Pacific, where land is a scarce resource. This is something that may be taken for granted by the tourists who visit that destination and do not consider how that land was used before tourism or how it might otherwise be used. Yet, as an economic activity, tourism must create enough jobs to compensate for lost jobs in agriculture or other traditional activities. New jobs must provide sufficient wages for local people to purchase from external suppliers the food that they need to support their families to compensate for the loss of the food they once produced for themselves.

Additionally, various types of construction resources are consumed in the development of tourism infrastructure. Local lumber resources may be used in hotel/

resort construction. As such, this resource is no longer available for local construction or other uses (e.g., as a source of fuel). Sand may be mined from beaches to make concrete, for building construction, or for construction of roads and airport runways. In both cases, trees and sand may additionally have the potential to serve as a tourism resource. If they are removed from the environment to be used as a construction resource, this also means that they will no longer contribute to the tourism base.

Water is a resource that is used in high quantities in tourism. The accommodation sector accounts for some of the greatest water demands at tourism destinations. Water may be used on the hotel/resort property in decorative fountains and to maintain green vegetation and flower gardens, even at destinations that have a dry climate. Likewise, water may be used in swimming pools and spa facilities. Guests account for a proportion of accommodations' water usage. Western tourists, in particular, have reportedly high water uses with baths, long showers, and in some case multiple showers per day. Guest services also constitute a source of water consumption. This includes kitchen and restaurant facilities but is primarily accounted for with laundry facilities, as the guests' sheets and towels are typically washed daily (figure 10.4). In addition, tourism attractions and activities utilize varying quantities of water resources, including water parks, golf courses, botanical gardens, and ski resorts that rely on snowmakers. Given the economic importance of tourism, these activities may receive priority access to water resources. This means that less water will be available for other local economic activities and residential uses.

Figure 10.4. We are used to seeing signs in tourist accommodations asking us to consider the environment as we decide how frequently we want our linens laundered. This sign displayed in a Caribbean vacation rental puts a different spin on the issue. ***Source:*** **Velvet Nelson**

The situation is similar for local fuel and/or energy resources. Accommodations and tourism attractions may require high electricity consumption to power their operations. This can put a strain on the destination's electricity-generating capacity. To prevent shortages that would affect tourists and tourism activities, thereby creating dissatisfaction with the destination experience, power may be cut to local business and/or residential customers at peak times.

In cases where tourism is already developed, local resources may be exploited for tourists' consumption. Economically, a high demand for locally made souvenirs would be considered a positive. However, to meet this demand, local residents may mine coral to make craft items and jewelry or remove trees and grasses to make boxes, baskets, and mats, and so on.

POLLUTION

Although tourism has been described as a "smokeless industry,"[6] it can either contribute to or directly result in all types of pollution at both the local and global scales. Water and air pollution are considered the most severe problems, but other types of pollution—such as noise or visual pollution—can also be a concern for tourism destinations.

Because so many tourism activities are located in coastal areas or on lakes and rivers, water pollution is one of the most significant types of pollution associated with tourism. Untreated sewage is typically the largest source of water pollution from tourism. This is generally attributed to the fact that many international destinations around the world have either no sewage treatment and therefore discharge the sewage directly into the water supply, or an inadequate system and only a portion of sewage gets treated. Tourism growth may take place at a faster rate than infrastructure growth; consequently, existing systems cannot keep up with the seasonal increases in population. Cruise ships also constitute a significant source of tourism-related water pollution. Again, sewage is the main source but others include fuel leakages and the illegal dumping of solid waste and chemicals from onboard activities. Other sources include fuels from recreational boats, chemical fertilizers, herbicides or pesticides used on resort properties and golf courses that leach into the groundwater or run off into water supplies, and even lotions and oils on the skin of tourists swimming in the water.

As discussed earlier, poor-quality water can increase the risk of waterborne diseases for both the local and tourist populations. Likewise, water pollution can contaminate the food supply. More than just an aesthetic detraction, it can also have a negative impact on the tourism resources for the destination. For example, the discharge of untreated sewage into water causes eutrophication, which is a process of nutrient enrichment. This stimulates algae growth which can be unattractive, have an unpleasant odor, and cause ecosystem damage, such as the suffocation of coral reefs.

Air pollution is the type that currently receives the greatest attention due to the issue of global climate change (box 10.3). However, air pollution may also be a localized problem with distinct consequences for destinations. In particular, one of the fundamental components of tourism is travel, which is dependent on the transportation that is available based on present technology. Where there is a dense concentration of vehicles for tourism, vehicle exhaust contributes to poor air quality. Interestingly, urban areas may be less likely to see a direct correlation between tourism and air pollution. These areas may already experience high numbers of vehicles, and they may have higher restrictions on cars (e.g., emission controls) and traffic (e.g., taxes or permits for vehicles in inner-city zones or traffic-free zones). In addition, there may be strong incentives or disincentives (e.g., limited parking areas, high parking fees, or confusing traffic patterns) for tourists to use public transportation systems, such as buses or trains, to get around in urban destinations. In contrast, natural places like the U.S. National Parks have experienced increasing problems with pollution during high seasons. Tourists need personal cars to reach and get around at these destinations. In recent years, new policies have been put into place to limit traffic congestion in the parks. Visitors may be required to park in designated areas and use buses for transport within the park. Moreover, parks such as Yosemite have replaced traditional diesel buses with lower emission biodiesel and hybrid buses.

Box 10.3. In-Depth: Climate Change Mitigation in Tourism

The relationship between global climate change and tourism has been gaining attention in recent years. This has become a key issue because the tourism industry is considered to be highly sensitive to the effects of climate change, as was discussed in chapter 6. Yet, it is recognized among researchers and stakeholders that a two-way relationship exists. According to the UNWTO, tourism is both a victim of and a contributor to climate change.* We have already looked at the first part of this relationship in chapter 6; now we will look at the second part and how tourism can mitigate its effect on climate change.

Research estimates that the tourism industry is responsible for approximately 5 percent of the world's total greenhouse gas emissions. Thus, tourism is a smaller contributor to the problem than heavy industry, but it is a contributor nonetheless. Approximately three-fourths of these emissions come from transportation to, from, and at the destination. Although the proportion of emissions is divided relatively evenly between air transport and auto transport (40 percent and 32 percent, respectively), air transport accounts for a much smaller proportion of tourist trips (approximately 17 percent). The next largest source of emissions is accommodations, with approximately 21 percent (table 10.1). Considering the phenomenal growth in the global tourism industry, tourism-related greenhouse gas emissions are projected to increase 152 percent between 2005 and 2035 if no mitigation measures are taken.†

Table 10.1. Distribution of Greenhouse Gas Emissions in Tourism

Sector	Percent
Transport	**75%**
Air	40%
Auto	32%
Other	3%
Accommodations	**21%**
Activities	**4%**

* United Nations World Tourism Organization, "FAQ—Climate Change and Tourism," accessed July 14, 2016, http://sdt.unwto.org/content/faq-climate-change-and-tourism

‡ Daniel Scott, Bas Amelung, Suzanne Becken, Jean-Paul Ceron, Ghislan Dubois, Stefan Gössling, Paul Peeters, and Murray C. Simpson, *Climate Change and Tourism: Responding to Global Challenges, Summary* (Madrid: World Tourism Organization and United Nations Environment Programme, 2007),14 and 18.

Climate change mitigation is defined as the technological, economic, and sociocultural changes that can lead to reductions in greenhouse gas emissions. Although one solution to reducing energy consumption and greenhouse gas emissions is to reduce travel, this is clearly unrealistic in the modern world. Thus, four mitigation strategies have been proposed for reducing greenhouse gas emissions from tourism: reducing energy consumption (e.g., increasing the use of mass transport over private cars; installing hotel key-card systems to ensure that lights, appliances, and in-room air conditioning units are not running when the guests are not in the room), improving energy efficiency (e.g., using hybrid or electric vehicles, relying less on air conditioning in buildings that utilize natural ventilation patterns), increasing the use of renewable energy (e.g., using biofuels in vehicles, using solar water-heating systems), and offsetting carbon emissions through the development of renewable energy projects or the planting of trees to act as carbon sinks (e.g., fees paid by tourists based on the amount of emissions incurred during travel, or investment made by tourism businesses to offset the emissions incurred by their operations). This last strategy—offsetting emissions—is the most controversial, and critics have argued that it provides wealthy tourists with a means of easing their guilt without actually having to change their behavior.

Because the tourism system is international in nature, includes both public and private sector interests, and encompasses a range of different types of businesses and services, efforts to coordinate and regulate mitigation efforts are and will continue to be difficult. Moreover, the implementation of these mitigation strategies requires a financial investment that may be beyond the means of local tourism stakeholders. Yet, because tourism is perhaps more vulnerable to the effects of climate change than other industries, many stakeholders feel that it is vital to take a leading role in making changes. There has been a precedent set by some tourism industry associations and entrepreneurs who have voluntarily adopted mitigation strategies. However, for many small tourism operators around the world, there are still significant barriers, including a lack of sufficient knowledge about mitigation strategies, the expense associated with implementing strategies (e.g., purchasing renewable energy infrastructure, retrofitting facilities with energy efficient technology), and even tourist resistance to strategies.[‡]

Research has repeatedly shown that awareness of climate change does not mean that tourists will change their behavior to mitigate effects.[§] For example, tourists who are able to take one vacation per year are often unwilling to sacrifice their trip on the basis of environmental effects. Moreover, they will make choices to get the most from their experiences on vacation, like flying to get to the destination faster (versus bus or train) or leaving the air conditioning on in the room to ensure a cool and comfortable environment upon return. This behavior may even be contradictory to the tourists' behavior at home, but it is justified as short-term and out-of-the-ordinary.[**] As a vital part of the system, tourists must also be willing to make changes. Education about destination-specific climate change effects may help influence tourists' attitudes, but new policies may be needed to act as a catalyst for change.

Discussion topic: Find an example of a tourism venture that has implemented one or more of the climate change mitigation strategies. Identify what actions have been taken and discuss the potential benefits of these actions.

‡ Velvet Nelson, "Investigating Energy Issues in Dominica's Accommodations," *Tourism and Hospitality Research* 10 (2010): 353.

§ Peter Burns and Lyn Bibbings, "Climate Change and Tourism," in *The Routledge Handbook of Tourism and the Environment*, ed. Andrew Holden and David Fennell (London: Routledge, 2013), 413–4.

** Scott A. Cohen and James E.S. Higham, "Contradictions in Climate Concern: Performances at Home and Away," in *Tourism, Climate Change and Sustainability*, ed. Maharaj Vijay Reddy and Keith Wilkes (London: Routledge, 2013), 259.

Even when air pollution is not the direct result of tourism, it can have a negative impact on destinations. Air pollution adversely affects the health and quality of life of local residents. Similarly, tourists, particularly those with preexisting respiratory conditions such as allergies, asthma, or emphysema, can be affected by short-term exposure to environments with poor air quality. In addition, air pollution can adversely affect the quality of tourism resources. For example, air pollution has been cited as one of the greatest threats to ancient archaeological sites, such as the Parthenon in Athens, Greece. Related to air pollution, acid rain has the potential to damage forest resources, as has occurred in Germany's well-known Black Forest region.

Noise pollution takes place in areas with a dense concentration of tourism facilities and infrastructure, such as airports, roads, or entertainment districts. This type of pollution can generate annoyance and dissatisfaction among tourists. Tourists who are looking for relaxation, peace, and quiet will be frustrated with their experience if the destination is populated by large numbers of families with loud and boisterous children or young adults having loud parties late into the night. Of course, this issue may be avoided by thorough research resulting in an appropriate choice of destination and/or resort in the pre-trip stage. Likewise, tourists visiting natural and/or sacred sites where a reverence for the environment is expected will be unhappy with the levels of noise from excessive numbers of tourists, air or road traffic, and others. As with air pollution, this has been a growing problem in some of the popular U.S. National Parks. For example, noise levels have risen as a result of the increase in air traffic for scenic flights over the Grand Canyon. Noise pollution from tourism can also adversely affect local residents. For these people, the noise generated from tourism may be more than an annoyance; it may contribute to a general decline in their quality of life or contribute to serious physiological and psychological health concerns.

Visual pollution results in a decline in the aesthetic quality of an environment. This may occur when landscapes are changed by tourism development. The construction of tourism infrastructure may be considered visual pollution if it seems out of place in that particular environment. This can refer to the location of a hotel on an otherwise undeveloped beach or a ski lift or ski slope on an otherwise forested mountainside. Visual pollution may also occur as landscapes are degraded by tourist activities. One of the most commonly cited examples is the trash generated on mountain treks in remote destinations such as the Andes or the Himalayas. Early tourists and tour guides in these areas were not always careful about the waste that was produced during the course of the trip. As it was left behind, it became an unexpected source of visual pollution for future generations of tourists expecting a more pristine environment. Awareness of this issue has led to cleanup efforts and stricter regulations concerning waste disposal on such excursions. Finally, visual pollution can also result from tourism facilities that have been constructed without consideration for local environments, materials, and architectural styles.

LANDSCAPE DESTRUCTION

The various costs of tourism development and tourist activities can ultimately contribute to, or result in, the destruction of landscapes. In addition to changing the fundamental nature and appearance of the landscape, this can contribute to further environmental problems such as disruption of habitats, fragmentation of ecosystems, and reductions in biodiversity.

Much of the world's tourism development has taken place in coastal areas. There are several unique and specialized ecosystems in these areas—including sand dunes, coastal wetlands and mangroves, and coral reefs—all of which have been threatened by tourism development. Each of these ecosystems has a high level of biodiversity and helps protect the coastal land area from erosion and the potentially damaging effects of storm waves and tidal surges. However, sand dunes have been leveled and wetlands drained for beachfront hotel/resort development. This can lead to problems with erosion and beach loss, as well as an increased amount of silt in coastal waters, which will smother coral reefs. Tourists themselves may trample dunes and damage coral by touching it, standing on it, or taking pieces as souvenirs. The nature of the land may be lost to the construction of tourism infrastructure and facilities. In a forested environment, this contributes to local and global problems associated with deforestation, ranging from increased erosion to increased carbon dioxide in the atmosphere.

Finally, these environmental impacts can destroy habitat and disrupt the species that inhabit them. The destruction of an ecosystem can contribute to a loss of biodiversity, while the destruction of parts of the landscape may fragment the wider habitat and affect species' migration patterns. The encroachment of tourism activities into these ecosystems may bring species into closer contact with people, both tourists and tourism industry workers. As this can affect eating and/or breeding patterns, it can have long-term effects on the species.

Factors in Environmental Effects

As with the other effects, there are some clear examples of how tourism directly affects the environment. This is particularly applicable when tourism activities are developed in environments where few or only small-scale human activities otherwise occur. For example, the development of a ski resort in an undeveloped area involves considerable changes to the landscape: the removal of trees and boulders, the recontouring of the landscape to create runs, and the construction of roads, lifts, accommodation facilities, and more. This development alone—without considering the potential effects of operation—may contribute to or result in deforestation, habitat destruction, loss of biodiversity, destabilization of the slope, erosion, an increased risk of landslides and avalanches, and visual pollution. Yet, in many cases, it may be difficult to separate what effects directly result from tourism and what would have occurred as a result of residential and industrial activities undertaken by the local population. This is particularly the case when tourism activities occur in already densely populated, highly urbanized, and/or industrialized areas. These destinations may experience problems with water or air pollution, yet tourism's contribution may be indistinguishable from that of other local industries.

The specific environmental effects of tourism at a destination, and the extent of these effects, will likely vary widely. The factors that may determine these effects can include the quantity of tourists that visit the destination, the carrying capacity of the destination, the seasonality of tourism, the type of destination, the level of infrastructure, local environmental policies and regulations, and the nature of the environment at the destination.

For example, as with social effects, it is often assumed that the larger the quantity of tourists, the greater the environmental effects. This can be true when the number of tourists visiting that destination within a relatively short period of time exceeds its carrying capacity. Carrying capacity is a widely used concept in environmental geography, as well as related fields such as biology, to indicate the size of a species or population that an environment can support and sustain. Adapted and applied for our purposes, **tourism carrying capacity** refers to the number of tourists a destination or attraction can support and sustain. This helps the destination/attraction to understand its ability to withstand tourist use. Likewise, the destination must recognize that if the carrying capacity is exceeded, it is likely to result in varying degrees of damage that can diminish tourist satisfaction.

The tourism season can also be a factor in the nature and extent of effects. Large quantities of tourists during the high season may put extreme amounts of pressure on local resources. If the carrying capacity is greatly exceeded during this time, it may cause irreversible damage from which the environment of the destination will not be able to recover. Provided the damage is not irreversible, however, most destinations have a low or off season during which the site will receive few visitors. This may provide enough time for the environment to recover.

Similarly, the type of destination and the level of infrastructure will also play a role in what environmental effects may occur. For example, mass tourism is associated with higher quantities of tourists; thus, the potential for negative environmental effects may be multiplied in comparison with small-scale niche tourism. Indeed, many long-standing, popular mass tourism destinations have experienced some of the worst environmental effects of tourism. Over sixteen million tourists visit Myrtle Beach, South Carolina each year.[7] The destination has battled reputation problems due to reports of high levels of bacteria in the water, and during the 2016 season, temporary swimming advisories were issued for several sections of the beach. Yet, other well-planned and developed mass tourism destinations may have the infrastructure in place to handle such quantities of tourists and strict regulations to control negative impacts.

In contrast, when a new and/or developing destination starts to receive more than just a few drifters, the infrastructure simply may not be in place yet to handle these numbers, even though they are still small compared to large-scale mass destinations. Moreover, even small numbers of tourists can have a negative impact on the destination's environment. For example, hikers in backcountry areas can cause considerable damage when they stray from prescribed paths, leave ruts or scars, disturb wildlife, pick plants, fell trees for firewood, light campfires carelessly, or improperly discard waste.

The nature of the environment at the destination can also determine the extent of effects from tourism. Fragile ecosystems, such as mountains, rain forests, or coral reefs, may be more vulnerable than others in that they are less able to withstand human use and recover from overuse. Likewise, historic and prehistoric sites are also vulnerable and need to be highly regulated to ensure that they are not adversely affected by increased exhaust from car traffic, wear and tear from foot traffic, dust and debris deposits, and careless or malicious behavior (e.g., vandalism and theft). Each of these environments has lower tourism carrying capacities. In some cases, the benefits of tourism may be negated when more of the visitor entrance fees must go toward combating the problems generated by tourism rather than restoring and/or preserving additional sites.

Knowledge and Education

Education—of tourists, tourism industry workers, and local residents at a tourism destination—is often one of the simplest and easiest-to-implement means of preventing the negative environmental effects of tourism. In the case of tourists, ignorant and careless behavior can have a direct impact on the environment of the places they visit. Yet, tourists have little connection to these environments, and given the short-term nature of their experiences in these environments, they may not see the consequences of their behavior. For example, some tourists feel that they are paying for the services that a hotel or resort provides.[8] Thus, they will use the facilities as they see fit—whether it is having their linens laundered on a daily basis or leaving the room's lights, air conditioning, and/or appliances on when they are not in the room—without considering the implications of their wasteful resource consumption. Tourists may give little attention to their waste without considering that items left behind may alter the eating habits of local wildlife, cause some species to fall ill or die, or attract predators (figure 10.5). While only a small subset of tourists travel specifically for educational

Figure 10.5. This sign reminds visitors that deer on Miyajima Island, Japan, should be treated as wildlife and remain undisturbed. ***Source:*** **Velvet Nelson**

purposes, the potential nonetheless exists for tourists to learn about the places they visit and to understand the consequences of their actions at that place.

Tourism industry workers play an important role in the implementation of mitigation strategies. A destination may have good intentions in devising a code of conduct or a sustainable development policy (see chapter 11); however, these strategies will fail if tourism industry workers are not properly informed of it and do not understand its rationale. For example, most tourists have been made aware of water consumption issues use by hotel placards informing them of the destination's water resources and requesting that guests elect to reuse their linens. Many are willing to support this policy on the basis that they would not change their linens daily at home. However, it is too often the case that tourists find that their linens have been replaced by the housekeeping staff regardless of their decision. Similarly, tourists may be requested to separate their trash into designated bins for recycling, only to see staff dumping the bins together as waste. These tourists may become frustrated with the lack of follow-through and therefore ignore such requests in the future.

Finally, local residents must also understand the pertinent environmental issues of the destination and the strategies that are being undertaken to maintain its resources. Again, the best efforts of tourism stakeholders to develop activities with minimal environmental costs can be undermined by unsustainable activities undertaken by the local population. For example, the destination may seek to establish policies to conserve its resources—say sand or trees. However, if people in the local community have a basic need for these resources, or if they have little direct stake in tourism but can profit from the extraction of these resources, they will use them. This, of course, erodes the basis for tourism at the destination and contributes to a decline in the environmental quality and general quality of life in that place.

Conclusion

Environmental geography is a vital component in the geography of tourism, as it represents the intersection of people and environment. Although there are certainly exceptions, much of tourism involves some type of interaction between tourists and the environments of the places that they visit. As with all human–environment interactions, this will have an impact on the environment. While these impacts are more likely to have negative consequences, concerted efforts can be made by stakeholders at all scales (i.e., locally, nationally, and globally) to maximize the benefits that tourism can have for the environment at the destination and to minimize the costs.

Key Terms

- climate change mitigation
- environmental geography
- protected area
- tourism carrying capacity

Notes

1. UNESCO, "Cornwall and West Devon Mining Landscape," accessed June 29, 2016, http://whc.unesco.org/en/list/1215
2. Visit Baltimore, "Baltimore Harbor," accessed July 14, 2016, http://baltimore.org/neighborhoods-maps-transportation/inner-harbor
3. Columbus Zoo and Aquarium, "The Wilds—History," accessed July 14, 2016, https://thewilds.columbuszoo.org/home/about/about-the-wilds/history
4. Serenity Holidays Ltd., "Mandina and Makasutu: Our Story," accessed July 14, 2016, http://www.mandinalodges.com/our-story
5. Kalemani Jo Mulongoy and Stuart Chape, "Protected Areas and Biodiversity: An Overview of Key Issues," *United Nations Environment Programme*, accessed July 14, 2016, https://portals.iucn.org/library/sites/library/files/documents/2004-011.pdf
6. Andrew Holden, *Environment and Tourism*, 2nd ed. (London: Routledge, 2008): 67.
7. Myrtle Beach Area Chamber of Commerce, "2014 Annual Marketing Report," accessed June 29, 2016, http://www.myrtlebeachareamarketing.com/docs/preso/2014AnnualMarketingReport.pdf
8. Velvet Nelson, "Investigating Energy Issues in Dominica's Accommodations," *Tourism and Hospitality Research* 10 (2010): 353.

CHAPTER 11

Tourism and Sustainable Development

Over the past several decades, tourism has experienced extraordinary growth. With the sheer volume of people traveling internationally and domestically, stakeholders around the world have been forced to look beyond the positive effects of this tourism to consider the present and future negative effects. At the same time, the concept of sustainable development has evolved into both a way of understanding the world and a way of working to solve key issues in the world. Thus, sustainable development has been used to understand how all of the effects of tourism already discussed—economic, social, and environmental—are interconnected, and stakeholders are increasingly considering how tourism can be used as a vehicle for sustainable development.

In this process, adequate planning, development, and management of tourism are vital. While these issues may seem as though they would be best approached from a business perspective, the geographic perspective has been incredibly influential in developing concepts and strategies to ensure the best possible interactions between tourists and the places they visit.

This chapter considers the relationship between tourism and sustainable development as well as the strategies used to promote sustainable tourism. These strategies play an important role in determining the nature of a destination, the type of tourists that will visit, and the experiences they will have there. First, we need to look at the concept of sustainable development and how it has been applied in tourism. Then we will discuss some spatial management strategies that may be applied at destinations to maximize the economic, social, and environmental benefits of tourism and minimize its costs, to ensure long-term sustainability.

Sustainable Development

Although now more than thirty years old, sustainable development has been described as one of the most important concepts of our time.[1] The 1987 World Commission on Environment and Development Report titled *Our Common Future*, often referred to as the Brundtland Report after the chairperson of the commission, is widely recognized for its importance in shaping the concept and securing its place on the international political agenda. As defined in the report, sustainable development should be "development that meets the needs of the present without compromising the ability of future generations to meet their own needs."[2] This intergenerational idea of sustainable development resonated with many people throughout the world. It forced people to consider that the decisions made and the actions undertaken now may impact their ability, and their children's ability, to survive and thrive in the future. This prompted them to think about new ways of doing things.

The Brundtland Report laid the foundation for the United Nations Conference on Environment and Development that came to be known as the "Earth Summit" in Rio de Janeiro, Brazil in 1992. The purpose of this conference was to identify the principles of an agenda for action toward sustainable development, which continued to be viewed as development that would not threaten the needs of present and future generations. In the years that followed, however, this idea—while still important—was critiqued as being just an idea, not a practical approach that would help guide actions. By the 2002 United Nations World Summit on Sustainable Development in Johannesburg, South Africa, the focus had shifted to a holistic approach that linked economic development, social development, and environmental sustainability.

A typical conceptualization of this shows three overlapping circles, one each for economy, social (community), and environment. The middle of the diagram, where all three circles intersect, represents development that is sustainable (figure 11.1). That is the ideal we are working toward. In reality, development is often unbalanced, with emphasis placed on one or possibly two spheres at the expense of the other(s). However, sustainable development reminds us that they are all ultimately connected. Economic development cannot be maintained with a deteriorating environmental resource base, and the quality of the environment cannot be maintained in the face of poverty. The Rio+20 Summit further expanded upon this idea of interconnectedness. Sustainable development was viewed as a process of promoting sustained, inclusive, equitable economic growth and social development that would reduce inequalities and raise basic

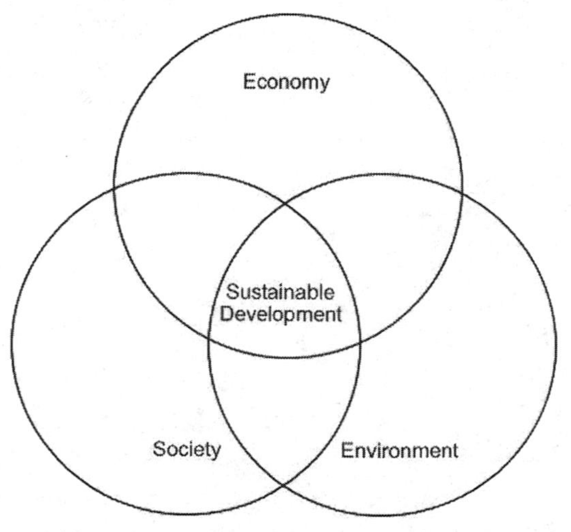

Figure 11.1. This graphic representation illustrates the three overlapping components of sustainable development: economy, society, and environment.

standards of living as well as promoting integrated and sustainable management of natural resources and ecosystems.[3]

The idea of sustainable development has been criticized as vague and open to different interpretations by various governmental agencies, NGOs, business interests, and researchers.[4] Some of the strictest interpretations seek to limit most forms of economic development on the basis that sustainability and development are mutually exclusive. Other interpretations take a far more relaxed stance and seek to allow all but the most destructive forms of development. In addition, critics of sustainable development have argued that, as an idea, it lacks practical applications that would guide the processes of development in light of the specific circumstances of places around the world.

In today's world, however, sustainable development is increasingly accepted as a means of trying to understand and work toward key issues. As such, various stakeholders are pushing the concept from an idea to an approach. At the global scale, world leaders took the opportunity to build upon the efforts started in 2000 with the creation of the Millennium Development Goals (MDGs). The MDGs were a set of eight goals (table 11.1), with related targets, primarily aimed at reducing extreme poverty in the less developed countries of the world over the course of a fifteen-year period.[5] As this period came to an end in 2015, the Sustainable Development Goals (SDGs) were established to focus efforts toward sustainable development over the next fifteen years all around the world, not just rural areas or less developed countries.[6]

Table 11.1. Millennium Development Goals (MDGs) and Sustainable Development Goals (SDGs)

MDGs	
Goal 1:	Eradicate extreme hunger and poverty
Goal 2:	Achieve universal primary education
Goal 3:	Promote gender equality and empower women
Goal 4:	Reduce child mortality
Goal 5:	Improve maternal health
Goal 6:	Combat HIV/AIDS, malaria, and other diseases
Goal 7:	Ensure environmental sustainability
Goal 8:	Develop a global partnership for development
SDGs	
Goal 1:	End poverty in all its forms everywhere
Goal 2:	End hunger, achieve food security and improved nutrition, and promote sustainable agriculture
Goal 3:	Ensure healthy lives and promote well-being for all at all ages
Goal 4:	Ensure inclusive and equitable quality education and promote lifelong learning opportunities for all

(continued)

Table 11.1 (*continued*)

Goal 5:	Achieve gender equality and empower all women and girls
Goal 6:	Ensure availability and sustainable management of water and sanitation for all
Goal 7:	Ensure access to affordable, reliable, sustainable, and modern energy for all
Goal 8:	Promote sustained, inclusive, and sustainable economic growth, full and productive employment, and decent work for all
Goal 9:	Build resilient infrastructure, promote inclusive and sustainable industrialization, and foster innovation
Goal 10:	Reduce inequality within and among countries
Goal 11:	Make cities and human settlements inclusive, safe, resilient, and sustainable
Goal 12:	Ensure sustainable consumption and production patterns
Goal 13:	Take urgent action to combat climate change and its impacts
Goal 14:	Conserve and sustainably use the oceans, seas, and marine resources for sustainable development
Goal 15:	Protect, restore, and promote sustainable use of terrestrial ecosystems, sustainably manage forests, combat desertification, halt and reverse land degradation, and halt biodiversity loss
Goal 16:	Promote peaceful and inclusive societies for sustainable development, provide access to justice for all, and build effective, accountable, and inclusive institutions at all levels
Goal 17:	Strengthen the means of implementation and revitalize the global partnership for sustainable development

Sustainable Tourism

The applications for sustainable development in tourism have generated considerable interest, as well as additional debate. There is no standard definition of sustainable tourism,[7] although many organizations have proposed their own ideas. The UNWTO definition states that sustainable tourism is "tourism that takes full account of its current and future economic, social and environmental impacts, addressing the needs of visitors, the industry, the environment and host communities."[8] This definition is expanded into three points. Sustainable tourism should be a viable, equitable, long-term economic venture offering stable employment and reducing poverty. It should respect and conserve local cultural heritage and values and promote intercultural understanding and tolerance. Finally, it should use environmental resources efficiently, maintain ecological processes, and conserve natural heritage and biodiversity.

Essentially, sustainable tourism should maximize the potential benefits discussed in the previous chapters.

Despite the general support for the ideas laid out above (and elsewhere), sustainability in tourism has been criticized as another buzzword used in government rhetoric and corporate public relations to generate a positive perception of the industry. Moreover, there have been different interpretations about how to apply these ideas in practice. A recent critique discussed three interpretations of the concept: the ability to maintain tourism over time, an approach to small-scale niche tourism, and a tool for sustainable development.[9]

In the first interpretation, sustainable tourism is approached from the typical definition of sustainable as "able to be maintained." Tourism stakeholders clearly have an interest in maintaining the industry over time. This perspective is criticized as placing an undue emphasis on economic interests. While social and environmental issues may be a lower priority, they must be considerations in the planning and maintenance of tourism nonetheless. As we have already seen, negative social and environmental effects from tourism have the potential to erode or destroy the foundation for tourism at a destination. Thus, to maintain the tourism industry, stakeholders must employ the holistic approach of sustainable development that encompasses economic development, social development, and environmental sustainability.

In the second interpretation, sustainable tourism is considered a response to the negative experiences of rampant mass tourism development. The concept is correlated to niche tourism, which is typically presented as an alternative to large-scale organized mass tourism. This is based on assumptions that small-scale niche tourism operators are more sensitive to social issues or environmental limits. Indeed, there are many scenarios that support this assumption. For example, local entrepreneurs who are members of the community and have a better understanding of, and relationship with, the social and natural environment of that place have a vested interest in working toward the betterment of that community. Likewise, ethically driven entrepreneurs are willing to invest in social and environmental initiatives that support their beliefs, even if it reduces their profit margin. However, there are also many scenarios where small-scale tourism operators do not have the time to devote to anything other than the day-to-day activities involved in running a business. They may not have the knowledge needed to properly plan and manage tourism or the capital needed to invest in sustainability strategies.

This perspective is also criticized because equating sustainable tourism with niche tourism implies that mass tourism cannot be undertaken sustainably. As an approach to tourism rather than a product, like those discussed in chapter 3, sustainable tourism has the potential to be extended to all forms of tourism. In fact, it may be argued that it is particularly important for sustainable practices to be implemented at mass tourism destinations because of the potential for negative consequences to be magnified by the large scale at which tourism takes place. There are certainly examples of unsustainable mass tourism development fueled by large, profit-driven corporations. However, in today's business environment, many companies have corporate responsibility programs that consider community relations and environmental impacts. Some even use a "triple bottom line" accounting framework that considers not only profit but also the social and environmental consequences of company policies and activities. In addition, corporations often have more resources to assess potential impacts and make investments in the appropriate strategies to minimize these impacts.

In the third interpretation, sustainable tourism is viewed as a tool that can be used to contribute to the goals of sustainable development. While this perspective is relatively well supported, the primary criticism is a lack of discussion or guidelines that would be of practical use in the development of such a sustainable tourism. Thus, in recent years, there has been much discussion of tourism's potential to contribute to the specific goals and targets laid out in the MDGs.[10] Using the framework of the goals, UNWTO resources offered suggestions for tourism organizations to consider in their corporate responsibility programs.

Three goals, in particular, presented the clearest opportunities for tourism. To support MDG 1—eradicate extreme poverty and hunger—businesses are encouraged to recruit workers and source products from the local community, support fair trade, and contribute to humanitarian efforts. For MDG 3—promote gender equality and empower women—organizations should promote education and training opportunities for women, from hosting craft skills workshops to providing technical training opportunities and supporting scholarships for managerial development. Businesses could also consider family support structures, such as flexible hours or in-house childcare, which would provide women with greater opportunities to participate in the workforce. For MDG 7—ensure environmental sustainability—stakeholders should be aware of and comply with environmental policies or recommendations, work toward greater resource efficiency, use renewable energies, recycle wastes, and support wildlife and ecosystem preservation.[11]

Preliminary research suggests that tour operators have a low-level awareness of the MDGs.[12] Thus, the potential for tourism to contribute to these goals will go unrealized. As the deadline for the MDGs came at the end of 2015, research into the full impact of these goals, and tourism's role, will continue in the coming years. In the meantime, the 2030 Agenda for Sustainable Development has already begun with the SDGs. Sustainable tourism is addressed in the targets for three of these goals, including SDG 8: promote sustained, inclusive, and sustainable economic growth, full and productive employment, and decent work for all; SDG 12: ensure sustainable consumption and production patterns; and SDG 14: conserve and sustainably use the oceans, seas, and marine resources for sustainable development. The UNWTO is already working to increase awareness of the goals and their relationship to tourism among stakeholders,[13] and we can expect to see more research on this topic in the future.

Planning, Developing, and Managing Destinations

Sustainable tourism depends on good planning, development, and management to minimize the negative and maximize the positive effects of tourism. Many strategies, borrowed from the topical branches of geography and other fields, have been adapted and proposed as means of developing the spaces of tourism in ways that will achieve this goal. Among others, some of these include participatory planning, construction regulations, land management strategies, carrying capacity analysis, and codes of conduct.

PLANNING

Planning is fundamental to the development and management of sustainable tourism. Through this process, stakeholders must identify both goals (i.e., what they want as well as what they do not want from tourism) and strategies that will allow them to achieve those goals. This provides a framework to direct the efforts of various stakeholders not only to facilitate the tourism experience but also the sustainability of the industry. The planning process can be used to coordinate with stakeholders in other industries as well. Resources may need to be managed for multiple uses (e.g., maritime resources used in tourism and fishing) to prevent competition or conflict that could undermine one or both industries. Planning provides opportunities to create linkages that can benefit other industries as well. While planning is important prior to tourism development, it is important to remember that tourism is a dynamic system. Planning needs to be conducted periodically to reflect changing circumstances.[14]

Planning alone is not sufficient to maximize the benefits of tourism and to minimize its costs. It is important to consider who is involved, and how they are involved, in the planning process. For example, outsiders (e.g., corporate representatives, public sector officials, or consultants) may take control of the process because they have knowledge and/or experience with tourism planning. However, they will not have the community's knowledge about the specific place in which the project will be developed. If only a select group of local stakeholders are consulted or are given a voice in the process, the needs and perspectives of the remaining stakeholders may be ignored. In either scenario, the project is unlikely to be sustainable. Although the answers to these questions—who and how—will be contingent upon community dynamics and sociocultural norms, in general, the poorest segments of the population, minorities, and women are the ones who are most commonly excluded from the planning process. For tourism to support the goals of sustainable development, though, these are the groups who must be involved in planning to ensure that they are able to experience the benefits of tourism.

Participation in planning can occur to varying degrees. At a minimum, participation involves information sharing. This means that potential stakeholders should be informed, in advance, about a planned project. Conversely, stakeholders may be asked to provide information that will be used in the planning process. Taken a step farther, this information sharing can lead to consultation, where stakeholders are invited to participate in discussions about the proposed project and its implications. This gives stakeholders an opportunity to voice their opinions, but if the project is still controlled by outsider experts, or powerful insiders, they may have little influence over the process. The next step involves collaborative decision making. In this scenario, stakeholder groups are given the opportunity to use their local knowledge to determine what strategies are most likely to yield the greatest benefits. Finally, empowerment is the highest degree of participation, in which stakeholders assume a leading role in the planning process as well as project development and management.[15]

Participatory planning is intended to engage all potential stakeholders in the process, especially those whose voices have not previously been heard. This helps "flatten" power relations and promote equality. Participation allows people to share their knowledge and gain confidence about their role in the community and/or project. At the same time, tourism planners and developers gain insight into the local circumstances as well as what, exactly, are the needs and priorities of the community. Incorporated into the planning process, this will contribute to the sustainability of the project. Working together at this stage helps communities and tourism operators establish relationships, which will be more likely to translate into benefits later (e.g., hiring from the local community or creating linkages with local suppliers). Finally, stakeholders become invested in and assume ownership over the project and are motivated to work toward its success.[16]

This type of participatory planning can be time consuming and possibly even contentious. Stakeholders will have different perspectives that may not be reconciled. Even efforts to try to engage all stakeholder groups in the process may not be successful. Existing power structures or social norms in the community may be too strong to be overcome to allow everyone to have a voice. Stakeholders may not trust each other to listen and to act with the community's best interests in mind as opposed to their own. Indeed, there are many challenges involved in the planning process, but the potential consequences are too great for it to be neglected.

DEVELOPMENT

Various policies can be used to limit the extent or to shape the character of tourism development. For example, construction regulations on many island destinations around the world mandate that buildings must be no taller than or even shorter than the height of surrounding vegetation, typically palm trees. This equates to two or three stories. This type of regulation seeks to restrict the potential negative impacts of mass tourism development. Essentially, it prevents the development of the high-rise megaresorts that dominate the coastal landscape of many popular 3 and 5S destinations. This effectively places limits on the numbers of tourists in that space, as well as the amount of resources used and waste generated by those tourists. Similarly, development companies may be required to submit architectural designs for tourism infrastructure to ensure that it does not conflict with local styles and become a source of visual pollution.

Spatial zoning is a land management strategy that designates permissible uses of an area based on its resources and/or character. In tourism, zoning determines what tourism activities may be undertaken where. Typically, governmental regulatory agencies identify the resources within particular areas of a destination, as well as the demand for tourism opportunities in those areas. Then, officials determine which areas have the most appropriate resources to allow those activities. GIS is increasingly being used as a tool in this process. Each zone permits an increasing amount of human activity. This includes primary conservation areas with strictly controlled access, natural areas with minimal facilities, and recreation areas with the greatest access and opportunities for hiking, fishing, camping, picnicking, and more.

Zoning may be used to either spatially concentrate tourists or disperse them. **Preferred sites** are typically planned locations that attract visitors through advertisements

Box 11.1. In-Depth: Enclave Tourism

Enclave tourism refers to geographically isolated and spatially concentrated tourism facilities and activities. We tend to associate enclave tourism with large-scale, all-inclusive resorts catering to the S tourism market. Indeed, the Dominican Republic—the Caribbean's largest destination—is often cited as a primary example of enclave tourism, where foreign tourists spend their entire vacation inside the walls of a resort compound. In fact, these resorts have been described as "concentration camps of leisure."* However, enclave tourism can be associated with other tourism products as well. These facilities can be located in remote rural areas based on different, primarily physical, resources.

Enclave tourism most commonly emerges in less developed countries that are using tourism as an economic development strategy. These almost exclusively foreign-owned facilities may develop as a function of constraints in the local infrastructure. For example, small, poor, rural communities will not have the extent of infrastructure—whether it is good-quality roads, reliable sources of power, or access to clean water—that is necessary to provide the base for tourism. A multinational company comes in and develops what they need within a specific geographic area. These self-contained spaces are separate and closed off from any existing communities in the area. This creates a form of spatial segregation. Local people will not receive the benefits of the newly developed infrastructure, and they will no longer have access to any resources that exist in that area. The company may also import food, equipment, and even personnel; therefore, economic leakages are high, and the local people receive few benefits from the newly developed industry. As a result, they may come to resent, and possibly even undermine, tourism. In addition, the resort will have standardized facilities and an appearance that has little relation to the wider characteristics of the place in which it is situated.

Foreign visitors constitute the primary market for such places. Enclave tourism typically reflects a package purchase, where visitors pay a single fee for everything from transportation to accommodation, meals, drinks, sightseeing excursions, and more. This fee is paid up-front to an agent located in the tourists' country of origin. For mass tourists, this is considered an attractive option because it is easy. The agent has already done the research, negotiation, and logistical arrangements of the trip.† For certain demographics of tourists, such as families with small children or senior citizens, it may also be considered attractive because it reduces or eliminates any potential hassles associated with patterns of movement at the destination. For example, the package may include dedicated transportation from the airport to the resort so that tourists will not have to deal with things like navigating unfamiliar roads, negotiating with potentially unscrupulous drivers, physically managing luggage, and so on.

Especially in the case of less developed nations, enclave tourism can present an attractive option for foreign tourists who have a demand for certain places but have doubts or concerns about the actual experience.‡ These enclaves are predicated on providing the expected level of quality in accommodations in otherwise materially poor areas and assurances that standards of sanitation and hygiene are at the level tourists demand. They provide reliable transportation services to/from the resort and activities outside of the resort, as well as whatever security is deemed necessary to ensure that tourists are safe, even in potentially volatile environments.

---

* Alastair Reid, "Reflections: Waiting for Columbus," *New Yorker*, February 24, 1992, 75.

† Wineaster Anderson, "Enclave Tourism and Its Socio-Economic Impact in Emerging Destinations," *Anatolia—An International Journal of Tourism and Hospitality Research* 22, no. 3 (2011): 362.

‡ Anderson, "Enclave Tourism," 363.

The S tourism enclave resorts are designed to be autonomous; nearly all of the facilities and/or services that tourists might desire are included in the initial price paid and can be obtained on site. Tourists visiting these resorts have little need or incentive to leave. They see little of the wider destination that they have traveled to, and there is little interaction with local people. Moreover, little additional money is spent at the destination and virtually none of it outside the resort. However, enclave tourism based on other products, such as nature tourism, will experience a slightly different variation on this pattern. These tourists will leave a clearly demarcated spatial area to have the intended experience, such as wildlife viewing, but the structure of the package keeps them isolated from the place nonetheless.

For example, the Okavango Delta in Botswana is one of the country's most important tourism destinations, as it is recognized as a wetland and wildlife habitat of international significance. There are approximately 122,000 people living in this area, which receives an average of 50,000 foreign tourists a year. This number has been growing in recent years as the Delta has come to be known as one of the world's new and exotic destinations. Yet, the tourism infrastructure remains poorly developed in Botswana. Thus, a form of enclave tourism was developed to meet the specific demands of international tourism without developing a comprehensive tourism infrastructure. These tourists arrive in the country's primate city (the city that is larger than all others in the country, usually in less developed countries) and are transported by the tour company directly to their resort. The tourists have virtually no interaction with the local population.

The majority of the resorts and tour companies operating in the Delta are foreign owned, and few of the goods and/or services used by these companies are obtained from the local population. Poverty levels among this population have been high, and it has only increased since tourism has come to the area. These people were not included in decisions about tourism, and they have received few benefits from the industry. Moreover, they no longer have access to certain areas, and they feel increasingly disconnected from the place. Given these factors, the local population has come to resent both foreign tourists and the tourism industry. According to one study, "interviews with community leaders and household representatives in the Okavango indicate that there is a general assumption that the Delta has been taken from them by government and given to foreign tour operators. . . . They believe that their resources have been usurped from them by foreign tourism investors."[§]

Amidst these criticisms, enclave tourism has been developed in destinations around the world. Recently, scholars have called for new research to explore options that will help extend the benefits of tourism beyond the enclave in these destinations.[**] Although there will be significant challenges in this process of changing the enclave model, it is viewed as vital to the promotion of sustainable tourism development.

Discussion topic: Describe a model of enclave tourism that you think would support the goals of sustainable development.

Tourism on the web: Botswana Tourism Organization, "Okavango Delta," at http://www.botswanatourism.co.bw/destination/okavango-delta

---

[§] Joseph E. Mbaiwa, "Enclave Tourism and Its Socio-Economic Impacts in the Okavango Delta, Botswana," *Tourism Management* 26 (2005): 163.

[**] Perunjodi Naidoo and Richard Sharpley, "Local Perceptions of the Relative Contributions of Enclave Tourism and Agritourism to Community Well-Being: The Case of Mauritius," *Journal of Destination Marketing & Management* 5 (2016): 19; Robin Nunkoo and Haywantee Ramkissoon, "Stakeholders' Views of Enclave Tourism: A Grounded Theory Approach," *Journal of Hospitality & Tourism Research* 40, no. 5 (2016): 558.

Figure 11.2. Managed natural areas may provide limited facilities for activities like camping, such as this space in Acadia National Park, Maine. This prevents users from environmentally destructive behavior as they attempt to create their own spaces. However, the majority of users will look for camping alternatives outside of the park that have more amenities such as bathrooms, showers, grills, refreshment stands, convenience stores, recreation areas, and more. *Source:* Kim Sinkhorn

and promotions; they have facilities like parking lots, restrooms, refreshments, picnic areas, designated paths, and/or information and interpretation centers. These sites spatially concentrate general visitors to ensure that their needs are met and to limit the effects of tourism to one particular area that is designed to handle it. The experience of these places may be enough for many visitors who do not feel the need to venture into other zones with less infrastructure and more fragile ecosystems (figure 11.2). In contrast, tools like planned scenic drives or tourist routes may be used to disperse tourists. These routes take people away from pressure points and spread them out over a wider area so as to not exceed the carrying capacity in one particular place.

MANAGEMENT

Stakeholders must take an active role in destination management to prevent the negative effects of tourism. This involves monitoring and regulating the use of attractions and facilities as well as managing patterns of visitor behavior. Carrying capacity analysis has been one technique for monitoring use. For the purposes of tourism, the carrying capacity concept has been used in different ways. Physical, environmental, perceptual, and social carrying capacities are some examples of the concept that are particularly useful in the geography of tourism.

Physical carrying capacity represents a somewhat literal interpretation of the concept in that it refers to the limits of a particular space. This may include things like the number of cars that a tourist site's parking area will hold or the actual number of people that the site can reasonably contain. As such, it is fairly straightforward and allows explicit restrictions to be put in place. **Environmental carrying capacity** refers to the extent of tourism that can take place at a site before its environment experiences negative effects. This can be more difficult to understand because it may not be based simply on the number of tourists but also on the type and accumulation of tourism activities. **Perceptual carrying capacity** refers to the extent of tourism that can take place at a site before tourist dissatisfaction occurs. This carrying capacity will be reached when tourists decide that a site is too crowded and choose to go elsewhere.[17] Perceptual carrying capacity can also be difficult to determine because the perceived level of crowding is primarily based not only on individual preferences but also on cultural conventions (box 11.2). Finally, **social carrying capacity** refers to the extent of tourism that can take place at a site before the local community becomes dissatisfied.[18] As with perceptual carrying capacity, social carrying capacity is contingent upon personal and social factors.

Box 11.2. Experience: Perceptual Carrying Capacity in Andean Peru

Perceptual carrying capacity can be difficult for stakeholders to manage because it is not only individual but also contextual. Those top-tier attractions, like Machu Picchu, are highly sought after. Consequently, most sites now have restrictions based on environmental carrying capacity. For many people, their perceptual carrying capacity is lower still. Nonetheless, they are willing to accept other visitors as part of the experience at these places in today's world, even if they are not yet ready to embrace them. In other places, or with other experiences, people may be less willing to push the limits of their perceptual carrying capacity. In addition, it can be hard to adjust our perceptual carrying capacities as places change. Foregoing the well-traveled Inca Trail, Ann elected to hike to the remote ruins of Choquequirao and encountered only a handful of other people along the way. That experience, however, may be threatened by new tourism developments.

Peru was one of my "bucket list" destinations. Like most people, I had seen amazing pictures of Machu Picchu, and I wanted to see it for myself. But I had also read stories about how the site was becoming increasingly overrun with visitors. Even the Inca Trail had gotten so crowded that the government had to put restrictions on how many people could hike it at any given time. It's not that I expected to see a place like Machu Picchu by myself, but I didn't want my experience of it to be overwhelmed by sheer numbers of people. So I made the decision to visit Peru sooner rather than later.

Since visitors are required to hire a guide to hike the Inca Trail, I started researching tour companies. In the process, I learned about some lesser-known routes through the Sacred Valley. In the end, I decided to skip the Inca Trail. Of course, I still went to Machu Picchu. Along with countless other people, I left Cusco on a bus to catch the train to Aguas Calientes and then another bus up the winding road to the site entrance. Still, I was fortunate that it was the low season, so I was able to experience the place without being crowded or rushed. It was as awesome as I had hoped, and I'm glad I went when I did. But it was the hike that turned out to be a truly extraordinary experience.

Described as the sister city to Machu Picchu, Choquequirao immediately caught my attention. The site is actually larger than Machu Picchu, but less than a third of it has been excavated. There are no roads or railroads to get there, only footpaths. On the first day, I hiked down about 4,000 feet to a camp near the Apurímac River. I saw one other couple who also camped there that night. The old bridge had collapsed with a landslide, so early on the second morning, I climbed into a basket and was carried across the rushing river by a hand pulley system. Then I had to hike up another 5,000 feet or so—to almost 10,000 feet—to the site. I live in a very flat region and had little opportunity to condition for hiking in the Andes, so that took most of the day.

It was late in the afternoon by the time I got to the entrance of Choquequirao. When I signed the visitor log, my name was the third on the list. In fact, for that whole week, the highest number of visitors per day was seven, and that included a family of five. I don't think I ever saw the other two visitors while I was at the site; I literally had the place to myself. As exhausted as I was, I climbed up even higher to get the bird's eye view of the site. The ruins were amazing, but it was the stillness of that place high in the mountains that took my breath away (well, that and probably the altitude). Then, after a few minutes, the clouds rolled in, settled over the mountain, and concealed the ruins. I think that made it even more impressive for me, that I had gotten a glimpse of something rare and special.

Since it took another two days to return the same way, I had plenty of time to talk to my guide. He was a kid from the little farming village just below the ruins. He told me that the government was planning to build a cable car up the mountain to transport visitors to Choquequirao. I know his view, but I don't pretend to understand all of the potential positives and negatives of such a project for the people who live in that area. Still, the idea of it made me sad. I later read an article that suggested the site could eventually get as many as three thousand visitors per day. I cannot even begin to describe the ways in which the experience would be changed. But things do change. I am sure someone was once saddened at the thought of a train being built that would one day take me to Machu Picchu. Perhaps it just makes us appreciate those rare experiences all the more.

—*Ann*

Figure 11.3. In the Peruvian Andes, less than one-third of the Incan site of Choquequirao is excavated. *Source:* Barret Bailey

Aside from the physical carrying capacity, it can be extremely difficult to quantify capacity limits. There are often many different economic, social, and environmental factors that influence the carrying capacity in a particular place. Instead, carrying capacity analysis is considered one part of ongoing management strategies. Destination stakeholders should monitor sites for potential environmental impacts and pay attention to both tourist and community attitudes about the extent of visitation through surveys or social media reactions. Ultimately, setting capacity limits will involve a value judgment based on the available information.

After identifying impacts and limits of tourism in particular places, stakeholders may need to regulate the usage of these places through various entry restrictions and/or fees. Only a certain number of visitors may be permitted at a site at any given time, or certain areas may be deemed off limits. Visitors may be required to travel with a guide. High fees may also be imposed to keep visitor numbers low. For example, Galápagos National Park is a highly attractive tourism destination but one that has a distinct need for controlling both tourist numbers and behavior. The park has an adult foreign tourist entrance fee of US$100, and all visitors are required to travel with a certified guide. Not only does this promote a high-quality visitor experience with interpretation of the islands' natural features, it also ensures that park rules and regulations are enforced.

Similarly, governmental agencies, NGOs, tourism industry associations, and even entrepreneurs have proposed guidelines or **codes of conduct** to help mitigate the negative effects of tourism. These codes may be targeted at any number of tourism stakeholders, including the industry, the local community, and tourists themselves. They help raise awareness about various issues, such as preservation of natural or cultural resources, with the aim of influencing patterns of behavior (figure 11.4).

Figure 11.4. The coastal waters off British Columbia and Washington State offer excellent opportunities to see whales, orca, dolphins, seals, sea lions, and more. Tour operators must follow strict guidelines about their interactions with these species, such as how close they can get for viewing. *Source:* Barret Bailey.

In one example, the well-known environmental NGO World Wildlife Federation (WWF) developed a set of codes specifically to minimize the negative environmental effects of tourism in the Mediterranean destination region. The overarching categories of their code of conduct for tourists include (1) support integration between environmental conservation and tourism development; (2) support the conservation of biodiversity; (3) use natural resources in a sustainable way; (4) minimize your environmental impact; (5) respect local cultures; (6) respect historic sites; (7) local communities should benefit from tourism; (8) choose a reputable tour operator involved in environmental protection with trained, professional staff; (9) make your trip an opportunity to learn about the Mediterranean; and (10) comply with regulations.

Correspondingly, the categories for their code of conduct for the tourism industry (defined as tour operators, hotels, and airlines) include (1) support integration between environmental conservation and tourism development; (2) support the conservation of biodiversity; (3) use natural resources in a sustainable way; (4) minimize consumption, waste, and pollution; (5) consider local cultures and attitudes; (6) respect historic sites; (7) provide benefits to local communities; (8) educate and train staff to support sustainable tourism; (9) ensure that tourism is educational; and (10) comply with regulations.[19]

Codes of conduct such as these have limitations in that the information outlined often takes the form of broad principles rather than specific policies. Consequently, the target audience of such codes may not have the necessary knowledge, or perhaps even the capacity, to carry out their recommendations. For example, in the first code, tourists are encouraged to choose a reputable tour operator. While this seems fairly obvious (after all, what tourist intentionally chooses a disreputable operator), tourists to a place may have little means of distinguishing between operators. In addition, the adoption of codes is voluntary rather than mandated. There may be little means of monitoring practices to ensure that the tourists, or tour operators, are adhering to the codes. Because of this concern, codes have been criticized as little more than marketing ploys that lead tourists to believe they are participating in sustainable tourism.[20]

Finally, as discussed in previous chapters, education is a valuable destination management tool.

The Evolution of Destinations

Over thirty years ago, Richard Butler proposed the **tourist area life cycle** (TALC) model, sometimes also referred to as the resort life cycle.[21] Ever since, it has been widely discussed and applied to cases of tourism development around the world. Butler has argued that there was (and perhaps still is) a need to challenge the prevailing ideal that once a place was established as a tourist destination, interest in and visits to it would be maintained indefinitely.[22] In reality, few destinations can remain unchanged over time. Tourism is a dynamic system. Transportation innovations have increased the accessibility of places around the world, and new destinations are developing all the time. With greater freedom to explore new, different, and unknown destinations, modern tourists have less place loyalty. A destination that does not respond to market trends will be perceived as outdated and unfashionable. Consequently, it will lose competitiveness in this highly competitive industry.

As a geographer, Butler's initial idea focused on the spatial implications of growth and development of tourism destinations. Using concepts such as carrying capacity, he argued that there are limits to growth. A destination that does not manage its tourism resources in light of the demands being placed on it will experience a decline in quality. This leads to a corresponding decline in the quality of tourists' experience and ultimately a decline in tourist visits. The TALC model provides a means of thinking about the development and evolution of destinations over the course of a series of stages. It describes changes in the character of the destination as well as in the types of tourists visiting and the nature of the effects from tourism there.

The first stage in the model is exploration. In this stage, tourists begin to be attracted to the destination for its inherent physical and/or human resources. Therefore, the primary attractions are most likely to be natural or human (not originally intended for tourism). The first tourists to "discover" the destination are typically adventurous, most likely categorized at the drifter end of the spectrum. With only a small number of tourists, the effects of tourism—positive or negative—are generally minimal. Given both the undeveloped nature of the destination and the type of tourists visiting in this stage, however, there is often a high level of interaction between tourists and local people.

The second stage is involvement. Following the arrival of the first tourists, local people begin to recognize the demand for tourism and develop new facilities. Subsequently, the public sector may offer some support, such as infrastructure development. The new stakeholders may begin to advertise the destination to encourage visits. Consequently, the destination experiences an increase in tourist numbers, more of whom would be characterized as explorers, and characteristics of the tourism industry in that place, such as season, become clearer. In this stage, the destination may experience some of the positive economic, social, and environmental effects of tourism.

The third stage is development. The number of tourists continues to increase, and more development occurs. Control over the tourism industry begins to pass from small business owners and local offices to a national governmental agency and large-scale, possibly even multinational, companies. With this continued development, individual mass tourists may begin to arrive. In this stage, the destination begins to experience many changes and an increase in the negative effects of tourism. For example, leakages will likely increase, tensions may build between locals and outsiders, and the overuse of resources may become apparent.

The fourth stage is consolidation. The tourism industry has become firmly established in that place, and the destination has become characterized by major multinational chain hotels and restaurants. A distinct central tourism district has emerged with a dense concentration of infrastructure and activities. However, many of the earliest facilities have become dated and may need to be upgraded. Tourism is the main economic contributor at the destination. With organized mass tourists visiting the destination, numbers are at a high, but the rate of increase begins to slow. In order to attract new markets, the destination may undertake widespread promotional campaigns.

The fifth stage is stagnation. The original natural and/or human attractions have been replaced by human-designed and artificial attractions. Consequently, the tourist area becomes divorced from the character of the place in which it is situated. The infrastructure has continued to deteriorate, and it experiences greater economic, social, and/or environmental consequences from the tourism industry. Although it is well known, it suffers from a poor reputation. The peak in tourism has been reached, and thus major promotion efforts must be undertaken. In addition, substantial discounts may need to be offered to maintain visitor numbers; however, cheap vacation packages will attract a new demographic of tourist that will further discourage earlier generations of tourists from returning.

The sixth stage can consist of decline, stabilization, or rejuvenation, depending on the decisions of destination stakeholders. If no action is taken, the tourist area will enter a period of decline. It may be immediate and drastic or slow and prolonged, but tourists will move on to other, newer destinations. The area will receive a smaller number of tourists from a more limited geographic area for weekend or day-trips. However, if minor adjustments are made and/or efforts are undertaken to better protect tourism resources, the tourist area may stabilize or perhaps see a limited amount of growth. If significant redevelopment projects are undertaken, the area can rejuvenate. This will involve investment in new facilities or upgrading existing ones. It may also involve creating new (human-designed) attractions, finding new ways to utilize previously untapped natural or human (not designed) attractions, and/or trying to attract new markets. If successful, this will create a new wave of growth.

TALC was created to be a general model of the tourism process. While it was intended to be applicable to destinations in various contexts around the world, not all will progress through the stages in the same way. The model has been criticized as being descriptive in nature, meaning that it is most useful in describing the process of development after it has occurred. Indeed, it can be difficult to identify and analyze the stages of a destination's development as it is taking place. Nonetheless, it can be predictive in the sense that it identifies what will happen if destinations are not appropriately planned, developed, and managed from the beginning. Although the concept of sustainable development had not been established when this model was proposed, the two are very closely related. As with sustainable development, TALC requires the acceptance of limits to development and necessitates a long-term perspective to minimize the negative effects of tourism.

Box 11.3. Case Study: The 2016 Olympic Games and Concerns about Sustainable Development in Rio de Janeiro

Cities around the world compete to host the Olympic Games. These mega-events hold the potential for improving the host country's international standing, boosting economic development, increasing tourism, promoting urban development, and more.* However, events also bring unprecedented global scrutiny of the host city and country not only during the event but before and after. This can generate criticisms of preexisting concerns in the host environment as well as issues emerging as a result of the games. For example, events have been increasingly criticized for negative environmental impacts, including greenhouse gas emissions associated with transportation to and from the site as well as from activities during the event, large quantities of waste generated and the lack of recycling, poor use of renewable

* John R. Gold and Margaret M. Gold, "Introduction," in *Olympic Cities: City Agendas, Planning and the World's Games, 1896-2016*, 2nd ed., ed. John R. Gold and Margaret M. Gold (London: Routledge, 2011): 6.

energy and material sources, the poor incorporation of new facilities into the surrounding landscape, and so on.[†]

In the wake of the Brundtland Report and the Earth Summit, the International Olympic Committee (IOC) began to consider issues of sustainability. In 1994, the IOC announced that it would consider the environmental consequences of candidate cities' proposals.[‡] In 1996, the Olympic Charter was amended to show that one of the roles of the IOC is "to encourage and support a responsible concern for environmental issues, to promote sustainable development in sport and require that the Olympic Games are held accordingly."[§] In 2005, the IOC established the Olympic Games Global Impact Project, which requires candidate/host cities to look at the economic, social, and environmental impacts of the games. Vancouver, host of the 2010 Winter Olympic Games, was the first city to go through the full process from proposal to post-games evaluation.[**]

Figure 11.5. With the 2016 Olympic Games, the world will become familiar with the sight of Rio de Janeiro, Brazil—but also with many of the country's economic, social, and environmental issues. *Source:* Heather Camacho-Cole.

At the time of writing, the opening ceremony of the 2016 Summer Olympic Games in Rio de Janeiro, Brazil, was two weeks away, and the media attention on the city was intense. Some commentators pointed out that Rio was not the first host city to face challenges prior to the start of the games.[††] Others, however, painted a far bleaker

[†] Arthur P.J. Mol, "Sustainability as Global Attractor: The Greening of the 2008 Beijing Olympics," *Global Networks* 10, no. 4 (2010): 516.

[‡] Gold and Gold, "Introduction," 3.

[§] International Olympic Committee, *Olympic Charter*, accessed July 20, 2016, https://stillmed.olympic.org/Documents/olympic_charter_en.pdf

[**] Gold and Gold, "Introduction," 5.

[††] Catherine E. Sciochet, *"Is Rio Ready? Other Olympic Host Cities Faced Problems, Too," CNN*, July 20, 2016, accessed July 20, 2016, http://www.cnn.com/2016/07/20/world/olympics-problems/

picture: Brazil has been facing economic recession with rising unemployment rates and inflation; the Zika epidemic; the impeachment of President Dilma Rousseff; and crime, corruption, and environmental crises. The state of Rio de Janeiro was recently forced to declare financial disaster and cut funding to many public sectors, including health care, education, and security. The latter is of particular concern; 85,000 police and soldiers were expected to provide security for the games, but there were police protests due to salary delays and poor working conditions.[‡‡]

While it was hoped that the infrastructure development associated with the games would create new opportunities for economic development, critics argued that the wealthiest segments of the population were the ones who stood to gain. In addition, many social and environmental issues remained unaddressed. For example, the number of people displaced by development projects associated with the games was contested, with numbers ranging from hundreds to tens of thousands, and the impacts of displacement were unknown (e.g., Where were people relocated? Were they compensated? Who would benefit from the new use of property?). Communities that were not displaced but affected by development complained that there was no dialogue with the government. Many lacked basic services, such as running water, safe electricity, sewage, and waste disposal, and some faced increased costs of living. Promised environmental cleanup efforts fell short of their targets. One water treatment facility was built but was not operational, while another facility was never built. Guanabara Bay, site of Olympic sailing events, remained highly polluted. In addition to the visual and olfactory detraction of this pollution, there were harmful levels of viruses and bacteria in the water.[§§]

The Olympic Games is not the cause of many of Rio's economic, social, and environmental problems. However, there was little indication that preparation for the games had done much to help some of these problems, and preparations may have exacerbated other problems. The effect of the games on the city and the country will not be known for some time to come. At the time of writing, though, an opinion poll showed that two out of three Brazilians believed that hosting the games would cause the country more harm than good.[***]

Discussion topic: Do you think the Olympic Games could be used as a tool for local sustainable development, or should the focus be on the sustainability of the games (i.e., minimizing negative impacts)? Explain.

Tourism on the web: City of Rio de Janeiro Tourism Authority, "Visit Rio," http://visit.rio/en/home-2/

[‡‡] "Majority of Brazilians Think Rio Olympics Will Cause More Harm Than Good," *Associated Press*, July 19, 2016, accessed July 20, 2016, http://olympics.cbc.ca/news/article/majority-brazilians-think-rio-olympics-will-cause-more-harm-than-good.html

[§§] Jonathan Watts and Bruce Douglas, "Rio Olympics: Who Are the Real Winners and Losers?" *The Guardian*, July 19, 2016, accessed July 20, 2016, https://www.theguardian.com/cities/2016/jul/19/rio-olympics-who-are-the-real-winners-and-losers

[***] "Majority of Brazilians."

Conclusion

Although sustainable development did not originate in the field, geography offers a natural approach to the concept. Geography integrates knowledge from both the physical and social sciences to understand the interrelationships that exist between people and their environment. Thus, geographers are already familiar with the holistic approach linking economy, society, and environment promoted in sustainable development. From this background, tourism geographers have been among the leaders in research on sustainable tourism. This research continues to search for tourism planning, development, and management strategies to maximize the benefits and minimize the costs of tourism. These strategies are vital in sustaining tourism at the destination, thereby extending its "life," as well as supporting sustainable development in that place.

Key Terms

- code of conduct
- enclave tourism
- environmental carrying capacity
- perceptual carrying capacity
- physical carrying capacity
- preferred sites
- social carrying capacity
- spatial zoning
- sustainable tourism
- tourist area life cycle

Notes

1. Jeffrey D. Sachs, *The Age of Sustainable Development* (New York: Columbia University Press, 2015), 1.
2. World Commission on Environment and Development, *Our Common Future: Report of the World Commission on Environment and Development* (Oxford: Oxford University, 1987), 41.
3. Sachs, *The Age of Sustainable Development*, 5–6.
4. Andrew Holden, *Environment and Tourism*, 2nd ed. (London: Routledge, 2008), 169.
5. United Nations, "United Nations Millennium Development Goals," accessed July 1, 2016, http://www.un.org/millenniumgoals/
6. United Nations, "SDGs: Sustainable Development Knowledge Platform," accessed July 1, 2016, https://sustainabledevelopment.un.org/sdgs
7. Edmund A. Spindler, "The History of Sustainability: The Origins and Effects of a Popular Concept," in *Sustainability in Tourism: A Multidisciplinary Approach*, ed. Ian Jenkins and Roland Schröder (Wiesbaden: Springer Gabler, 2013), 29.
8. United Nations World Tourism Organization, "Sustainable Development of Tourism," accessed July 5, 2016, http://sdt.unwto.org/content/about-us-5
9. Jörn W. Mundt, *Tourism and Sustainable Development: Reconsidering a Concept of Vague Policies* (Berlin: Erich Schmidt Verlag, 2011), 121–2.
10. Kelly S. Bricker, Rosemary Black, and Stuart Cottrell (eds.), *Sustainable Tourism & the Millennium Development Goals: Effecting Positive Change* (Burlington: Jones & Bartlett Learning, 2013); Jarkko Saarinen, Christian M. Rogerson, and Haretsebe Manwa (eds.), *Tourism and the Millennium Development Goals: Tourism, Local Communities and Development* (London: Routledge, 2013).
11. United Nations World Tourism Organization, "Tourism & the Millennium Development Goals (MDGs)," accessed July 5, 2016, http://icr.unwto.org/en/content/tourism-millennium-development-goals-mdgs
12. Marina Novelli and Alexander Hellwig, "The UN Millennium Development Goals, Tourism and Development: The Tour Operators' Perspective," *Current Issues in Tourism* 14, no. 3 (2011): 211.

13. United Nations World Tourism Organization, "Tourism and the Sustainable Development Goals," accessed July 5, 2016, http://www.e-unwto.org/doi/pdf/10.18111/9789284417254

14. Stephen Williams, *Tourism Geography* (London: Routledge, 1998), 128.

15. Peter Rogers, Kazi F. Jalal and John A. Boyd, *An Introduction to Sustainable Development* (London: Earthscan, 2008), 230.

16. Heather Mair, "Trust and Participatory Tourism Planning," in *Trust, Tourism Development and Planning*, ed. Robin Nunkoo and Stephen L.J. Smith (London: Routledge, 2015), 58.

17. Williams, *Tourism Geography*, 116.

18. Holden, *Environment and Tourism*, 188.

19. Simone Borelli and Stefania Minestrini, "WWF Mediterranean Programme," accessed July 14, 2016, http://www.monachus-guardian.org/library/medpro01.pdf

20. Peter Mason, *Tourism Impacts, Planning and Management*, 3rd ed. (London: Routledge, 2016), 199.

21. R.W. Butler, "The Concept of a Tourist Area Cycle Evolution: Implications for Management of Resources," *Canadian Geographer* 24, no. 1 (1980): 5–12.

22. Richard Butler, "The Resort Cycle Two Decades On," in *Tourism in the 21st Century: Lessons from Experience*, ed. Bill Faulkner, Gianna Moscardo, and Eric Laws (London: Continuum, 2000), 288.

Part IV

TOURISM AND PLACE

Destinations are the places of tourism. Just the idea of them is enough to captivate our imagination and create a demand for our experience of them. We formulate an idea in our minds of what we think it will be like and then, if we can, we try to turn these daydreams into reality. Given this opportunity, there are many factors that will shape our trip—from our expectations to the ways we choose to experience the destination. As tourists, we would never conceptualize this in geographic terms. However, as geographers, we know that our basic concepts help us better understand the patterns of the world, including those of tourism.

Geography is described as the study of places. The concept of place is used as a means of understanding the character of parts of the earth's surface as well as the ways in which people think about and interact with them. Adapted for the geography of tourism, we can use place as a tool to help us understand the character of tourism destinations, the ways in which people think about those destinations, the ways in which they interact with them, and how they represent them to others. Earlier in the book, we used the topical branches of geography to help us better understand the context of and key issues in tourism. In this final section, we will use place to explore tourism as a geographic phenomenon. Chapter 12 discusses how various representations—from those in literature to social media—create ideas about places as well as the experience of those places. Chapter 13 examines the tools of place promotion and branding in tourism, including the use of Travel 2.0. Finally, chapter 14 considers the ways in which tourists experience the places they visit.

CHAPTER 12

Representations of Place and Tourism

Place is a way of understanding the world. It refers to the parts of the earth that have been given meaning. Our understanding of places to which we have never been is shaped by the ways they are represented through media. The proliferation of media allows more places to be "experienced" than ever before. In fact, there are few parts of the world that we have not been exposed to in one form or another and therefore have an impression of what we think that place is like. Representations cannot replace direct experience. It may be easy to take these representations as accurate portrayals of reality, especially visual representations such as photographs and videos. However, as vivid as these images may be, they are nonetheless partial and selective. The audience becomes a passive observer who sees only what someone else has chosen for them to see. Still, representations of place are recognized to be extraordinarily important in tourism. As put by tourism scholar Dean MacCannell, "Usually, the first contact a sightseer has with a sight is not the sight itself but with some representation thereof."[1]

In this chapter, we will examine **place representations**. This describes the ways places are summarized and portrayed to an audience that then creates ideas and images about those places. This has implications for our discussion of the geography of tourism, as these ideas and images factor into tourists' decisions to visit a place and shape their expectations for their experiences there. The first part of the chapter introduces ideas about place representations and their relationship to tourism, while the second part provides a brief discussion of the types of popular and social media that contribute to our ideas about places and our expectations for tourism experiences in those places. Finally, the chapter concludes by examining some of the consequences of representations of places and their peoples through travel and tourism.

Types of Representations

As long as people have traveled, they have represented the places they have experienced through written descriptions and visual illustrations. These representations have proliferated exponentially since those times, as all parts of the world have become more accessible, more people have had the opportunity to travel, and new forms of media have allowed us to vicariously experience places in a multitude of ways. Many of these representations have no overt connection to tourism and are not explicitly intended to encourage visitation to the place depicted. Yet, these representations must be considered along with the less ambiguous promotional representations (chapter 13). Both have the potential to create distinct impressions of places in the minds of their audience, which factors into demand for travel as well as destination choice. Moreover, there may be an explicit relationship between popular media and place promotion. A film may highlight a place as a means of advertising, similar to product placement, or a place may draw upon popular literary, film, or even music references in its promotions.

POPULAR MEDIA

Literature—including plays, poetry, and prose—was one of the earliest representations of other places to popular audiences. This medium has been credited with creating some of the most powerful ideas of places discussed above. For example, far-off and exotic tropical islands frequently played an integral role in literature, not only providing the setting but also distinctly shaping the events of the story. William Shakespeare's *The Tempest* (early seventeenth century) was written at a time in which reports were coming back from parts of the world that were being "discovered," and his story contributed to the mythology of these places. Although it is unclear whether the setting for the play is an island off the coast of Africa or in the Caribbean, the specific location is less important than the idea of the place represented to audiences who would never have any direct experience with the tropical island environment. These ideas have persisted over time in works such as *Robinson Crusoe* (Daniel Defoe, 1719), *Treasure Island* (Robert Louis Stevenson, 1883), and *Lord of the Flies* (William Golding, 1954). Although these novels may not be read as widely as they were in the past, the stories and themes nonetheless remain familiar to us today. In fact, they are often updated and given modern twists, such as in Alex Garland's *The Beach* (1996).

Literature has also played a role in shaping ideas about distinct places and, consequently, in creating a demand for experiences there. For example, the Lake Poets (i.e., William Wordsworth, Samuel Coleridge, and Robert Southey) played an instrumental role in popularizing England's Lake District during the time period in which domestic tourism was expanding. In particular, Wordsworth's poetry is considered to be intimately connected to the region. Having lived most of his life among the lakes, he was often inspired by the landscape, which was represented in his work. In one of his most well-known works, "I Wandered Lonely As a Cloud" (1804), he reflected on a sight he encountered along Ullswater Lake. The readers of such works often became interested in experiencing these places for themselves, and new developments in tourism at the time, such as the railroad, increasingly allowed them to do so. Similarly, in the American context, authors such as Washington Irving and James Fenimore Cooper helped popularize the Hudson River Valley. Also at a time in which tourism was developing in the region, stories such as Irving's "Rip Van Winkle" (1819) and Cooper's *Last of the Mohicans* (1826) described real places that their readers might know or be able to experience for themselves.

Literature continues to represent places and shape ideas of them. Indeed, many avid readers will argue that written descriptions constitute some of the most powerful conceptions of place because they work in concert with their imagination (box 12.1). However, the rise of visual media has, to some extent, superseded the importance of literature. While novels still represent places, it is often the film adaptations of those novels that reach the widest audience. That the visual is extraordinarily important is evidenced by the fact that new versions of such novels are produced with images from the film replacing the original cover art. In the example of Alex Garland's *The Beach* discussed above, the novel was a best seller in Europe but had relatively little impact on the American market. Yet, the story is still well known today among American audiences from the movie version starring Leonardo DiCaprio (2000). After more than a decade, the film continues to inspire tourism, and rather than seeking the places that might have inspired Garland to write the story, tourists have flocked to the place in which the movie was filmed (figure 12.1).

Figure 12.1. Maya Bay, on Thailand's Phi Phi Leh Island, served as one of the film locations for The Beach. Today, the official tourism website for the province notes about Maya Bay: "Beautiful it is, secluded it isn't—thousands of people visit each day."* *Source:* Wesley Mills Eskonen

* Krabi Tourism, "Maya Bay Krabi—Thailand," Krabi Tourism, accessed July 11, 2016, http://www.krabi-tourism.com/phiphi/maya-bay.htm.

Box 12.1. Experience: Literature and Travel

Those people who are not avid readers might not understand why I included a section on literature in a book on geography and tourism. But avid readers will know. They will tell you how powerful words can be in igniting our imagination and appealing to our emotions. These words take us to new places, or back to ones we know well, and stimulate the demand to actually go there. Whether it is poetry and French literature, as Pamela describes below, or a popular young adult novel, there is a distinct relationship between literature and travel.

Armchair travel has long been associated with the comforts of a room of one's own, upholstery, and coziness. With an open book in hand, the body remains sedentary while the imagination roams. While reading can be a form of shoestring travel, it also inspires us to plan itineraries that engage our senses and, like pilgrims, experience sites of our beloved authors and characters. Travelogues are but one literary genre that inspire travel. Poetry has the power to stir us and draw us out of the chair.

Poetry compels me to travel. William Butler Yeats immortalizes the beauty of Glencar Falls with its lake in his poem "The Stolen Child": "Where the wandering water gushes/From the hills above Glen-Car,/ In pools among the rushes." Landscape's power to move us is the muse for art. As we walked and approached the falls, I could conceive of the young Yeats' fascination with the force of the water and its mist; lasting childhood impressions are imprinted on the adult poet who draws on Irish folklore to imagine the faeries as beings that inhabit the natural world. Standing in the place where the poem's speaker warns the child against being lured by the faeries, "Come away, O human child!" I experienced the conjoined spiritual power of poetry and the natural world. And so this poem was one of many literary works that shaped an itinerary to Ireland.

Fiction too has inspired travel, or shaped opinions about travel. A student of the French language, Hugo's *Notre-Dame de Paris*—or *The Hunchback of Notre Dame* in its English iteration—beckoned me to this European capital. A crowd of Parisians open the novel in a celebration of the Feast of Fools; today, you may be disappointed by the crowds of tourists flocking to climb the towers for a glimpse of gargoyles and the chimera who kept the outcast Quasimodo company. But those bells! One reads about Quasimodo's dutiful bell-ringing, but nothing—I mean, nothing—prepares you for the sound of those bells. Each time I hear them, in the tower or on the ground, I am filled with sound and transported back to a medieval Paris.

In grade school, Antoine de Saint-Exupéry's *Le Petit Prince* introduced me to the baobob tree. Later, my studies came to include Francophone literature of former French colonies such as Ken Bugal's *The Abandoned Baobob*. I read about baobab trees, cassava, harmattan winds, Wolof, and bright textiles. It was not until I arrived in Senegal traveling between Dakar, Saint Louis, and Thiès would I lay eyes on these ubiquitous trees. However, the most impressive—the Brobdingnagian tree—was on Gorée Island where a group of school children took shade under its limbs. The tree's grandeur and age left me speechless. I imagined such a tree had borne silent witness to the island's history in the slave trade. Senegal engaged my senses in so many ways. To taste a variation of a rice-based peanuty-spiced dish called *Tiébou* (Ceebu) means to sit around the platter with others, who included fellow U.S. travelers, our polyglot guides, and even beggars from the street hanging about the mosque. Depending on the hosts, one might do a quick rinse before eating, but one almost always eats with one's hands what is in one's vicinity. The male host might break up the meat and toss it your way. *Teranga*, or Senegalese hospitality, is not something you experience in a novel.

And finally, for when I want to return to the stillness and beauty of the Rocky Mountains or the desert of the Southwest, I know to which books I may turn. I pull out Edward Abbey's *Desert Solitaire* and imagine I am back hiking in Arches or Canyonlands. Picking up Ivan Doig's *The Whistling Season* brings back memories of traveling through the open spaces and mountain ranges of Montana. For me, armchair travel includes a return to the familiar book and an escape into the descriptive prose where I can go home in spirit.

—Pamela

In addition to influencing the way we think about places, films may also encourage or discourage us from visiting these places for ourselves. This has become such an important representation that tourism promoters will often seek to capitalize on the publicity that films generate. For example, following the release of Baz Lurhman's film *Australia* (2008), the country's national and regional tourism associations began to use the film images of Australian actors Nicole Kidman and Hugh Jackman in their promotions.

In some cases, the place plays such an important role in the film that tourists seek to re-create the characters' experiences by visiting featured sites. For example, visitors to New York City often seek out iconic spots such as Tiffany & Co., prominently featured in the classic *Breakfast at Tiffany's* (1961), or the American Museum of Natural History, made famous by *Night at the Museum* (2006). In other cases, epic films highlighting dramatic landscape scenery often create a demand for experiences in such places, even though tourists do not expect to replicate the experiences of the film. The *Lord of the Rings* trilogy (2001–2003), for example, continues to stimulate a demand for tourism to New Zealand based on the unique natural landscapes that provided the setting for the action of the film.

Interestingly, movies can also create an interest in and a demand for tourism in a place, even when the movie was not actually filmed at that location. For example, the *Twilight* series (2008–2012) is set in the town of Forks, Washington. Although the movies were not filmed in the town, it has nonetheless seen a boost in tourism, and the Forks Chamber of Commerce even offers a "Forks Twilight Map" for visitors. Parts of the first three *Pirates of the Caribbean* (2003–2007) movies were filmed on islands in the region and contributed to the already powerful imagery of the tropical island theme. The fourth (2011) and the soon-to-be-released fifth (2017) installments were, in fact, not filmed in the region but in Hawaii and Australia, respectively. However, as with *The Tempest,* the actual location of the tropical island in the film becomes less important than the ideas created about the place, and the title of the series continues to promote the Caribbean regardless.

At the same time, an unflattering representation of a place in a film can change viewers' ideas of that place and convince them that they have no desire to visit. Such movies often portray tourists getting caught up in a violent local conflict (e.g., *No Escape* [2015], in which an American family faces a violent uprising in an unspecified Southeast Asian nation) or finding themselves—deservedly or not—stuck in the miserable conditions of a foreign prison (e.g., *Brokedown Palace* [1999], in which a pair of young American tourists are accused of drug smuggling and imprisoned in Thailand). In other examples, the stark portrayal of a place in a film might contrast with generally glorified representations of that place. For example, Paris is typically associated with romantic imagery of lovers walking down the historic boulevards or kissing on a bridge over the Seine. However, the action film *Taken* (2009) shows a much darker side of the city—and a much scarier idea of traveling abroad—when the young American tourists are kidnapped to be sold in the sex trade.

Serialized television shows can serve a similar purpose in creating ideas about general place types, such as the tropical island theme re-created in the fantasy series *Lost,* as well as specific places, such as scenes from Portland, Oregon featured in the supernatural series *Grimm* (2011–2017). While these shows may not be the primary motivation for tourists to visit that particular place, they may still shape patterns of tourism. In the case of Hawaii, regular viewers of *Lost* (2004–2010) might visit the show's film locations, such as the beach containing the "wreckage" of Oceanic flight 815. In the case of Portland, viewers can find a host of resources online to help them visit film locations, such as Nick's house, the police department, and the spice and tea shop.

Finally, travel-themed television programs blur the boundaries between these types of popular media, the more specific travel-related media, and explicit place promotions. For example, *Samantha Brown's Asia* (2010) highlights different aspects of destinations throughout the region. In addition, the Travel Channel hosts many programs that have become increasingly specialized to focus on certain components of destinations, such as food in Andrew Zimmern's *Bizarre Foods* (2006—present). Although most shows are primarily intended for entertainment purposes, there may also be an instructive aspect in which prospective travelers to the featured destination might be able to use the programs to help them plan their own trip.

Representations of place in any of these media have the potential to create a demand for tourism. However, much of the audience for these media will not have the opportunity to visit the places represented and will therefore only experience them vicariously through representations. As such, this will contribute to the creation of suppressed demand (see chapter 2). Of course, it is also worth noting that unpleasant representations of place in these types of popular media can also create negative feelings toward that place, whether those associations are deserved or not.

TRAVEL LITERATURE

Few written records have been preserved from the earliest eras of Western travel; however, starting with the Age of Exploration in the Elizabethan Era, explorers and adventurers kept logs and journals of their journeys, some of which were later published. This practice continued with the next wave of scientific travelers. In fact, their "authority" was often contingent upon their published accounts of the places they visited.[2] The primary purpose of published travel journals was to convey information and descriptions about the new places being explored. One of the most well-known examples of this type of account is Charles Darwin's *The Voyage of the Beagle: Journal of Researches into the Natural History and Geology of the Countries Visited During the Voyage of H.M.S. Beagle Round the World* (1839). In this text, Darwin described the geography, geology, biology, and anthropology of the places that he visited, including Brazil, the Galápagos Islands, New Zealand, Madagascar, and St. Helena.

As travel and tourism continued to evolve, a new type of text also evolved, with elements of other genres. Travel literature might include the "objective" descriptions of earlier scientific accounts, as well as stories reminiscent of the adventure novels discussed above, and the advice found in the tourist guidebooks that also emerged in the early nineteenth century. Travelogues became a popular genre of English-language literature at this time, particularly in England but also in the United States.

Some of the first examples of travel literature came from established authors of both fiction and nonfiction, including the likes of James Fenimore Cooper, Charles Dickens, Henry James, Robert Louis Stevenson, and Mark Twain. The well-known writers of this period had the time and the means to travel when both were still scarce for much of the population. They also had connections with publishing houses and were household names that would generate interest in and ultimately sell their books. Such writers sought out new places to experience and were among the first tourists to visit these places. Still intended to be informative about other places, these books were also clearly written to be entertaining. They helped fulfill the readers' sense of adventure and satisfy their curiosity about places they might not have had the opportunity to visit themselves.

Yet, these narratives were also extraordinarily influential in generating new attitudes toward travel. The written accounts of early tourists' journeys were one of the first ways in which the majority of the population was exposed to the idea of travel for pleasure, which helped tourism to come to be seen as a normal activity. As tourism began to open up to a wider market, these written accounts of what other places were like encouraged travel. Although not strictly intended to be promotional, they had the power to influence what destinations tourists chose, what activities they participated in, and what expectations they had for their experiences.

During the prolific career of Anthony Trollope, one of the Victorian Era's most recognized English novelists, he also wrote travel narratives based on his experiences across the Americas, Australia and New Zealand, Africa, and Iceland. He described the places he encountered and his experiences—good and bad—with these places. His personal writing style gave readers the impression he was speaking directly to them as they planned a (real or imagined) trip of their own. In *Central & South America with the West Indies* (1858), Trollope wrote the following about hiking in the Blue Mountains of Jamaica to view the sunrise: "As for the true ascent—the nasty, damp, dirty, slippery, boot-destroying, shin-breaking, veritable mountain! Let me recommend my friends to let it alone."[3] After publication, this work was frequently cited in other books on the region. In particular, one of the early published travel guides, *The Pocket Guide to the West Indies* (1910), cited Trollope's narrative as one of the "volumes which, in the opinion of the writer, should prove most useful and interesting to those contemplating a visit to the West Indies."[4]

With this precedent, the new generations of tourists who traveled for pleasure also wrote letters, kept journals, and made sketches over the course of their journey, for their own record as well as for the purpose of publishing their own travel narrative. As a result, tourists increasingly wrote these books for other tourists.[5] Because the places visited were becoming more familiar to the readers, there was less need for these writers to provide the same extent of detailed, descriptive information. Instead, they gave greater emphasis to their *experience* of place. As writers highlighted the specific sights they saw and provided advice based on what they did, potential tourists could essentially follow the itineraries that had been laid out by those who went before them.

At the same time, publishing companies began to produce explicit guidebooks to cater to the burgeoning tourist market. At the forefront of this industry, Thomas Cook began to put together a guidebook for his expeditions to describe the places that would be encountered and the sights seen during the course of the journey. On this side of the Atlantic, Gideon Minor Davison has been credited with producing the first American guidebook. It described a specific route—termed the "fashionable tour"—that Davison intended readers to follow. One of the unique characteristics of this new book, however, was that it was small and cheaply printed. As such, it was intended to be portable for the duration of the trip and disposable after the trip was completed.[6]

By the late nineteenth century, magazines were publishing not only special articles by travel writers but also dedicated travel columns that also provided advice to potential tourists. For example, the British women's magazine *Queen* was one of the first to make "The Tourist" a featured column. As travel was just starting to become more accessible to women, many were uncertain about what to expect from their first experience, especially in foreign countries. The column offered practical advice on travel arrangements, suitable accommodations, expected patterns of dress, and etiquette, among other topics. This was intended to give women the confidence to travel and the ability to experience new places with pleasure rather than fear or anxiety.[7]

The lines between this travel literature and early place promotion began to blur as some of the first tourism industry stakeholders began to "sponsor" writers. Railroad or steamship companies would hire writers to undertake trips or provide them with complimentary trips using their services, which the tourists would then write about. For example, in the preface to the travel narrative *Back to Sunny Seas* (1905), Frank Bullen wrote:

> But I want to make it perfectly clear that I was the guest of the great Royal Mail Steam Packet Company, whose hospitality to me was more generous and farther-reaching than I could ever have dreamed of receiving. Yet I would like to make it clear too, if possible, that I have subdued my natural bias in favour of the Company, so that I have written only what I believe to be literally and exactly true.[8]

These companies were interested in creating and maintaining demand for their services. To some extent, that also meant creating and maintaining a demand for experiences in the places that they served. As the tourism industry continued to grow and destinations began to compete for tourists, they, too, began to offer incentives for authors to visit and write about their places.

Travel writing continues to evolve, with new trends in both tourism and media. As a genre of literature, travel writing has experienced a marked decline since its rise to popularity in the nineteenth century. Of course, this may at least partially be attributed to the same reasons literature in general has experienced declining readership. Nonetheless, narratives of travel continue to appear in search results for travel books alongside modern guidebooks. These stories are still written by established authors, such as V.S. Naipaul, recipient of the Nobel Prize for Literature (e.g., *A Turn in the South*, 1989), and recently by other well-known personalities as well, such as chef and Travel Channel host Anthony Bourdain (e.g., *No Reservations: Around the World on an Empty Stomach*, 2007). As in the past, these individuals are more likely to have the flexibility and the means to undertake the extensive trips to unusual places that become the subject of such books. In fact, some of the most widely known travel writing today is based on the type of extended stays in a place that stretch the limits of the definition of tourism, such as *Under the Tuscan Sun: At Home in Italy* (Frances Mayes, 1996) or *Eat, Pray, Love: One Woman's Search for Everything Across Italy, India, and Indonesia* (Elizabeth Gilbert, 2007). The primary function of these books is entertainment; the instructive function of these books has all but disappeared, as few readers would have the ability or the interest to replicate the writers' experiences.

The more typical experience of modern tourism does not lend itself as well to travel writing. The majority of tourists travel to prominent destinations visited by thousands if not millions of other tourists each year. Although more people are able to participate in tourism than ever before, their trips are of a far shorter duration than earlier generations of tourists. Americans, for example, receive an average of two weeks' vacation time per year, and few take a single trip for even that duration. Given the extent of information about the places visited and the compressed time frame, modern tourists have fewer opportunities for extensive noteworthy experiences. No one writes letters to family and friends back home during these trips; few tourists even send postcards, which notoriously arrive well after the tourists' return. Instead, with increased access to the Internet, even at foreign destinations, many tourists today choose to participate in Travel 2.0 (see below).

Guidebooks have remained a popular source of information for tourists (box 12.2). Companies such as Baedeker's, Fodor's, Frommer's, Insight Guides, Lonely Planet, Michelin Green Guides, Rick Steves, and Rough Guides produced books covering most destinations—at least at the regional scale. With the ubiquity of mobile devices, like iPads, e-readers, and smartphones, some travelers opt to download the e-text version of these guidebooks as a portable reference to be used throughout the trip. However, many travelers are now supplementing the guidebook with, or even bypassing it in favor of, user-friendly social media sites. These sites provide the most up-to-date information available and advice from "average" tourists like themselves as opposed to a professional travel writer.

Box 12.2. In-Depth: Tourism Guidebooks

In the modern era of tourism, guidebooks became a mass cultural phenomenon. As more tourists sought individualized and independent tourism experiences all over the world, guidebooks became an important means of helping them confidently make their own travel decisions, prepare for their experiences, and navigate the destination efficiently. While the nature of tourism information is changing, guidebooks (paperback, e-text, or web) continue to serve as a pre-trip resource that influence tourists' perceptions of places. This shapes tourists' expectations of the destination and ultimately their satisfaction with its experience. In addition, guidebooks also influence decisions about where to go, what to see and do, and how to experience the destination.*

In contrast with other forms of travel literature—from the subjective personal travel narratives discussed in this chapter to the persuasive place promotion discussed in the next

* Chak Keung Simon Wong and Fung Ching Gladys Liu, "A Study of Pre-Trip Use of Travel Guidebooks by Leisure Travelers," *Tourism Management* 32 (2011): 618.

chapter—readers expect guidebooks to be more comprehensive, more practical, and more objective.† The reputation of the publisher is fundamentally dependent on its ability to provide information that is consistent with tourists' experiences at the destination.‡ In addition, the use of clear, precise language and an authoritative tone to describe places gives the impression of objectivity and accuracy. However, guidebook descriptions are not a straightforward depiction of reality but partial and selective representations.§

Guidebooks are specifically intended to make different, possibly foreign, places accessible to a diverse set of readers (i.e., different national or ethnic backgrounds, different demographics, different types of tourists, etc.). To achieve this, they reduce the complexity of places and present only certain aspects of them to avoid confusing tourists by presenting them with too much information. Guidebook authors/editors make the decisions about which sites to include, which attractions are worthy of seeing and experiencing. This is partially a reflection of preexisting patterns in tourism, where representations reinforce the reputation and popularity of known attractions. It is also a reflection of the authors'/editors' perspectives and preferences, yet guidebooks typically provide little insight into who these authors or editors are, what their backgrounds are, and what their priorities might be.

Guidebooks also have an incredible power to positively or negatively represent places. They can encourage visits to a place, identify new places, or re-create ideas about places that would allow them to be seen by potential tourists in a new light. Conversely, they can actively discourage visits or simply ignore places. With this type of passive representation, places may remain unknown, or potential tourists will infer that they are uninteresting or unattractive.** This not only shapes tourists' experiences but possibly even destination development.

As with other types of representations of place in tourism, guidebooks have an influence over the ways in which potential tourists think about, and actual tourists experience, places. Although these tourists are more likely to accept guidebook representations at face value, we must still consider them—and their effects—critically.

Discussion topic: Do you trust guidebooks to provide you with information about a destination? Are you more willing to trust guidebooks or traveler reviews? Explain.

Tourism on the web: Fodor's Travel, "Fodor's Travel Guides," at http://www.fodors.com; Lonely Planet, "Lonely Planet, at https://www.lonelyplanet.com/#; Rough Guides, "Travel Guide and Travel Information," at http://www.roughguides.com

---

† Rudy Koshar, "'What Ought to Be Seen': Tourists' Guidebooks and National Identities in Modern Germany and Europe," *Journal of Contemporary History* 33, no. 3 (1998): 326.

‡ Scott Laderman, "Shaping Memory of the Past: Discourse in Travel Guidebooks for Vietnam," *Mass Communication & Society* 5, no. 1 (2002): 93.

§ Deborah P. Bhattacharyya, "Mediating India: An Analysis of a Guidebook," *Annals of Tourism Research* 24, no. 2 (1997): 376.

** Malin Zillinger, "The Importance of Guidebooks for the Choice of Tourist Sites: A Study of German Tourists in Sweden," *Scandinavian Journal of Tourism and Hospitality* 6, no. 3 (2006): 231–2.

USER-GENERATED CONTENT (UGC)

In the era of Travel 2.0, tourists have once again become key producers of information about the places and experiences of travel. This is a part of the changes in the tourism system from a business-to-consumer model to a peer-to-peer one discussed in chapter 2. Travelers 2.0 use the descriptions, reactions, reviews, advice, photographs, and videos posted by other tourists to get ideas in the pre-trip stage and to plan their trip as well as during the movement/experience stage to make additional decisions. These tourists then have the opportunity to contribute their own ideas, experiences, and content throughout their trip and after. Some of this content is intended for select audiences, such as friends and followers, but much of it is made publicly available for audiences around the world to view, now and in the future.

As an expression of travel experiences, UGC has become increasingly popular. Many tourists post photographs, videos, and/or short descriptions of destinations or experiences on social networking sites. A growing number of amateur writers find an outlet for the stories of their travels in travel blogs. This can be seen on any general blog hosting site as well as specialized sites like TravelBlog, which cites over 200,000 members and an average of 100 new members a day (last updated February 2013).[9] As with earlier generations of travel writers, bloggers often provide descriptive information about the places visited and the peoples encountered as well as practical advice for readers who might one day follow in their footsteps. However, blogs are also stories, personal narratives of their experiences, their thoughts, and their emotions.[10]

For many tourists, the primary objective of writing a blog while traveling is to keep family and friends up to date on their activities (figure 12.2). Others privilege the process of identity construction that comes from having new experiences and trying to

Figure 12.2. Florinda Klevisser created this blog to keep her friends and family updated during the course of a three-month-long trip. Upon her return home, she used this blog as a basis for writing a travel narrative that was published in Italian—her native language—called *Viaggia Con Me*. *Source:* Florinda Klevisser

make sense of them (box 12.3). Still others seek the "social capital" they will gain by demonstrating their newly acquired knowledge or worldly experiences. Regardless of the motivation, these blogs are also publicly available. As a result, these sites are often encountered through keyword searches and therefore become part of the representations of a place that make up a potential tourist's pre-trip information search.

Box 12.3 Case Study: Tourist Identities in Travel Blog Narratives of Madeira's Levadas

As people, we transform our lives into stories. This allows us to organize and make sense of our experiences, to create meanings, and to construct identities. While much of this takes place in our own heads, we also use narratives to represent ourselves and our experiences to others. Travel and tourism often provide us with opportunities to learn about ourselves, from how we handle unfamiliar and possibly unexpected situations to how we understand our encounters with different peoples and places. Narratives help us make sense of this and to reflect not only on who we are but also on who we want to be.* From this, we tell our stories to others, whether it is a face-to-face conversation or online in the form of a travel blog.

Narratives are made intelligible through commonly accepted structures and discourses. For the former, tourists' stories generally involve "setting the stage" by establishing the context, setting, and actors. Then, they lay out selected events that lead to a climax. For the latter, today's tourists are familiar with the discourses of travel. This means that, subconsciously, they model their stories after the stories told by earlier generations of travel writers.† For example, in the discourse of travel as adventure, risks and challenges to be overcome (e.g., discomfort, illness, danger, etc.) are key events. While some risks are real, most are dramatized for the sake of a good story and for the social capital one gains by having such exciting experiences. Nonetheless, as tourists learn about themselves through these adventures, they have the opportunity to (re)define their identity.‡ Researchers examine travel blog narratives to gain insight into travel experiences as well as tourists' identities.

This research used travel blog narratives for Madeira. Madeira Island is part of an archipelago of volcanic islands that are located in the Atlantic Ocean west of the African coast and comprise an autonomous region of Portugal (map 12.1). Madeira is a popular European destination for approximately one million tourists a year, many of whom are repeat visitors.§ These tourists are primarily interested in rest and relaxation on a beach or at a hotel in a place with a pleasant climate. Still, most tourists will make an excursion to explore the island via the levadas, or water channels. The maintenance pathways along the levadas provide a highly unique opportunity for nature walks amidst a dramatic natural landscape (figure 12.3). Within this network of levadas, some are easily accessible from the main tourist zones, while others are in remote locations; some are easy walks, while others are challenging hikes. This variation allows the majority of Madeira's tourists to have the "levada walk" experience, regardless of physical ability.

---

* Stephen Wearing, Deborah Stevensen, and Tamara Young, *Tourist Cultures: Identity, Place and the Traveller* (Los Angeles: Sage, 2010).

† Chaim Noy, "This Trip Really Changed Me: Backpackers' Narratives of Self-Change," *Annals of Tourism Research* 31, no. 1 (2004): 92.

‡ Torun Eslrud, "Risk Creation in Traveling: Backpacker Adventure Narration," *Annals of Tourism Research* 28, no. 3 (2001): 598.

§ António Almeida and Antónia Correia, "Tourism Development in Madeira: An Analysis Based on the Life Cycle Approach," *Tourism Economics* 16, no. 2 (2010): 432.

Following the typical pattern, the bloggers began to set the stage for their narrative. They were aware of Madeira's image of a mass 3S destination, and, in fact, most were there for that very purpose. For example, Kate** started her vacation on the island with "a very relaxing three days with late starts, massages, great seafood and lots of time to read." However, to effectively construct an adventure narrative from their experience on the levada walk, the bloggers needed to establish an appropriate, but credible, setting. Liam wrote, "There's something very Jurassic Park about the Island which is awesome! (I think a lot of people think it's just a place for old people but there's plenty to do there! Lots of mountains and cliffs!)." This gives the reader an impression of a primordial place, at once exciting and dangerous. More specifically, the bloggers needed to ensure their readers had a sufficient understanding of the levada walk. For this purpose, Susan wrote:

> Although often cut high up in the hills [the levadas] offer a network of easy-to-follow, often mostly flat walks along the paths beside them which follow the contours. There' s a downside—it' s often a long way down. Many, if not most, of the levadas are cut high up into sheer rock faces and the paths can run out so you can find yourself walking on a 30 cm wide levada wall with water on one side and a sheer drop of hundreds of feet to the other. There are often—but not always—wire handrails at these points but many warnings for those who suffer from vertigo (my friend' s husband took her walking there a few years ago. It was her idea of hell). But the views are stunning.

Next, the bloggers positioned themselves as the main actor in the story. Although they had previously established themselves as tourists, their identity began to shift. They described their preparations for the experience, including hiking boots, backpacks, flashlights and head torches, GPS, and walking sticks. They contrasted themselves with the "tourist masses" on guided tours who were described as wearing all manner of unsuitable clothes and footwear (including a pair of white high-heeled shoes).

Then the bloggers began to construct their adventure, replete with challenges: narrow, muddy, slippery and/or rocky paths, damaged railings, fallen trees, dark tunnels, low overhangs, steep grades, and even stray dogs. While only one blogger actually reported suffering a fall, several bloggers claimed the ultimate challenge: facing death. One couple referred to their experiences as a "death walk." During their walk, they had a fight and continued on separately. First, Meredith wrote:

> The levada path got narrower and narrower and I found myself walking alone on a strip of concrete about a foot across with a drop of about 1,000 feet next to it. You just cannot see this horror in the photographs. Now I am absolutely terrified of heights. I was the kid who crawled across railway bridges . . . on my hands and knees. I had exactly the same sense of fear, only this time it was real, if I slipped I would die. No question.

Her partner, Chris wrote, "I knew nothing of all this drama. To begin with I was just having a sulk and trying to enjoy the scenery. But it didn't take long before I realise[d] that my life was at risk."

The core of such narratives is how the individual responds to these challenges, and in the end, the blogger always emerged triumphant. Meredith wrote, "At one point I can recall thinking that if I just hurled myself off the ledge at least I wouldn't have to suffer the surprise

** All names are pseudonyms.

of slipping. I cannot believe that I actually did this, I have never been braver." After she and Chris were reunited they declared, "We were so pleased to have survived the whole experience that we completely forgot our row. It is amazing how near death experiences help you to appreciate the small things in life." After this bit of adventure, the bloggers quickly reverted back to the tourist role. For example, Owen and his wife returned to their hotel, went to the pool, got massages at the spa compliments of their tour package, watched the sunset while drinking a glass of wine, and splurged on the most expensive meal of the trip because they felt they deserved it.

These bloggers were not looking for an "off-the-beaten-path destination" or profound life-changing experiences. Yet, travel has the potential to place tourists in unfamiliar and unexpected situations that may present (real or perceived) challenges. This can push the limits of their comfort zone, which allows them to reflect on who they are and how they want to represent themselves to others.[††]

Discussion topic: Using one of your own travel experiences, write an adventure narrative. What does this story tell us about (a) the place in which the experience took place, (b) the experience itself, and (c) you as the tourist and actor in the story?

Tourism on the web: Madeira Regional Tourism Board, "Site Oficial do Turismo da Madeira," at http://www.visitmadeira.pt/en-gb/homepage

Figure 12.3. The maintenance paths along Madeira's levadas—water channels—offer unique opportunities for nature walks. Some paths are wide and flat, while others are narrow and located on the edge of cliffs. *Source:* Velvet Nelson

[††] Velvet Nelson, "Tourist Identities in Narratives of Unexpected Adventure in Madeira," *International Journal of Tourism Research* 17, no. 6 (2015).

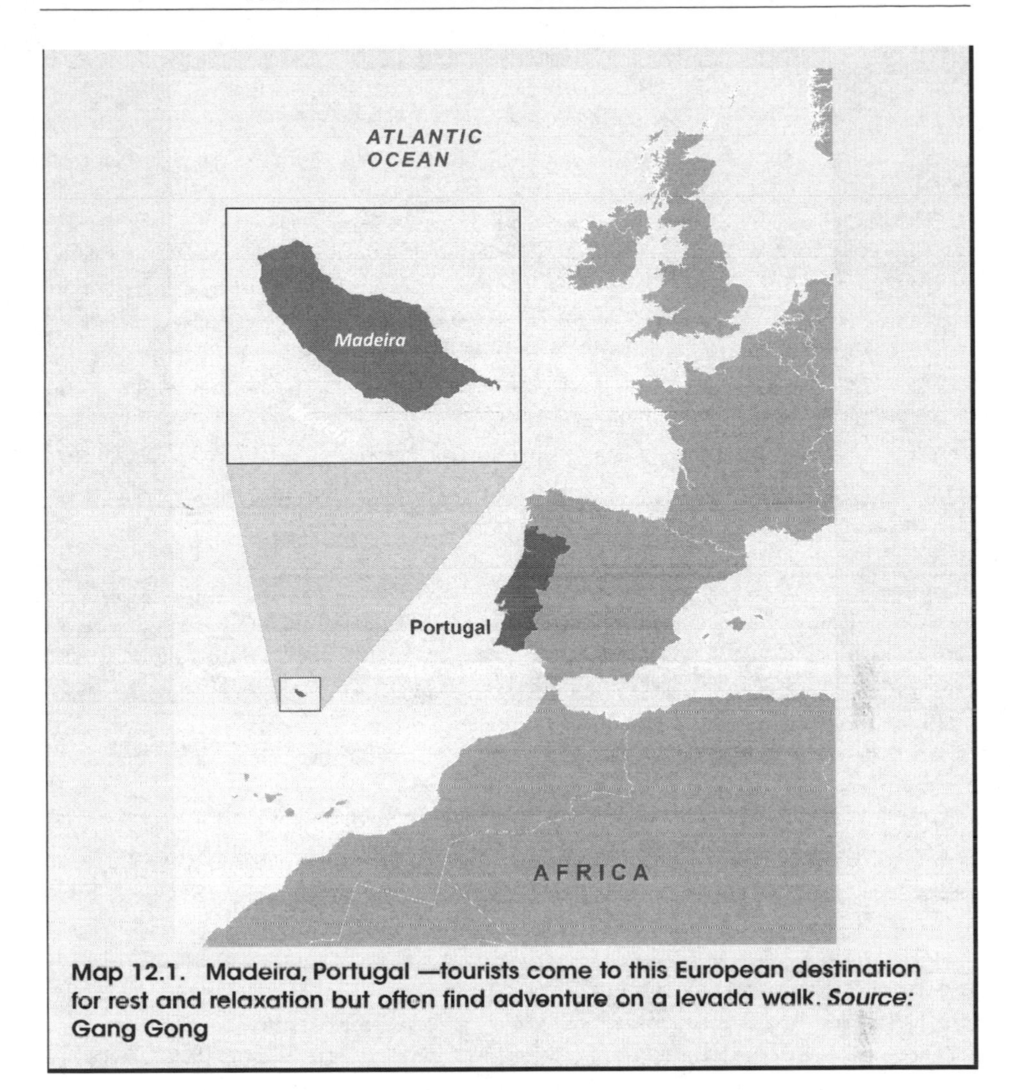

Map 12.1. Madeira, Portugal —tourists come to this European destination for rest and relaxation but often find adventure on a levada walk. ***Source:*** **Gang Gong**

As a source of travel information and inspiration, UGC—from these personal travel blogs to online review sites—has become extraordinarily significant. The risk associated with the destination decision-making process is considered higher than for other types of products because the purchase is made prior to the experience, and the experience cannot be returned if it is not satisfactory. **Word-of-mouth** (WOM) has long been recognized as an important source of information for trip planning. Traditionally, WOM involved person-to-person communication, and individuals were likely to rely on that person's opinions, based on his or her experiences, to help them make an informed pre-trip decision.

Travel 2.0 represents **electronic word-of-mouth** (eWOM or "word of mouse"). eWOM gives individuals access to the opinions of more tourists and about more places than what they would be able to obtain from their immediate social circle.[11] Thus, individuals will not know online reviewers, but recent research shows that social media content is often perceived to be more trustworthy when compared to place promotion intent on selling a product.[12] In recognition of this, companies that earlier produced travel guidebooks are not simply reproducing this content online in today's market but are establishing digital travel communities that allow consumers to be contributors.[13]

Nonetheless, it is important to consider whose perspectives or viewpoints are being conveyed. Studies have shown that, although many of today's tourists are interested in consuming UGC such as blogs and reviews, a much smaller proportion actually contribute to content.[14] While Web 2.0 technologies have often been considered democratizing, participation is still not universal. The digital divide has narrowed in recent years, especially with the use of mobile technologies, but the availability and cost of Internet access can still constitute barriers.[15] The generational divide has also narrowed, but lack of knowledge, comfort, and trust among older users who are less accustomed to such technologies can also hinder participation.[16] Other factors include culture, personality, and, quite often, free time.

In addition, there has been growing concern about the veracity of UGC. The news media has reported stories of business owners who pose as customers to write positive reviews of their services or negative reviews of their competitors.[17] In response, some sites claim to have verified reviews; for example, a hotel booking site may only post reviews from customers who made reservations through their site and had a confirmed stay. However, this policy still does not preclude reviews from individuals with a vested interested in the business. Reports have also questioned whether sites filter or suppress reviews, thereby offering a selective sample of customer opinions.

It is also important to keep in mind the various factors that might influence tourists' representations. In the past, tourists were more likely to recount their experiences during the post-trip stage. Although stories told in this stage were based on memory (and forgetting), tourists still had time to process their thoughts and to contextualize them within the circumstances of the particular experience and of the trip as a whole. Today, the real-time nature of mobile technologies has the potential to affect tourists' representations in different ways. Those who post an immediate reaction or review online may wish to revise their opinions upon reflection. For example, tourists who are tired, hungry, or intoxicated are likely to react differently to events than they would otherwise. Likewise, tourists who are unprepared for their experiences or have expectations that are initially unmet may think differently when the trip is over. Tourists' in-the-moment attitudes can even be swayed by the reactions of those around them, whether it is that of their traveling companion or simply a vocal stranger.

Ultimately, these representations in print or online are an individual's interpretation of peoples, places, and experiences. As consumers of these representations, we exercise common sense about bias and give more weight to the overall image of a destination or service than individual opinions. However, we should also be aware that the representations of others—as well as our own—are not innocent and can have distinct, if not always intended, consequences.

Consequences of Representations

Representations of place can be extraordinarily powerful. These are not accurate or objective portrayals of reality but partial and selective ideas about places based on individual perceptions, personal attitudes, and cultural contexts. Scholars have argued that the early travel writers discussed above cannot be disconnected from empire. These writers were familiar with colonial discourses, and they reproduced them in their travel narratives. This refers to the ways people thought and talked about colonies and colonial peoples. For example, travel writers used the colonial rhetoric of conquest and achievement as they described overcoming the challenges of traveling in foreign places, from the inefficiency of local systems to the inferiority of the infrastructure.[18] Places and peoples were viewed as the "other," and they were compared to those with which the travelers were familiar. This helped them to try to make sense of what they encountered, but it was nonetheless framed within preexisting ideas of what was right and superior.

Critics argue that the writing in travel blogs today has changed little from these early narratives (see blogger Bani Amor's "Dispatches" for a variety of perspectives on this issue).[19] Travel writers continue to use authoritative voices to describe the other places and peoples, but it is based on their often-limited experiences as well as their own background and perspectives. From a critical perspective, this is predicated on the assumption that those peoples are unable to represent themselves. Indeed, for many destinations, the only representations audiences around the world receive come from outsiders (e.g., tourists or marketers).

The ethnicity, culture, and/or daily life one comes from are viewed as the norm, while everything else is the "other." Narratives of othering use descriptive words to highlight differences and emotional words to demonstrate surprise, confusion, or delight. These narratives are used to show readers that they had an out-of-the-ordinary experience and, accordingly, to gain social status from having had such experiences.[20] However, in the modern world, there are fewer opportunities for truly extraordinary experiences. Thus, even mundane aspects of life in that place can be romanticized and made exotic by the visitor. For example, local people on the Caribbean island of Grenada take buses every day for routine purposes, but for foreign tourists the brightly painted buses traveling winding roads at breakneck speed using Morse code horn signals can constitute an adventure and a story to be told.

Authenticity is measured by the extent of difference. Tourists may consider those places that have been exposed to global culture not "authentic" (figure 12.4). This not only diminishes the value of that place, its people, and its culture, but it also leads to a search for places further "off the beaten path." This search for the new and different is a process of discovery (and conquest) compared to that of earlier generations of explorers. However, such tourists must consider what effect they have on the "authenticity" of such places. They may be responsible for bringing global culture to that place, even though they specifically came there because of its absence, or they may contribute to the process of museumization (chapter 9) as local people try to maintain tourist interest in their "traditional" culture.

Figure 12.4. Tourists may deem places that display elements of global culture "not authentic." For example, the Tunisian woman on the left is wearing Western clothing, while the woman on the right is carrying a plastic bottle of water. *Source:* Velvet Nelson

Finally, representations of places through travel have been relatively narrow due to a general lack of diversity in the media. Travel personalities have typically been white, heterosexual males. Thus, the media continues to reproduce limited perspectives, which influences others' ideas about places and their peoples. In addition, it perpetuates ideas about barriers to travel for other individuals. The tools of social media are helping to create a shift in this norm. Producers of online videos and contributors to blogs—of all genders, ethnicities, and sexual orientations—have been able to add their experiences and perspectives to the growing conversation about travel and the modern world. They provide examples for others who identify with them, which may give these individuals the confidence to make their potential demand for travel effective demand. Although these voices continue to remain outside of the mainstream media, they are growing and gaining attention. As more readers and viewers are exposed to diverse ideas and viewpoints, we would hope to see a shift away from the colonial discourses of the past to a more self-critical awareness of how we understand, engage with, and represent the places and peoples of our travels.

Conclusion

Place is one of the key concepts in geography, and it is equally important in the geography of tourism. Tourism is an inherently place-based geographic phenomenon. Yet, we must first understand the ways in which people think about places before we can begin to examine ways in which people interact with and experience places through tourism. Our most concrete ideas about places come from firsthand experience, but we also have ideas about all kinds of places we have never visited. In a sense, we "experience" these places through representations. These representations can be extraordinarily powerful in shaping our perceptions of places, which plays a role in determining whether or not we wish to visit that place for ourselves. However, the impact of representations does not end with the destination decision-making process. The ideas of places created by representations also shape the way we experience places, which will be examined in chapter 14.

Key Terms

- electronic word-of-mouth
- place representation
- word-of-mouth

Notes

1. Dean MacCannell, *The Tourist: A New Theory of the Leisure Class* (New York: Schocken Books, 1976; reprinted with foreword by Lucy R. Lippard; Berkeley: University of California Press, 1999), 110. Citations refer to the California edition.
2. Felix Driver, "Distance and Disturbance: Travel, Exploration and Knowledge in the Nineteenth Century," *Transactions of the Royal Historical Society* 14 (2004): 77.
3. Anthony Trollope, *The West Indies and the Spanish Main*, 4th ed. (London: Dawsons of Pall Mall, 1968), 50.
4. Algernon E. Aspinall, *The Pocket Guide to the West Indies*, 2nd ed. (London: Duckworth & Co., 1910), 42–3.
5. Mimi Sheller, *Consuming the Caribbean: From Arawaks to Zombies* (London: Routledge, 2003), 60.
6. Richard H. Gassan, *The Birth of American Tourism: New York, the Hudson Valley, and American Culture, 1790–1830* (Amherst: University of Massachusetts Press, 2008), 73–5.
7. Jill Steward, "'How and Where to Go': The Role of Travel Journalism in Britain and the Evolution of Foreign Travel, 1840–1914," in *Histories of Tourism: Representation, Identity, and Conflict*, ed. John Walton (Clevedon: Channel View Publications, 2005), 44–5.

8. Frank T. Bullen, *Back to Sunny Seas* (London: Smith, Elder & Co., 1905), vii.

9. TravelBlog, "About TravelBlog," accessed July 8, 2016, http://www.travelblog.org/about.html

10. Ana María Munar and Jens Kr. Steen Jacobsen, "Motivations for Sharing Tourism Experiences Through Social Media," *Tourism Management* 43 (2014): 47.

11. Daniel Leung, Rob Law, Hubert van Hoof, and Dimitrios Buhalis, "Social Media in Tourism and Hospitality: A Literature Review," *Journal of Travel & Tourism Marketing* 30 (2013): 8.

12. Stella Kladou and Eleni Mavragani, "Assessing Destination Image: An Online Marketing Approach and the Case of TripAdvisor," *Journal of Destination Marketing & Management* 4 (2015): 187.

13. Stephanie Hays, Stephen John Page, and Dimitrios Buhalis, "Social Media as a Destination Marketing Tool: Its Use by National Tourism Organizations," *Current Issues in Tourism* 16, no. 3 (2013): 212.

14. Eduardo Parra-López, Desiderio Gutiérrez-Taño, Ricardo J. Díaz-Armas, and Jacques Bulchand-Gidumal, "Travellers 2.0: Motivation, Opportunity and Ability to Use Social Media," in *Social Media in Travel, Tourism and Hospitality: Theory, Practice and Cases*, ed. Marianna Sigala, Evangelos Christou, and Ulrike Gretzel (Ashgate: Surrey, 2012), 177; Kyung-Hyan Yoo and Ulrike Gretzel, "Use and Creation of Social Media by Travellers," in *Social Media in Travel, Tourism and Hospitality: Theory, Practice and Cases*, ed. Marianna Sigala, Evangelos Christou, and Ulrike Gretzel (Ashgate: Surrey, 2012), 198.

15. Ulrike Gretzel, "Introduction to Part 3," in *Social Media in Travel, Tourism and Hospitality: Theory, Practice and Cases*, ed. Marianna Sigala, Evangelos Christou, and Ulrike Gretzel (Ashgate: Surrey, 2012), 167.

16. Ana María Munar and Jens Kr. Steen Jacobsen, "Trust and Involvement in Tourism Social Media and Web-Based Travel Information Sources," *Scandinavian Journal of Hospitality and Tourism* 13, no. 1 (2013): 3.

17. Raffaele Filier, Salma Alguezaui, and Frazer McLeay, "Why Do Travelers Trust TripAdvisor? Antecedents of Trust Towards Consumer-Generated Media and Its Influence on Recommendation Adoption and Word of Mouth," *Tourism Management* 51 (2015): 175.

18. Mary Louise Pratt, *Imperial Eyes: Travel Writing and Transculturation*, 2nd ed. (London: Routledge, 2008), 146.

19. Bani Amor, "Dispatches," accessed July 12, 2016, https://baniamor.com/dispatches/

20. Carmela Bosangit, Juline Dulnuan, and Miguela Mena, "Using Travel Blogs to Examine Postconsumption Behavior of Tourists," *Journal of Vacation Marketing* 18, no. 3 (2012): 214.

CHAPTER 13

Place Promotion in Tourism

"Beautiful beaches," "Spectacular sunsets," "Majestic mountains," "Fragrant flowers," and "Colorful costumes"—clichéd though they are, these are frequently used tourism terminologies to evoke vivid images of places in our minds. These ideas and images play an important role in shaping both our demand for and our experience of tourism destinations. In tourism studies, these issues are frequently approached from a marketing perspective. However, as places are the product to be sold in tourism, the geography of tourism is an equally relevant framework. In the previous chapter, we discussed how meanings of place are created indirectly through popular media and user-generated content (UGC); in this chapter, we will look at how meanings are explicitly produced through the practice of place promotion.

We will consider some of the issues associated with promoting places for tourism. These representations are intended to communicate specific meanings about a place to shape ideas about that place and ultimately to influence the decision to visit that place. In the first part of the chapter, we will discuss the concepts of place promotion, place branding, and place reputation management, while in the second part we will look at some of the strategies used to convey meanings about places. Next, the chapter reviews the types of promotional media that have been used in the past in tourism and highlights the current significance of social media. Finally, the chapter concludes by examining some of the consequences of place promotions in tourism.

Place Promotion, Branding, and Reputation Management

In addition to the representations of place discussed in the previous chapter, some are deliberately created with specific meanings to "sell" a place to external audiences. Tourism is a highly competitive global industry; therefore, the success of modern tourism destinations depends on the creation and promotion of clear and ultimately distinctive ideas. **Place promotion** is the deliberate use of marketing tools to communicate specific and selective ideas and images about a place to a target audience for the purpose of shaping perceptions of that place and ultimately influencing decisions. It is a form of advertising and can be deceptive. Because place promotion draws selectively upon the real nature of places and presents only those elements that will appeal to the target market segment, there may be many different representations of a destination. Each representation will highlight a different aspect of the destination, draw upon a different theme, and utilize different images to attract specific types of tourists.

Not all destinations use place promotion in the same way. For many of the world's largest tourism destinations—such as France, the United States, or Australia—there is little need to promote the destination as a whole. Ideas about these places already abound in representations, and there is a preexisting suppressed demand (i.e., many people already believe they want to visit these places if/when they have the opportunity). It may be more important that these destinations promote specific regions or places, as tourists are likely to only visit a part of the country during their trip. This can help provide more detailed information about the unique resources and experiences available in different places. However, for smaller and emerging destinations, place promotion is vital in raising awareness about the destination and what it has to offer

Branding was initially associated with consumer goods, but the concept was eventually applied to places as well. As such, "brand" is used as a metaphor for the ways places compete with each other for a variety of purposes in the modern, global world.[1] The idea of place branding has been much debated. Some scholars were concerned that place branding would lead to the commodification of places, much like the commodification of culture discussed in chapter 9. Others dismissed the idea because there are significantly more challenges to branding a place than branding a product. Places are complex, and such a brand would have to encompass a diverse set of products and services. Moreover, there are countless public and private stakeholders who would have to be involved in the process of producing, promoting, and supporting the brand.[2]

Credited with originating the term "place brand," Simon Anholt has also been one of the concept's greatest critics. In the introduction to his book, *Places: Identity, Image and Reputation*, Anholt argues that places (i.e., nations, regions, cities, etc.) have "brands" in the sense that they have reputations.[3] **Place reputation** refers to the composite of ideas and impressions held by external audiences.[4] These reputations are cultural phenomena that cannot be entirely controlled by a destination marketing organization or local stakeholders; all of the representations discussed in the previous chapter have an influence on reputation. Nonetheless, reputation plays an important role in the development and success of that place. Destination marketers will try to create a brand that identifies and communicates the core characteristics of the place. If they are successful, this will be reflected in a positive reputation.

Poorly known destinations—often new or remote destinations—are perhaps the easiest to brand and promote. These destinations lack a reputation; essentially, potential tourists have few ideas—positive or negative—about these places. Stakeholders have the opportunity to create and promote a desired brand identity to shape potential tourists' ideas about that place. However, research indicates that reputations are remarkably stable. Once a reputation is established, it can be very difficult and time consuming to change people's minds about a place.[5] A poor reputation may be based on serious past or present issues or events in the country or region of the country, including conflicts, political upheaval, acts of terrorism, human rights violations and/or atrocities, problems associated with the drug trade, crimes against tourists, and so on. Likewise, a poor reputation may be based on the destination's tourism industry, such as a poorly developed infrastructure, an unfriendly or hostile local population, and even an overdevelopment of tourism in which mass commercialization supplants the local character of the place or the sheer volume of tourists overwhelms the experience of place. Even places that have experienced positive changes may continue to receive less attention because of a poor reputation plagued by persistent stereotypes.

Place reputation management involves efforts to adjust the reputation so that it is closer to how stakeholders would like the place to be perceived.[6] Effective place reputation management requires a clear understanding of both perceptions of the place and the place itself. Stakeholders should be involved in the process of (re)defining a clear brand identity to ensure that they support the idea and reinforce it in their activities.[7] This identity must reflect the reality of the place. For tourism destinations, it is vital that potential tourists have appropriate expectations for that place to prevent dissatisfaction with the experience.

Some destinations with a poor reputation attempt to ignore the issue entirely in the creation of a new brand identity. For example, Detroit, Michigan, arguably had the worst reputation of any city in the United States based on reports of financial crisis, corruption, a decaying infrastructure, high crime rates, and more. The Detroit Metro Convention & Visitors Bureau's website includes a section titled "Comeback Stories" and briefly acknowledges that the city is experiencing revitalization. However, these stories highlight attractions such as the Motown Museum and the Detroit Institute of Arts rather than showing readers what progress has been made with regards to those issues that gave the city its bad reputation.[8] Others choose to address it head-on in their promotions to show that it has been resolved. In the case of Houston, Texas, discussed in box 13.1, media representations and promotions confirm the negative stereotypes about the city with which readers would likely be familiar. Then they challenge those stereotypes and present readers with a different, but realistic, view of the city.[9]

Box 13.1. Case Study: Place Reputation Management in Houston, Texas

"Okay Houston, we've had a problem here."

Astronaut John Swigert, Jr.'s misquoted and clichéd statement from the ill-fated 1970 Apollo 13 moon mission has become such a significant stereotype that it is often the first (and sometimes the only) thing that comes to mind when people think about Houston, Texas (figure 13.1). Other stereotypes for the city are based on wider Texas (i.e., cowboy) culture, heat and humidity, its association with the oil industry, and its notorious lack of formal zoning laws. If you are unfamiliar with any of these, consider that a *Business Week* article began: "Petrochem capital of the Americas, sprawl capital of the universe: Houston can seem like a city you cannot escape fast enough."* Clichés and stereotypes, whether they are true or not, affect the way we think about places.† Thus, despite its status as the fourth largest city in

* Lisa Grey, "A Free Day in Houston, Texas," *Business Week*, April 22, 2010, accessed May 6, 2014, http://www.visithoustontexas.com/about-houston/in-the-news/

† Simon Anholt, "Competitive Identity," in *Destination Brands: Managing Place Reputation*, 3rd edition, eds. Nigel Morgan, Annette Pritchard, and Roger Pride (Florence, KY: Routledge, 2011): 22.

Figure 13.1. Houston, Texas, is working to change its negative reputation, and stakeholders are using food as a means to generate new attention for the city. *Source:* Scott Jeffcote

the United States boasting an economy that would be among the top twenty-five countries in the world, Houston has had a reputation problem.

A negative place reputation can be very slow and difficult to change, but place reputation management seeks to adjust it so that it is closer to how stakeholders would like the place to be perceived.[‡] Scholars argue that this should reflect the real character of the place, or the quality of place (i.e., what is there, who is there, and what is going on there), that defines it and makes it attractive not just to residents or tourists but to both.[§] This helps generate a "buzz" about the place that can aid in the process of reputation management.

In the case of Houston, economic growth, urban regeneration, and immigration have all played a role in changing the dynamics of the city. However, one factor in particular has been getting attention: food. In 2015, Travel + Leisure media named Houston first among America's food cities. The article's author acknowledged that this ranking was considered a "Texas-sized upset," as the city surpassed well-known American food destinations like Chicago, New Orleans, New York, and San Francisco.[**] This increased recognition for Houston's locally driven culinary scene is providing a new attraction for the city, but it is also being used to change the city's reputation. In particular, culinary culture is used to highlight other aspects of quality of place.

Popular media articles on this culinary culture are primarily aimed at external audiences who are likely to have an ambiguous or negative image of the city. For example, prevailing ideas of the city based on sprawling freeways and generic urban infrastructure create a sense of placelessness. Articles encourage readers to look beyond the freeway infrastructure and corporate chain restaurants to find the local restaurants, from small neighborhood family-run

‡ Nigel Morgan, Annette Pritchard, and Roger Pride, "Tourism Places, Brands, and Reputation Management," in *Destination Brands: Managing Place Reputation*, 3rd edition, eds. Nigel Morgan, Annette Pritchard, and Roger Pride (Florence, KY: Routledge, 2011): 17.

§ Richard Florida, *The Rise of the Creative Class, Revisited* (New York: Basic Books, 2012), 280–1.

** Katrina Brown Hunt, "America's Best Cities for Foodies," *Travel + Leisure*, accessed May 26, 2015, http://www.travelandleisure.com/slideshows/americas-best-cities-for-foodies

ethnic restaurants to widely publicized concept restaurants developed by area chefs. These chefs are described as passionate and even obsessive about obtaining products locally and change their menus to take advantage of whatever is available at that particular time. Thus, readers are reminded that Houston is a port city, connected to the Gulf of Mexico via the Houston Ship Channel and Galveston Bay. This provides chefs with ready access to fresh seafood. Likewise, the area's subtropical climate allows access to fresh agricultural produce year round.

Prevailing ideas also characterize the city as having a provincial "redneck" culture. Again, food allows media articles to highlight the diversity of the city. After identifying communities of African Americans, Latin Americans, Eastern Europeans, Russians, Pakistanis, Indians, and Vietnamese, one author declares: "In fact, the Houston metro area has no single dominant ethnic group. None composes a majority, and none is likely to in the near future—a colorful aspect in which the city takes great pride. . . . This diversity also means the Houston area is a great place to enjoy Vietnamese, Mexican, Latin American (especially Brazilian) and Indian food. Then there's classic Southern soul food."[††] In addition, the city's young, talented, and innovative chefs are represented as emblematic of the city's growing creative culture.

Finally, the city's ambiguous reputation has contributed to the idea that it has little to offer visitors. The media focus on food presents cafés, restaurants, and bars as an integral part of "what's going on" in Houston. They are places to see and be seen. An article on the city's Midtown district describes a restaurant scene "where a microcosm of twenty-first-century Houston—African-Americans, Vietnamese, Hispanics, Anglos, college students in jeans, glad-handing politicians—mingles each morning with hungry out-of-towners." The same author notes that the neighborhood's renaissance is "not an artificial tourist trap." He cites the area's 18,500 residents, many of whom "are new to Houston; others are fleeing from, not to, the sprawling suburbs. They average 25 to 40 years old, and their vitality is evident day and night."[‡‡] As such, Houston is represented as a cosmopolitan and cutting-edge city, comparable to those with far better reputations.[§§]

Some cities are endowed with distinct amenities and a well-known positive reputation; however, many—like Houston—must work to establish or change their reputation to convince potential visitors of their attractiveness. As interest grows in food and beverage tourism, culinary culture will provide an attraction for Houston. This is tied to the unique circumstances of place, from its environment to its people and culture and the events taking place there. Reputations can be slow to change; thus, it remains to be seen whether this will be the catalyst the city needs to change its reputation.

Discussion topic: What factors of a place do you think have the greatest potential to change its reputation? Explain.

Tourism on the web: Greater Houston Convention & Visitors Bureau, "Houston Hotels, Events & Things to Do," at http://www.visithoustontexas.com

---

†† Eric Lucas, "High Energy Houston: Alaska Airlines' Newest City," *Alaska Airlines* (September 2009): 122.

‡‡ Harry Shattuck, "Midtown Houston," *Executive Travel* (October 2009): 69–70.

§§ Velvet Nelson, "Place Reputation: Representing Houston, Texas as a Creative Destination through Culinary Culture," *Tourism Geographies* 17, no. 2 (2015).

Destinations perceived to be similar to others present a challenge for place promotion. The Caribbean is one such example. People often see the islands of the region as a collective, and the fact that they have at times acted cooperatively to strengthen their position on the global market contributes to this perception. However, this becomes a problem when potential tourists think one island is the same as another. This not only affects the choice of specific destination but may also serve to discourage these tourists from returning and visiting another island because they think they have already had the "Caribbean experience." Consequently, each island tries to create and promote a distinctive place identity that plays up the resources that make them unique. St.

Lucia capitalizes on its iconic landscape, the Twin Pitons. These volcanic spires have UNESCO World Heritage status. Trinidad is known for hosting the biggest annual party in the region: its Carnival. One of the islands containing two countries—Dutch St. Maarten and French St. Martin—plays up the idea of two destinations in one.

Destinations at all geographic scales around the world have been involved in the processes of place promotion, branding, and reputation management. Tourism boards or convention and visitors' bureaus (e.g., Chicago Convention and Tourism Bureau) may be responsible for a local destination's image. Large destinations may have state- or regional-level agencies (e.g., Tourism Western Australia), while most countries now have some type of national tourism organization or association (e.g., Tourism Authority of Thailand). These organizations typically concentrate on creating a national tourism brand identity that can be promoted to an external or foreign audience, although there may be efforts to promote domestic tourism as well. In some cases, small and/or relatively similar destinations may work cooperatively through regional (supranational) tourism organizations to promote a specific destination region (e.g., Caribbean Tourism Organization, open to any country with a Caribbean coast). The resources available for place promotion vary widely based on the size of the destination, level of overall economic development, and the extent to which tourism development is a priority.

Promotional Strategies

The international tourism industry is highly competitive, and the need for destinations to create and promote a clear, positive, recognizable, and distinctive place identity to differentiate themselves from others is, perhaps, more important than ever. Yet, this task has become increasingly difficult. Destinations are no longer competing only with similar places in their region but with places around the world that claim the same kinds of attractions, such as beautiful landscapes or unique cultures.

Traditional marketing strategies represent the characteristics of a place in simplified form or stereotypes. This has been viewed as a concise means of identifying places that external audiences can easily recognize and remember. Destination slogans or "taglines" are used to condense the idea of the place represented to tourists into a short, memorable phrase. For example, the tagline for Malaysia is "simply Asia," at once situating the destination in the region and implying its authenticity. While such tools might be easily dismissed as superficial marketing ploys, they can be significant nonetheless (see, for example, the case on Dominica in chapter 10). Indeed, the ideas conveyed in a slogan can be so important that they may become controversial. For example, in 1991, the newly created European country of Slovenia was relatively unknown and needed to create a place identity for the purpose of stimulating tourism. Not only did Slovenia need to create a positive place identity for itself, but the country also needed to distance itself from the violence of the independence wars among the other Yugoslav states. As such, Slovenia adopted the slogan "The sunny side of the Alps" to capitalize on both the country's physical resources for tourism—particularly the attractive destinations in the Julian Alps like Lake Bled (figure 13.2) and the preexisting positive tourism imagery associated with the larger Alpine region. However, Italy

Figure 13.2. Slovenia's Lake Bled, surrounded by the snow-capped mountains of the Julian Alps, is one of the most popular tourism destinations in the country. Such iconic images are often used in tourism promotions. ***Source:*** **Tom Nelson**

objected to this slogan and the perceived implication that, as Slovenia's neighbor, it was not physically and/or metaphorically "sunny." Consequently, Slovenia was forced to abandon the slogan.

Additionally, traditional marketing has relied on visual images. Given the old adage "A picture is worth a thousand words," images are often considered the most important aspect of promotions. These representations are powerful in instantly creating an impression of a place in the viewer's mind. In fact, it has been argued that the image of a destination plays a key role in potential tourists' decision-making process.[10] Early tourism imagery included illustrations and black-and-white photographs; in today's digital media, it is dominated by vivid full-color photographs and video files. These images depict characteristics of the place, including attractive landscape vistas and iconic scenes, as well as tourists participating in activities and enjoying themselves there—such as relaxing on the beach, hiking in the forest, or dining at sunset.

Still, tourism is an experiential product that involves all of the senses. Destinations seeking to move beyond traditional strategies to create a more substantial brand have focused on the experience of, and the feelings associated with, the consumption of place. **Experiential marketing** is intended to more successfully engage potential tourists in the destination search. As it communicates the characteristics of a place, it appeals to readers' imagination,[11] inviting them to envision what their experience of that place would be like. In particular, the use of multiple senses in brand marketing is shown to clarify the brand identity, generate interest in the brand, shape consumer behavior, and ultimately create brand attachment.[12] Humans experience the world through their senses, which are linked to memories and can stimulate emotion.[13] They become more emotionally invested in the idea of the experience and therefore have a stronger desire to visit the place. This helps reduce the risk involved in the destination decision-making process. Thus, rather than simply informing readers of attractive destination attributes, experiential branding goes a step further by prompting readers to imagine they are actively having the experience described. Sights, scents, sounds, tastes, and touch all play a role in stimulating the imagination to create an overall feeling of the destination and an impression of the state of mind the visitor would have with such an experience.

The sense of sight has long been the most important sense in tourism promotions. In the era of modern technology, potential tourists will be exposed to vivid visual images. The use of sight in written descriptions must try to engage the reader. The narrator assumes the role of guide and points out details to the reader, as if he or she were there, such as the variations of brightly colored fruits at a tropical market. To further engage the reader, though, sight can be used more effectively in combination with other senses. Sounds can imply particular types of experiences. The narrator describes the array of sounds in an urban environment that encourages the reader to feel the activity and excitement of such a place, or the absence of sound in a natural environment that allows the reader to feel the stillness and peace.

Scents trigger memories and the emotions attached to those memories. Therefore, they can be used to create good feelings and even to reduce stress.[14] The narrator invites the reader to inhale the scent of fresh-baked bread or the aroma of fresh-brewed coffee. This has the potential to tap into the reader's memories, which can create pleasant feelings and associations. Moreover, it may prompt the reader to unconsciously inhale deeply, which has a calming effect. Taste is closely related to smell. As food and beverage experiences become a more important part of tourism, more destinations will promote the taste of place. In such cases, the narrator does not just identify key ingredients and distinctive dishes but describes the flavors, the sweetness, the crunchiness, etc. of these products. Finally, touch is used to stimulate the imagination. The narrator encourages the reader to imagine what it is like to feel powdery sand underneath his or her toes or the cool breeze brushing across his or her skin.

Combined, all of the senses can be brought together in a destination's promotional literature to illustrate the feelings, emotions, and states of mind the reader could have if he or she were to visit that place.[15] In theory, if the reader becomes emotionally invested in this imagined experience, it will stimulate a demand for the actual experience. Thus, he or she will choose that destination over all others.

There are some very obvious and standardized messages conveyed in place promotion (e.g., fun or relaxation), but there are also subtler messages embedded in them. Many of the themes used to convey ideas about places, generate interest, and play upon potential tourist motivations for travel have a long history in tourism representations (box 13.2). Some of these themes include excitement and adventure, tradition and timelessness, fantasy and romance, pristine and unspoiled, or exotic and different.

Box 13.2. In-Depth: The Cycle of Expectation in Caribbean Tourism Representations

Today, any reference to the Caribbean is likely to conjure up certain images: bright sunny skies, clear turquoise waters, soft white sands, and lush green palm trees gently swaying in the breeze. This is the ubiquitous imagery that populates the place promotions across the region's island destinations. Yet, these images associated with the modern tourism industry are hardly new. In fact, they have a long history that can be traced back to an earlier era of tourism in the region and the representations of the islands produced by the first generations of tourists.

As the Caribbean was one of the first colonial regions in the era of Western European colonialism, explorers, scientists, and plantation owners had long produced information about the region. However, it was not until the end of the Napoleonic Wars in 1815 that the Caribbean came to be seen as a potential tourism destination. At this time, the region was relatively free from conflict, largely devoid of a hostile native population, and increasingly accessible by transatlantic steamship routes. Many of the first tourists to the islands made sketches of the scenes they saw and kept journals of the experiences they had. Then, as tourism continued to increase, subsequent tourists relied on the accounts of those who had gone before them. They prepared for a journey by reading the available literature and even took these books along for reference. Thus, they already had an idea in their mind of the places they were going to see before they arrived. For the most part, they found that their own experiences lived up to their expectations, and they perpetuated these ideas in their own travel narratives.

Ultimately a circular relationship evolved between representations of place and experiences of place: Representations created preconceived ideas and images of the places to be

Figures 13.3 and 13.4. The waterfall is one of the most common features in tourism representations of landscapes, as shown in this image from 1887. Cascading through the center of the image, the waterfall is surrounded by jungle-like vegetation. The water collects in a pool at the bottom with large rocks in the foreground, and the people positioned at this pool marvel at the scene. Scenes highlighted in tourism promotions haven't changed much since then; modern tourism organizations often feature images such as this. *Sources:* (13.3) Froude, *English in the West Indies*, 72; (13.4) Velvet Nelson

visited; experiences in places tested those preconceptions. Because tourists generally felt that the experience lived up to their expectations, these ideas and images were reaffirmed and perpetuated through successive generations of tourists. As a result, traces of the past may be seen in modern representations of tourism in the Caribbean.* Certainly the nature of these representations has changed over time, but the legacy is nonetheless clearly seen. For example, in 1869, English historian and novelist Charles Kingsley wrote of Dominica, "The whole island, from peak to shore, seems some glorious jewel—an emerald with tints of sapphire and topaz, hanging between blue sea and white surf below, and blue sky and white cloud above."† A modern Caribbean Tourism Organization publication described St. Lucia as "a brilliant green jewel in the blue Caribbean . . . from the Pitons' majestic twin peaks rising above the southeast coast, to Mt. Gimie—the island's highest point—to its miles of pristine beach. . . ."‡ Although early visual representations were charcoal sketches, pen-and-ink drawings, or black-and-white photographs, similar scenes are depicted in color photographs of today (figures 13.3 and 13.4).

Discussion topic: Why do you think the same ideas and images that were used by travel writers in the past are still used today?

Tourism on the web: Caribbean Tourism Organization, "The Official Tourism Website of the Caribbean," at http://www.caribbeantravel.com

---

* Velvet Nelson, "Traces of the Past: The Cycle of Expectation in Caribbean Tourism Representations," *Journal of Tourism and Cultural Change* 5 (2007): 11.

† Charles Kingsley, *At Last: A Christmas in the West Indies* (New York: Harper & Brothers Publishers, 1871), 57.

‡ Caribbean Tourism Organization, *Caribbean Vacation Planner* (Coral Gables, FL: Gold Book, 2002), 53.

One of the most common tourism representations of place is an empty natural landscape—such as a deserted beach or an undisturbed forest—without people or evidence of people. This emphasizes the naturalness and authenticity of the destination; it implies an earlier time, when life was slower, simpler, and people had a closer connection to their environment. As such, it is targeted at tourists who live in places with the opposite character: namely, overcrowded urban areas with the fast pace of modern life. While this has typically been intended for the traditional major tourist-generating regions of northeastern North America and Northwestern Europe, it can now also apply to the emerging generating regions in parts of East and South Asia. Clearly, this plays upon the tourist inversions discussed in chapter 2 in terms of creating a sense of contrast with those places and activities that make up tourists' daily lives.

Moreover, this type of representation invites potential tourists to imagine themselves in that place. The setting is provided and some suggestions might be made about what could be done there, but everything else is left open. This allows viewers to tap into their own dreams and desires and fill in what they want from the experience. This helps create a demand for the imagined experience and encourages tourists to visit that place for the purpose of turning fantasy into reality.

Promotional Media

At least initially, tourism stakeholders relied on travelers' written accounts and word-of-mouth to promote both places and the services that would allow people to get to those places. Throughout much of the nineteenth century, rail companies in Britain did relatively little to advertise their services or the places they served. However, their American counterparts more quickly realized the value of generating tourism for the purpose of creating a steady market for the places they served. These companies, and later the British ones, produced abundant information and advertised in newspapers and magazines.

Both of these patterns were perpetuated in the twentieth century. Travel service companies—also including airlines by this time—and national/regional tourism organizations continued to produce pamphlets, brochures, and magazines to highlight the specific attractions of the destination(s). These materials could be distributed to potential tourists by mail (through targeted marketing or by request) or through travel agencies. In addition, these stakeholders continued to adapt to new forms of media by using television advertisements as well. To some extent, television advertising is still used, often by specific destinations or resorts. For example, Visit California has run a series of television ads on the "misconceptions" of life in California featuring a host of celebrities ranging from Jason Mraz to January Jones, Shaun White, Betty White, and more.

With the rise of Web 1.0, the Internet emerged as the most important medium for place promotion. Although Internet access is still not evenly distributed around the world, the areas that continue to have the least access are the least developed places and the ones that export relatively few tourists. For the major tourist-generating regions, the Internet became the way in which the majority of people learned about a destination, booked their travel arrangements, and formed expectations for their experiences. Electronic media is increasingly replacing print media for pre-trip research as well as during the movement and experience stages. While many destinations and attractions continue to make information available at the point of consumption, this is also changing with the dramatic growth of smartphone use.

The destination website, typically produced by national, regional, or local tourism organizations, has become an important source for place promotion. This includes raising awareness about the destination, providing contextual and logistical information, and creating a demand for tourism to that place. Websites were considered an improvement over print media in place promotion. They are more flexible in that they have the ability to provide a greater quantity of information. This information can be customized for the market (e.g., one representation of the destination for an American market when American/English is selected as the language and a different representation for a Chinese market when Mandarin is selected) or for the tourism product (e.g., sections on ecotourism, cultural tourism, food and beverage tourism, etc.). Websites also offer a greater quantity and variety of media than was possible in printed materials, including more images as well as sound and video files. In addition to these destination websites, tourism promotion on the web is also found in sites for specific hotels and resorts (e.g., Sandals.com), tour packages (e.g., VikingRiverCruises.com), and various types of tourism experiences (e.g., SkyAdventures.travel).

Social Media

As we saw in the previous chapter, tourists are increasingly representing the places they visit through social media, and these representations are becoming extraordinarily significant. The Travel 2.0 technologies are going to exert more and more influence over tourist decisions and behavior.[16] For example, UGC has been found to be "search engine friendly," meaning that tourists may encounter these representations prior to those produced by stakeholders. Additionally, UGC is increasingly perceived to be more credible than information intended to "sell" places, experiences, or services.[17] This presents a distinct challenge for place reputation management efforts, as stakeholders are no longer in complete control over the image of their destination and their tourism products and services. Yet, Travel 2.0 should also be viewed as an opportunity for stakeholders.

Travel 2.0 provides tourism stakeholders with unprecedented insight into tourists' perceptions, preferences, expectations, experiences, and reactions. Blogs offer in-depth information about the tourist process, from how they learned about the destination to how they arrived there, what they did, what they thought about it, and what they want to tell others. This allows destination organizers to understand their tourists and even to identify areas of potential within the destination environment (e.g., features that could serve as an attraction or services that may be needed) that they may not have previously considered. Reviews highlight those features that visitors like and dislike about a destination or business, which helps stakeholders to understand their strengths and weaknesses. Stakeholders might also consider what visitors like and dislike about their competitors to see how they compare. They might also look for potential differences between different visitor demographics (e.g., country of origin) to better understand their markets and demand.

Analyzing social media comments and conversation allows stakeholders to effectively "listen" to visitors and to make improvements—in service provision and/or place promotion—accordingly. This can increase visitor satisfaction, which will be reflected in social media, contribute to a positive destination reputation, and potentially promote visitor loyalty. Staying attuned to social media helps stakeholders stay responsive to changes, whether it is a new type of tourist to the destination or new patterns of behavior among today's tourists. Stakeholders can also use review sites as an opportunity to enhance customer relationship management. For example, they can express appreciation to those reviewers who provide positive feedback. At the same time, they might be able to mitigate the damage of a negative review if they respond promptly and courteously, provide explanations as appropriate, but most important, demonstrate that they are taking the complaint seriously and working to remedy any issues raised.[18]

In addition to being *consumers* of Travel 2.0 information, tourism stakeholders should be active *users*. Tourism stakeholders have been slow to understand the power of social media and to use it for their benefit.[19] These stakeholders may remain unconvinced of the potential of social media as a marketing tool or believe that social media users are not their target audience (e.g., social media is only used by young people who may lack the power to make travel decisions or the financial resources to travel). They may not know how to make social media work for them, or they may be unable to spare the human resources to devote the necessary time and energy into monitoring and using social media (box 13.3).

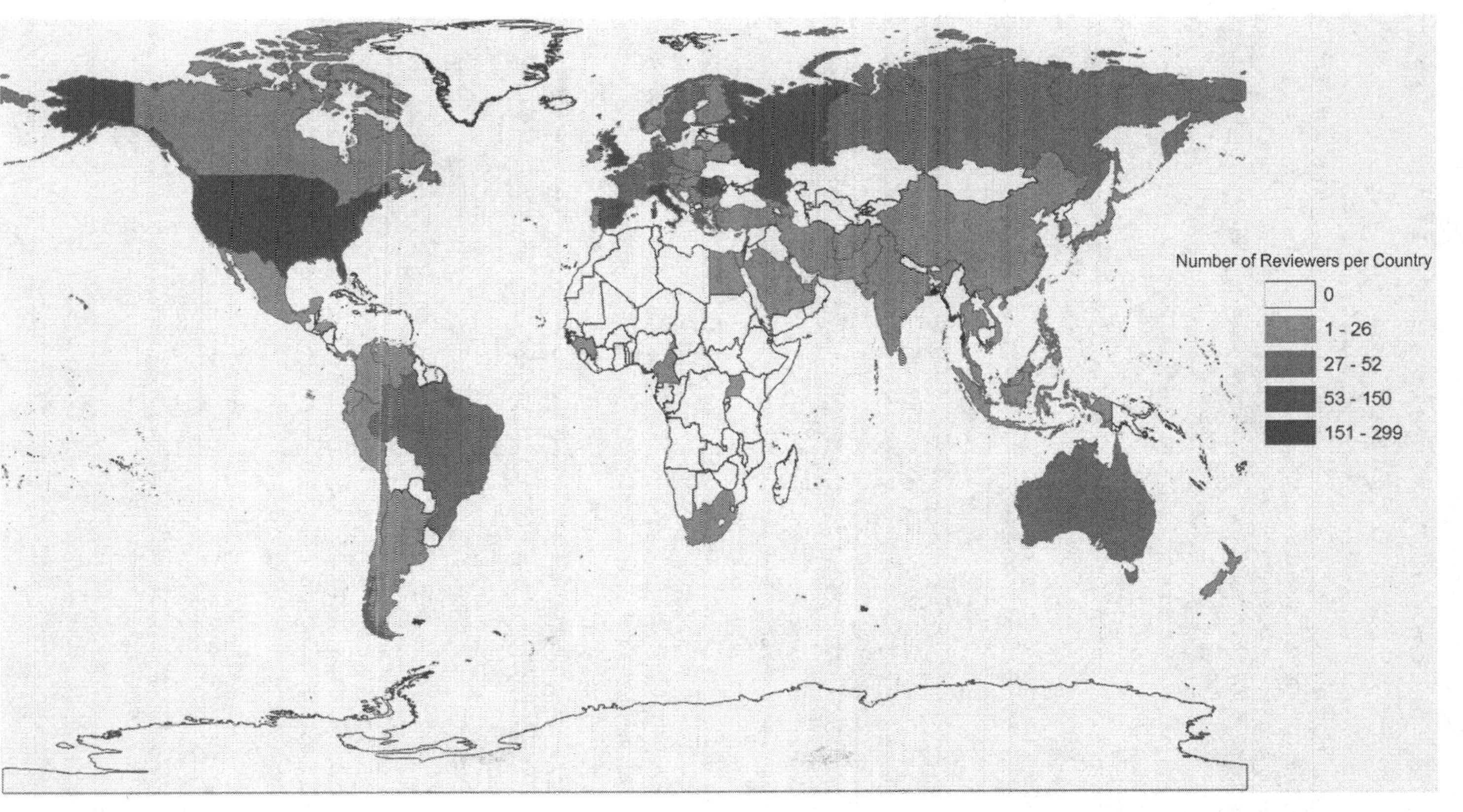

Map 13.1. Tourists all over the world now use review sites such as TripAdvisor, and some of the most popular attractions have tens of thousands of reviews. In the case of one attraction, Romania's Bran Castle—identified as Dracula's Castle on TripAdvisor—86 percent of reviewers (1,952 as of April 2016) identified their country of origin. Eighty-four countries were represented. Although the attraction itself does not collect country of origin data, stakeholders estimate their largest sources include Romania, the United Kingdom, the United States, Japan, Germany, Spain, Israel, Austria, and Italy. *Source:* Samuel Adu-Prah

Box 13.3. Experience: Marketing and Social Media at a History Museum

Travel 2.0 is creating new opportunities, as well as new challenges, for tourism destinations, attractions, and service providers. To maintain a competitive advantage, businesses and organizations increasingly need to use and monitor social media. This role often falls to someone traditionally responsible for marketing. For example, Megan has been the marketing coordinator for a small history museum for seven years. In that time, her work has increasingly moved online.

As an undergrad, I studied history. As a graduate student, I studied recreation, park, and tourism sciences. As the marketing coordinator for a history museum, I am able to combine my two areas of interest. One of my greatest challenges in this job is the fact that we receive such a wide range of visitors—from history enthusiasts to those who have no clue about our subject matter, from senior citizens to elementary school groups. It can be difficult to market a site to such a broad audience, but it's what makes my job fun. I am primarily responsible for all marketing for the museum, education demonstrations, guest speakers, and special events. Social media has been a useful tool in getting our name out there.

Social media has really taken off. When I first started working here at the museum, my supervisor had started a Facebook page. We have now expanded that to Twitter, YouTube, Instagram, and even Pinterest. I try to post regularly at least on Facebook, Twitter, and Instagram; I announce events, showcase items from our collection, or share articles related to our topic. Mostly we get likes and shares. I think it is a great way to show people that we are doing some really cool things here and to try to overcome some of the perceptions they might have of history museums as "boring" places. In particular, our followers seem to really like the "behind the scenes" posts.

I also monitor our TripAdvisor reviews and Groupon ratings. Overall, responses have been pretty positive. People comment that they didn't know we were here, they really learned a lot, and they enjoyed their visit. It can be frustrating to get a negative review, especially when the complaint is not really justified. For example, one person gave us a "poor" rating on TripAdvisor because the museum was closed when they were there. But the museum is always closed on that day, and that information is clearly posted in multiple places. That being said, negative reviews can also be useful in letting us know how we can improve.

It takes time for me to monitor all of the sites and a bit of experimentation to figure out what to post on social media and when. I have read the studies that make recommendations on these things, but in my experience, you cannot always predict what is going to generate the most interest. I have found that posts with pictures generally get more attention, and tagging fellow employees helps me reach more people because their friends see it in addition to our followers. I also think it is important to support (i.e., "like" or "follow") other museums and businesses in the area.

I like the immediacy of social media; I can post an update or a live video when things happen. The technology allows me to be creative and to interact with our followers in a way that wasn't possible before. It helps me to get our name out there. Most of our social media followers are history buffs and are generally from the local area, but I have also been able to reach people who wouldn't know about us otherwise. Recently people have started tagging in at the museum and posting pictures from their visit. This electronic word-of-mouth has become a really powerful tool for promoting places. Social media should not be underestimated for smaller sites like us. We simply don't have much of a budget for traditional marketing, but social media gives us the potential to reach an even wider audience. Overall, I think social media has been a great tool for us.

—Megan

The research is clear: Tourism organizations must incorporate social media into their marketing plans or they will lack a critical competitive advantage.[20] Stakeholders are increasingly recognizing this; however, some remain reactionary. They "get on" social media sites but essentially maintain existing marketing strategies. In other words, they post the same imagery and promotional taglines or phrases used on the website on various social media outlets. They may even post identical messages on various Web 1.0 and 2.0 sites to try to reach audiences in at least one place. However, such content is unlikely to generate the conversation that is essential in social media. Stakeholders need to find more innovative ways to not only communicate the desired message but also to effectively engage their audience.

For example, stakeholders might pose a question that encourages social media users to respond with their opinions about what the destination has to offer (e.g., What is your favorite place at destination X?). Stakeholders can learn from users' answers, but these answers can also promote the destination by identifying and highlighting aspects that other users might not be aware of. In addition, the activity (i.e., responses to both the original question and to users' answers) can be effective in generating a "buzz" about the destination.[21] Stakeholders might also ask questions about what types of experiences or services that potential and/or actual tourists would like to see provided at the destination. In this way, stakeholders have the opportunity to interact and develop relationships with visitors. This helps eliminate the conceptual "distance" between the two that was often perpetuated by the authoritative mode of "pushing" information to audiences for them to consume that was typical in Travel 1.0. Visitors can feel that their opinions are valued and perhaps even become more "invested" in the destination.

Most stakeholders take advantage of existing social media sites, such as Facebook or Twitter. Because they act as users of these sites, they do not have to pay for promotion, but they have access to audiences worldwide.[22] Because people interact with these sites regularly, destination content is viewed more frequently than just during the pre-trip information search. This increases awareness about the destination and helps to create a demand for its experience. In addition, users may be inspired to view the destination page based on teasers in posts and tweets, and one-click links make it easy for them to do so. To put this in perspective, one study found that, in 2010, VisitBritain's Facebook page was viewed fifty-three million times. By way of comparison, the official tourism website was viewed eighteen million times.[23] In general, the most effective approach is considered to be a balanced one; stakeholders should maintain an active presence to keep the destination in the minds of potential visitors without annoying users by bombarding them with content.[24]

Finally, stakeholders are also recognizing the power of blogs as a source of travel information and inspiration. As with the social media sites discussed above, blogs offer a low-cost means of connecting with potential first time and repeat visitors. Destinations such as Luxembourg include a blog on their official tourism website that provides information on specific topics, special events, or simply different perspectives.[25] For example, destination marketers may recruit local residents and businesses to write about their favorite things to do or what places mean to them to create a deeper sense of place. Likewise, they might reproduce visitor stories to give potential tourists an idea of what their experience might be like. With the first-person narrative style of blogs, this presents a more informal source of information that is intended to be more engaging than traditional place promotion.[26] The comments section of blogs also creates further opportunities for discussion.

Travel 2.0 has especially been seen as an asset for smaller destinations and tourism businesses.[27] Social media allows them to raise awareness among potentially limitless audiences, which would have been virtually impossible in the past. In addition, they can use the power of previous visitors' reviews to give potential visitors the confidence to take a chance on an unknown entity. As such, they have the potential to compete with larger destinations or corporations with well-known reputations and more resources.

Consequences of Promotions

One of the most prominent scholars on the geography of place, Edward Relph, criticized representations of place, particularly advertising and promotion. He argued that these representations create superficial ideas of places based on simplified, recognizable, perhaps even exaggerated concepts that may be readily accepted by an external audience. These representations have little to do with reality, the meanings that the place has for the people who live there, or the sense of place that can only be obtained through direct experience.[28]

Tourism promotions naturally represent places as something to be experienced and enjoyed by tourists. As such, aspects of the place, including its people, may be objectified as they are essentially turned into an attraction for tourists. The idea of place as playground discounts the daily activities, lifestyles, and livelihood patterns of local people. Moreover, promotions that represent places without people or nontourist activities can reinforce the segregation of local people and tourists. For example, destination organizers may prevent residents from undertaking activities in a place to ensure that the "natural" condition of the site that is used to attract tourists is maintained for their pleasure.

Place promotion must maintain a balance between tapping into generalized ideas of place that appeal to audiences and creating a sense of distinction among other destinations. Because some destinations appeal to the same tourist motivations as others, promotions from places around the world draw upon the same themes. These places then run the risk of becoming "placeless." **Placelessness** is described as a loss of identity, in which one place looks and feels like other places, often as a result of the superficial, stereotypical images circulated by the media.[29] As tourism destinations, these places essentially have the same experiences to offer. For example, the idea of a tropical island paradise is clearly important in tourism representations, but the stereotypical imagery associated with it can, in fact, describe places in many different parts of the world. In their search for a destination, potential tourists may seek this *type* of place, where they believe they will have the experience they desire, rather than a specific place. As such, a destination needs to be able to not only attract potential tourists' interest with these themes but also provide them with a reason to choose it over other destinations that may appeal to the same motivations or desires.

Perhaps most important, the representations in place promotion must maintain a balance between presenting the characteristics of a destination that are most likely to attract tourists and creating realistic expectations for the experiences they would have at the destination. The ideas and images created for promotional purposes are often simplified and generalized, and not all aspects of the destination will fit that mold. To some extent, destinations can mitigate the potential for conflict between expectations and reality by the way in which tourism is developed. In many destinations, the tourism infrastructure channels tourists into certain places that are most likely to fit the idea of the destination presented to them before they arrive. Likewise, they are kept away from those parts of a place that do not fit the image, such as an inner-city slum adjacent to a fashionable metropolitan district, a section of clear-cut forest near popular hiking trails, or a landfill just a few miles from a pristine beach. Because of this geographic separation, tourists' expectations for a place are often met or even exceeded.

However, if tourists encounter a reality that is vastly different from their expectations, this increases the likelihood they will be dissatisfied with their experiences. In the past, these tourists would express this dissatisfaction to family and friends in the post-trip stage. In the era of Travel 2.0, tourists can log on to social network and microblog sites to immediately vent their frustrations and to post more in-depth criticisms in reviews and blogs after they return home. One negative report will be unlikely to have much impact, but if certain themes become recurring it will affect the reputation of the destination or tourism service provider. People are skeptical of advertising. If it becomes apparent that promotional images and literature show one thing but tourists report something else, the latter can prove to be more powerful.

Conclusion

Stakeholders in places all over the world are engaged in promotion, branding, and reputation management to try to raise awareness among external audiences, create positive perceptions, and encourage people to visit their place over any other. However, in the modern world, these stakeholders do not have complete control over place reputation. While social media presents a distinct challenge for people in many places, it also offers many opportunities for tourism organizations. There is a learning curve, as organizations try to figure out how to make social media work for them, and research into these issues is also in the early stages. Travel 2.0 is a very interesting development in tourism and one that will be fun to follow in the coming years.

Key Terms

- experiential marketing
- placelessness
- place promotion
- place reputation
- place reputation management

Notes

1. Simon Anholt, *Places: Identity, Image and Reputation* (Houndmills: Palgrave Macmillan, 2010), 1.
2. Maja Konecnik and Frank Go, "Tourism Destination Brand Identity: The Case of Slovenia,"*Brand Management* 15, no. 3 (2008): 179.
3. Anholt, *Places*, 2.
4. Evan Cleave and Arku Godwin, "Place Branding and Economic Development at the Local Level in Ontario, Canada," *Geojournal* 80, no. 3 (2015): 325.
5. Nigel Morgan, Annette Pritchard, and Roger Pride, "Tourism Places, Brands, and Reputation Management," in *Destination Brands: Managing Place Reputation*, 3rd ed., ed. Nigel Morgan, Annette Pritchard, and Roger Pride (Florence, KY: Routledge, 2011), 1.
6. Morgan, Pritchard, and Pride, "Tourism Places, Brands, and Reputation Management," 17.
7. Simon Anholt, "Competitive Identity," in *Destination Brands: Managing Place Reputation*, 3rd ed., ed. Nigel Morgan, Annette Pritchard, and Roger Pride (Florence, KY: Routledge, 2011), 22.
8. Detroit Metro Convention & Visitors Bureau, "Metro Detroit—Tourism and Travel," accessed July 15, 2016, http://visitdetroit.com
9. Velvet Nelson, "Place Reputation: Representing Houston, Texas as a Creative Destination through Culinary Culture," *Tourism Geographies* 17, no. 2 (2015): 199.
10. Kelly J. MacKay and Daniel R. Fesenmaier, "Pictorial Element of Destination in Image Formation," *Annals of Tourism Research* 24, no. 3 (1997): 538.
11. Simon Hudson and J.R. Brent Ritchie, "Branding a Memorable Destination Experience. The Case of 'Brand Canada'," *International Journal of Tourism Research* 11 (2009): 218.
12. Dora Agapito, Patrícia Oom do Valle, and Júlio da Costa Mendes, "Sensory Marketing and Tourist Experiences," *Spatial and Organizational Dynamics* 10 (2012): 10.
13. Annica Isacsson, Leena Alakoski, and Asta Bäck, "Using Multiple Senses in Tourism Marketing: The Helsinki Expert, Eckerö Line and Linnanmäki Amusement Park Cases," *Turismos: An International Multidisciplinary Journal of Tourism* 4, no. 3 (2009): 171.
14. Isacsson, Alakoski, and Bäck, "Using Multiple Senses in Tourism Marketing," 171.
15. Velvet Nelson, "Experiential Branding of Grenada's Spice Island Brand," in *Travel, Tourism, and Identity: Culture & Civilization, Volume 7*, ed. Gabriel Ricci (New Brunswick, NJ: Transaction Publishers), 123.
16. Khaldoon Nusair, Mehmet Erdem, Fevzi Okumus, and Anil Bilgihan, "Users' Attitudes toward Online Social Networks in Travel," in *Social Media in Travel, Tourism and Hospitality: Theory, Practice and Cases*, ed. Marianna Sigala, Evangelos Christou, and Ulrike Gretzel (Ashgate: Surrey, 2012), 219.
17. Ulrike Gretzel, "Introduction to Part 3," in *Social Media in Travel, Tourism and Hospitality: Theory, Practice and Cases*, ed. Marianna Sigala, Evangelos Christou, and Ulrike Gretzel (Ashgate: Surrey, 2012), 167.
18. Daniel Leung, Rob Law, Hubert van Hoof, and Dimitrios Buhalis, "Social Media in Tourism and Hospitality: A Literature Review," *Journal of Travel & Tourism Marketing* 30 (2013): 10, 13–14.
19. Stella Kladou and Eleni Mavragani, "Assessing Destination Image: An Online Marketing Approach and the Case of TripAdvisor," *Journal of Destination Marketing & Management* 4 (2015): 189.

20. Stephanie Hays, Stephen John Page, and Dimitrios Buhalis, "Social Media as a Destination Marketing Tool: Its Use by National Tourism Organization," *Current Issues in Tourism* 16, no. 3 (2013): 213.

21. Hays, Page, and Buhalis, "Social Media as a Destination Marketing Tool," 221.

22. Uglješa Stankov, Lazar Lazić, and Vanja Dragiević, "The Extent and Use of Basic Facebook User-Generated Content by the National Tourism Organizations in Europe,"*European Journal of Tourism Research* 3, no. 2 (2010): 106.

23. Hays, Page, and Buhalis, "Social Media as a Destination Marketing Tool," 232.

24. Nusair, Erdem, Okumus, and Bilgihan, "Users' Attitudes toward Online Social Networks," 219.

25. Grand Duchy of Luxembourg, "Blog—Visit Luxembourg," accessed July 18, 2016, http://www.visitluxembourg.com/en/blog

26. Leung, Law, van Hoof, and Buhalis, "Social Media in Tourism and Hospitality," 11.

27. EvangelosChristou, "Introduction to Part 2," in *Social Media in Travel, Tourism and Hospitality: Theory, Practice and Cases*, ed. Marianna Sigala, Evangelos Christou, and Ulrike Gretzel (Ashgate: Surrey, 2012), 69.

28. Edward Relph, *Place and Placelessness* (London: Pion, 1976), 58.

29. Relph, *Place and Placelessness*, 90.

CHAPTER 14

Experiences of Place in Tourism

While popular and social media representations and promotions play an important role in the creation of place meanings, these meanings can also be individual. Overarching cultural conventions, personal preferences, and perhaps most importantly, direct experiences with places will shape these meanings. Although place is an important topic in geography, studies in the geography of tourism have been criticized for giving the experience of place in the context of tourism relatively little attention.[1] This is primarily due to the fact that the human geography traditionally has focused more on the meanings that come from experiences in places that are most familiar. However, the potential to draw upon this tradition in the geography of tourism to explore the experience of other places is clear.

Places are complex entities, and they become even more complex with the development of tourism. The character of a place may be changed as a result of tourism, and it may become more stratified as some areas of the place embrace the influx of outsiders while others remain reserved for locals. Likewise, the meanings associated with a place may be changed and new layers of meaning may be added, based on the experiences of outsiders in addition to those of insiders. This chapter further examines the relationship between place and tourism; in particular, how tourism shapes the character of places and how tourists experience the places they visit.

Places and Tourism

In the previous chapters, we saw how important representations of places are in shaping the ways in which people think about tourism destinations. While these representations are selective in the images that are offered to potential tourists, they must have some basis in the character of the place; otherwise, the destination runs the risk of tourist dissatisfaction when the experience does not match up with expectations. As such, the character of a place is important in attracting and maintaining tourism. Yet, the unique character of a place may ultimately be affected by tourism.

One of the most influential works on the geography of place has been Edward Relph's *Place and Placelessness* (1976). In this work, Relph defines a geography of places that are unique and full of meaning; these places create a world that is rich and varied. He contrasts this with a placeless geography. Non-places have few characteristics that situate them in their location or distinguish one from another, and they lack meanings beyond certain stereotypical ideas. Thus, in a placeless geography, the character of the setting is devoid of significant or unique features, and people do not recognize that places are different. Consequently, placelessness involves both a look and feel of sameness.

For example, the tropical beach has been described as one such non-place. These beaches feature the same, typically stereotyped characteristics (e.g., sunny skies, palm trees, white sands, clear waters, possibly umbrellas and lounge chairs), regardless of their actual location in the tropical world. In fact, even when such a place is visited, there may be few readily apparent features that would distinguish it from other, similar places or indicate the wider character of the place in which it is situated. These beaches are loaded with superficial meanings, such as fun, relaxation, and escape, but they often lack the depth of meaning associated with places that are unique (figure 14.1).

Relph is particularly critical of tourism and argues that it plays an integral role in creating placelessness: "Tourism is an homogenizing influence and its effects everywhere seem to be the same—the destruction of the local and regional landscape that very often initiated tourism, and its replacement by conventional tourist architecture and synthetic landscapes and pseudo-places."[2] In other words, tourism destinations are prone to becoming non-places. This is often attributed to the standardization of mass tourism. Multinational companies that build resorts and restaurants in the same style and offer the

Figure 14.1. Is there anything about this scene that might give you a clue as to where it is? *Source:* Scott Jeffcote

same services regardless of location tend to characterize mass tourism destinations. These multinational companies reflect the demands of organized mass tourists that the places they visit—even if they are foreign places—have at least certain elements of home that are familiar and comfortable. To maintain tourism, destination stakeholders seek to meet these demands. However, in the process, this fundamentally changes the character of that place, at least within certain areas (i.e., the tourist zone). Thus, there may be a certain sameness to mass tourism destinations in many parts of the world.

Another variation of placelessness refers to places that are artificial, contrived, and have little relationship to the history and/or reality of the places in which they are situated (box 14.1). Again citing tourism as a crucial contributing influence, Relph describes this as a process of "Disneyfication" in which the synthetic world of the theme park has begun to affect the character and development of other places.[3] Indeed, scholars have applied this concept to various places around the world, including existing places that have been subject to Disneyfication, such as New Orleans, as well as places that have been developed in this way, such as Las Vegas.

Box 14.1. Case Study: Washington State's Bavarian Town

Leavenworth is a small town in central Washington State. The area's first people were the Yakima, Chinook, and Wenatchi tribes. In the nineteenth century, Euro-American settlers began to arrive in search of furs, gold, and good farmland. The town's economy was largely tied to the logging industry and the railroad, with the headquarters of the Great Northern Railway Company. In the early twentieth century, however, the company left the town, the railroad was rerouted, and the sawmill closed. The town struggled until the 1960s. At this time, community leaders made the decision to remodel the town in an attempt to develop a tourism industry. With the area's natural backdrop of the Cascade Mountains, they decided to create a Bavarian village.* Bavaria is a state in southeastern Germany and the country's most popular tourist region. Scenic mountains and medieval towns are considered typical features of the landscape, while distinctive cultural traditions and festivals round out its characteristics.

Leavenworth was not intended to replicate a particular town in Bavaria; it was based on an idea of a Bavarian town. Theme towns are not uncommon in the United States, but they are usually based on a resident ethnic population. In this case, there was no connection with Germany beyond an impression of place. For this redevelopment scheme to be successful, all stakeholders had to buy into the idea. If centrally located businesses chose not to participate, it would destroy the appearance of "authenticity." Of course, Leavenworth is not an authentic Bavarian town. In this case, authenticity was used to refer to the extent to which the town looked like or appeared to be a Bavarian town. Once support for the project was secured, stakeholders had to formulate a plan and begin the transformation.† The first step was an extensive renovation project in the downtown area based on Bavarian architectural styles.

* City of Leavenworth, Washington, "History," accessed July 19, 2016, http://cityofleavenworth.com/about-leavenworth/leavenworth-history/

† Stephen Frenkel and Judy Walton, "Bavarian Leavenworth and the Symbolic Economy of a Theme Town," *The Geographical Review* 90, no. 4 (2000): 559, 565, 569, 574.

Stakeholders wanted this transformation to be more than just aesthetic, so they also adopted elements of Bavarian culture, such as traditional clothing and musical styles. They developed German restaurants, bakeries, and local craft breweries as well as souvenir shops selling products that one might find in Germany, such as nutcrackers, cuckoo clocks, beer steins, and music boxes. They opened a nutcracker museum (self-described as "probably the world's largest collection of nut-opening devices"‡), which became one of their biggest attractions. They launched their own versions of Bavarian festivals, including Maifest, Oktoberfest (one of the largest outside of Munich, figure 14.2), and a Christmas Lighting Ceremony and Christkindlmarkt (a Bavarian-style Christmas market).

We know that tourism development can have both positive and negative impacts on a place, but the reinvention of Leavenworth for the purpose of establishing a new economic base has been successful. The Bavarian identity of the town became a point of local pride, and the concept was even emulated in other places.§ The Leavenworth Chamber of Commerce's tourist website invites potential visitors to experience Bavaria, and nearly two million tourists come to the town each year.** None of these tourists believe that they are, in fact, experiencing Bavaria, but clearly the destination holds appeal for many people. Research could give us insight into what aspects of place attract visitors and what meanings Leavenworth holds for them.

Discussion topic: Do you think Leavenworth has created a distinctive place in the United States or a placeless destination? Explain.

Tourism on the web: Leavenworth Chamber of Commerce, "Leavenworth Washington Hotels, Lodging, Festivals & Events," http://www.leavenworth.org

Figure 14.2. German festivals, such as Oktoberfest, are popular in places around the United States. *Source:* Scott Jeffcote

‡ Leavenworth Nutcracker Museum, "NM About Us," accessed July 19, 2016, http://www.nutcrackermuseum.com/about.htm

§ Frenkel and Walton, "Bavarian Leavenworth," 568.

** City of Leavenworth, Washington, "History."

Although there is certainly some truth in the relationship between tourism and placelessness (both non-places and Disneyfied places), this type of blanket criticism of tourism is not entirely justified. There is a tremendous variety in tourism, and not all tourists are looking for a standardized experience. Tourists at the opposite end of the spectrum—the explorers and drifters—specifically avoid such destinations and instead seek out new and different places to experience. Although tourism will inevitably bring changes to these destinations, these changes can be made conscientiously to avoid destroying the character of that place.

In addition, the perceived homogenization of places in the modern world has stimulated a process of localization. In other words, in the face of standardization as a result of global processes, including tourism, some places have attempted to reassert local interests, traditions, and distinctiveness. This helps reinforce, and in some cases re-create, a unique sense of identity and character for places that might otherwise be lost. While local people may initiate this process of localization to protect their heritage, it also has the distinct advantage of giving that place a competitive advantage among tourists (or place consumers) who are looking for a unique experience of place.

Sense of Place

Sense of place refers to the association with and emotional attachment to places based on the meanings given to those places. It is one of the ways in which we are connected to the world and therefore an integral part of the human experience.[4] A sense of place is developed by experience in and a relationship with a place. In particular, geographer Yi-Fu Tuan has argued that "sense of place is rarely acquired in passing. To know a place well requires long residence and deep involvement."[5] Thus, the places of our everyday lives are those that hold the most meaning for us and therefore are the ones to which we are most attached. As such, these are the places, relationships, and meanings that geographers have been most interested in. Nonetheless, sense of place can contribute to our understanding of the geography of tourism.

For example, the relationship with and feelings we have toward the places of our everyday lives can play a role in the demand for tourism. In addition to feelings of affection and attachment, the familiarity of these places can generate feelings of complacency or even hostility if we begin to perceive that we are tied to or imprisoned there. Even though we know that these are the places to which we will always return, we may still feel the need for a temporary change of place. Tourism provides us with this opportunity.

The sense of place that we have for "our" places is instinctive and unconscious. In fact, it is something that we think little about. However, developing a sense of place can also be a conscious act. Although tourists will not develop a sense of a place equal to an insider's, they nonetheless have the potential to gain insight into a place if they are willing to be open-minded and sensitive to its nuances.[6] Taken a step farther, tourism becomes the means of experiencing new places and places in new ways. Unfamiliar places are experienced differently than familiar ones. While we take certain aspects of a familiar place for granted, everything in a new place is different and unknown.

We may have a greater sense of curiosity and excitement. Activities that seem mundane in our daily lives—driving from one place to another, taking a walk up the street, going to the store, finding something to eat—can suddenly turn into an adventure. As we have the potential to encounter new things, we tend to observe more carefully. We typically have a greater sense of security in the places that we consider our own; thus, to varying degrees based on the context, we may even be challenged to pay more attention to our surroundings in a new place to find what we need or to keep ourselves safe.

Because of widespread representations of places in today's world, there are few places for which we have no preconceptions or expectations. The meanings these places hold may be abstract and are most likely based on stereotypes. However, with every experience we have in places, we build upon these preconceived ideas. Over the course of a trip, we create more nuanced, personal meanings of place that constitute a type of sense of place.

Finally, experiences in new places often cause us to reflect on our experiences in those places most familiar to us, those that constitute the setting of our daily lives. In some cases, the sudden absence of those aspects of a place that we take for granted may prompt us to appreciate them more upon our return, at least for a little while. We may find that there are aspects of a place we would rather see changed, to be more like that of a place visited. In essence, experiences in other places may cause us to refine our sense of place.

Experience of Place

There are countless factors that can affect tourists' experience, ranging from poor infrastructure to the presence of pests. Perhaps one of the greatest factors that have a distinct impact on tourists' experiences is the weather conditions of a place at the time of a trip.[7] Nothing is likely to ruin a tourist's experience more than unexpected and undesirable weather conditions that prevent them from seeing or doing the things they had planned. While it may be an unusual—perhaps even unprecedented—occurrence for that place, it may be the only experience tourists have with that place. Of course, other factors may be unrelated to the place but will affect the tourists' experience of it nonetheless. For example, tourists who are sick during their vacation may not be able to participate in certain activities, and they are likely to enter into their experiences with far less enthusiasm than they would have otherwise.

Interactions with the people at a destination can play an important role in tourists' experience of that place. In a new place, we may not know where to go, how to act, or whom to trust. Tourists are often wary of being taken advantage of or cheated—in some cases, rightly so. Tourists may have little knowledge of how much things should cost or the way things work; consequently, they are at the mercy of tourism stakeholders and local people to deal with them fairly. Encounters with local people who are honest and friendly, or those who go out of their way to help strangers, can have an extraordinarily positive impact on tourists' experience. Conversely, encounters with even a few people at the destination who are dishonest, unhelpful, hostile, or harassing can ultimately shape the way tourists forever think about that place.

Personal factors, such as previous experiences and personality, play a role in the way an individual experiences a place. We approach experiences with different attitudes. Some tourists feel apprehension, anxiety, or even fear, perhaps from the very moment they leave home, at the unknown of experiencing a new place. This is most often the case among tourists who have had little experience with new and especially different places. However, this should lessen with time spent in the new place, as the tourist becomes more familiar and comfortable with the circumstances. In contrast, other tourists may experience a sense of euphoria at being in a place where everything is novel. This, too, can lessen with time as the novelty of the experience begins to wear off.

Tourists have different logistical options for experiencing a destination. The following sections discuss guided tours and independent travel as two options and how each shapes the experience of place.

GUIDED TOURS

Tourists can experience a destination through a guided tour. There is an endless variety of experiences that range from a complete package trip to a day-long excursion as one part of a larger trip, from a group with dozens of participants to a one-on-one experience. Tourists might choose this experience for a number of different reasons, reflecting the type of tourist, the desired tourism product, the choice of destination, or the motivations for the trip. Essentially, guided tours can serve different purposes and provide different types of experiences of place.

For example, organized mass tourists are often interested in the convenience of a package tour such as the "European Dream" eight-day coach tour spanning London, Amsterdam, Paris, Lucerne, Milan, Venice, and Rome. The itinerary is preplanned (e.g., what places to visit and for how long), and all of the logistical arrangements have already been made (e.g., how to travel, where to stay, where to eat). This creates a "worry-free" holiday for tourists who do not have the time or interest in planning a trip and do not want any surprises. These tours may be considered suitable for relatively inexperienced travelers who are anxious about traveling in an unfamiliar, especially foreign, destination where they may not know the customs or speak the language.

Although these tours are tremendously popular, they are commonly criticized for minimizing the experience of place. There is little need to come to the destination with any knowledge of the place, as all arrangements have already been made and guides provide necessary information along the way. With a set itinerary, there is little opportunity for exploration and interaction with the place or its people. The spontaneity of the tourism experience is eliminated, and tourists are reduced to passive observers of place through the windows of a climate-controlled bus.

In comparison, individual mass tourists are likely to visit the same, or similar, destinations as their organized counterparts (e.g., London or Rome); however, they will outline their own itineraries and/or make their own travel arrangements. This will require more research and planning in the pre-trip stage. At the destination, they may choose to do a combination of the guided tour and independent travel for their experience; for example, they might take a day tour or sightseeing bus to get information about the place and to see the highlighted attractions before exploring a bit more on their own.

Guided tours may also be used to facilitate certain types of special-interest tourism, such as those that require specific skills. For example, tourists may be interested in participating in an activity at the destination—such as rock climbing, scuba diving, or horseback riding—but have little previous experience with that activity. As such, a tour provides them with instruction, necessary equipment, and a guide to help them along the way and ensure their safety. Likewise, special-interest tourism may require in-depth local knowledge. Tourists interested in bird watching, wildlife photography, hunting, or fishing may require a local guide who will know when and where they will have the greatest opportunities for these activities.

Finally, guided tours may be necessary to allow tourists to visit places they would not otherwise know about or have access to (i.e., MacCannell's back region stages discussed in chapter 7). This can include places not generally made known to outsiders, such as an unmarked hiking trail, or those places not open to outsiders except on a tour because of logistical or safety reasons (e.g., the subterranean passages of the Seattle Underground). Some destinations impose such specific regulations on tourists that a guide is necessary to ensure that proper procedures are followed; some of the strictest controlled destinations actually require that tourists travel with a guide. This is the case in places like Tibet and Bhutan. Thus, while the types of tourists who visit these "off-the-beaten-track" destinations are fairly adventurous explorers and drifters looking for a unique experience, they must travel in a different manner than they would normally.

The existence of tour guides dates back to the earliest eras of tourism. Tour guides played a particularly important role in the Grand Tour era, before the tourism industry and infrastructure were developed. With little in the way of guidebook information, maps, signs, or other features that facilitate tourism, outsiders were dependent on guides. On one hand, the guides would literally guide tourists in places that were unfamiliar, inaccessible to outsiders, and in which they would be met with suspicion or hostility by the local population. On the other hand, these guides would metaphorically guide tourists in the process of personal development that was intended to accompany an experience such as the Grand Tour.[8]

Modern tour guides continue to serve a variety of functions (figure 14.3). In addition to taking care of logistical arrangements, guides are generally responsible for the safety and well-being of tourists during the course of the trip. At the same time, they are responsible for ensuring that tourists are familiar with and abide by local customs and policies. This is particularly the case in places where guides are required. For example, if a tourist does something in Tibet that is prohibited by the Chinese government (e.g., staying to travel independently after the tour or engaging in political activism), the tour guide and/or the travel agency that arranged the tour will be punished for the offense.

Tour guides are expected to have a good knowledge of the places visited. While some tour itineraries are preplanned, others are flexible, and it is up to the guide to determine the course of the trip/excursion to reflect the interests of tour participants. The guide may be expected to find routes that will yield the best opportunities to

Figure 14.3. This tour guide in St. Kitts (Carribean) transported his tour participants to a local hiking trail, provided information about the area's flora and fauna, and told stories about exploring the forest as a child growing up on the island. *Source:* Tom Nelson

encounter desired points of interest. This is particularly applicable on special interest tours where those points are moving targets (e.g., wildlife). In addition, guides must be able to convey their knowledge of a place to participants. This must be done in a way that is easily understood by visitors, which may require translating or interpreting things that might seem strange or unusual to outsiders. Yet, guides must also balance providing information about places and entertaining tourists on holiday (for a humorous take on this topic, see the 2009 film *My Life in Ruins*).

Tour guides may be required to have certain skill sets, such as fluency in multiple languages, and also some knowledge of the tourists' culture to understand what their interests are, what type of experience they want, and how to best represent the places visited. Some tourists may want information about the places, while others may be more interested in myths or personal stories. In some cases, tourists may not want any interpretation at all; the guide is simply intended to facilitate travel and highlight sights to be seen and/or photographed. Guides must be flexible and accommodating to meet the needs of tour participants as they arise.

Tour guides can have a highly important role to play in tourists' experiences of a place. They constitute another form of representation; they represent local peoples and places to tourists. As such, they have tremendous power in determining what is important and will be seen—and, conversely, what will not—and shaping what tourists think about those places and their experiences. Guides are considered an essential interface between tourists and the destination.[9] However, they are not entirely autonomous; they may be required to represent the aspects and stories of the place that are officially sanctioned. In other words, they may be limited by what places the government will allow guides to take tourists and what topics they can discuss (see box 4.2).

Tourists place a great deal of trust in their guides to be honest, give them accurate information, and generally deal with them fairly. Of course, this is not always the case. Guides may fabricate information, advise them to purchase inauthentic souvenirs, or require them to pay additional "fees" that line their pockets or those of their acquaintances. However, tour companies depend on their reputation, and social media allows dissatisfied customers to spread the word about any problems they had with a tour. Moreover, because tour guides play such an important role in representing the destination, governments frequently implement regulations and/or require licenses for tour guides. Nonetheless, unlicensed guides operating in the informal sector of the economy are common in many destinations around the world, and experiences with them vary widely.

INDEPENDENT TRAVEL

Many tourists—particularly explorers and drifters—prefer to experience a place on their own. Just as organized package tours are criticized for minimizing tourists' experience of place, independent explorations are often considered to provide the greatest opportunities for tourists to develop a sense of place. Of course, a deeper experience of a place does not necessarily translate into a positive one, and there can be both advantages and disadvantages of going it alone at the destination.

Those who prefer independent travel typically value the flexibility to set their own itinerary. Based on their interests and priorities, they can choose what places they want to visit and what sights to see. These may be primary attractions, but one of the advantages of independent travel is the ability to get off the traditional tourist track and experience more of the place than the front regions. For example, tourists come en masse to visit Piazza dei Miracoli, the main tourist complex in Pisa, Italy, featuring the famous Leaning Tower. This is a well-known primary attraction and one that all kinds of tourists to Tuscany are likely to see. However, those on a guided tour will likely only experience this part of Pisa, while those traveling independently have the potential to explore other parts of the historic city if they choose.

Likewise, tourists who travel independently are not bound by a strict schedule, unless they set it for themselves. They have the flexibility to linger at a site that they find enjoyable or interesting without feeling rushed, and conversely, they are free to move on to the next attraction if they decide they have done all they wanted in that place. Consequently, tourists can feel that they had the fullest experience of a place with little perceived wasted time (e.g., waiting at rendezvous points).

To some extent, those who travel by personal vehicle may be subject to the same criticisms as those who travel by tour bus: that their only experience of other places is from a distance and in passing. However, personal vehicles can be used as a means of getting *to* a destination but not the primary means of *experiencing* it. Independent travelers may choose to walk or use public transportation at the destination, which will provide opportunities for interactions with local people, access to back regions, and insight into the lived experience of the place.

This more flexible style of travel allows for greater spontaneity in the experience of place. Independent travel does not always go as planned; in fact, it frequently does not. Yet, for many tourists, their most memorable and rewarding experiences of a place are those that were stumbled upon by accident in the course of exploring—and, in some cases, getting lost—on their own. These are the experiences where they met interesting local people, found a great restaurant, saw places they never would have encountered otherwise, and observed or participated in a unique local event.

Of course, not all unexpected experiences are pleasant ones. For many tourists, the prospect of traveling without a guide and facing the unexpected alone are great sources of stress. Tour participants benefit from operators who scout out the best attractions, accommodations, or restaurants. Tourists who plan their own trips, on the other hand, must make selections from their best guess based on whatever information is available. Traveler reviews are seen as a means of avoiding the worst experiences, but they cannot prevent us from running into issues.

Tour participants benefit from dedicated transportation that takes them directly to points of interest. In contrast, those who rely on public transportation may be frustrated by restrictive schedules and an inability to get to tourist sites not served by transportation systems. Independent tourists also run the risk of getting lost. For some, this is an adventure and creates opportunities, but for others, it is a source of stress. These tourists face the potential frustration of not reaching the desired attractions or the anxiety of finding themselves in undesirable, possibly unsafe, locations.

Independent tourists may not always have access to the same extent of information about the places and attractions visited. Sites have varying degrees of information available to independent tourists. Some highly developed attractions have self-guided audio tours and site-specific guidebooks (both increasingly available via mobile phone, figure 14.4) and well-annotated displays or interactive monitors. However, less developed or local attractions with fewer resources may provide little information about the site or use only the local language in guides and displays. This information is also typically limited to basic facts without providing the level of detail or richer stories that a good tour guide might have to offer.

Finally, tour guides act as a middleman between tourists and local people. These guides should speak the local language and understand local customs and therefore be able to help tourists navigate foreign destinations. Even the most conscientious tourists who try to familiarize themselves with the local culture and speak some basic words of the language can run into problems with miscommunication and misunderstanding

Figure 14.4. Mobile phone tours are increasingly replacing text-based displays and recorded audio boxes to convey information to visitors at various types of attractions. This "maritimes voices tour" provides short descriptions about the ships at Hyde Street Pier, part of the San Francisco Maritime National Historical Park. *Source:* Velvet Nelson

when they have to manage various situations on their own. The stress, frustration, and/or dissatisfaction that arise from any of these aspects of a trip can affect tourists' attitudes toward the place and their experience of it.

SOLO INDEPENDENT FEMALE TRAVEL

Although it is recognized that tourists are not a single, homogenous group that will experience places in the same ways, tourism research has not always given these differences much consideration. However, supported by developments in feminist geography, more attention has recently been given to the factors that cause women to experience places differently than men. This is exemplified by the attitudes toward and patterns of solo independent travel (i.e., traveling without a companion or an organized tour group).

Although there are examples of female travelers throughout history, travel has generally been seen as the province of men. Throughout the early eras of tourism, only a small percentage of women had the time, money, and social standing to be able to travel, and very few traveled alone. The prevailing sociocultural attitude was that it was inappropriate for respectable women to travel extensively and simply unacceptable for them to travel by themselves. Those who did were viewed by society as eccentric at best; at worst, they faced a ruined reputation.

Opportunities for women in the more developed Western countries to travel gradually increased over the course of the twentieth century. More women entered the workforce (i.e., the public sphere), became more financially independent, and to some extent, experienced less social pressure to start a family at a relatively young age. The number of women traveling now is equal to that of men, and an increasing number of women are choosing solo independent travel. Yet, attitudes toward women traveling—especially abroad and alone—remain somewhat antiquated.

For the most part within modern societies, the idea that a woman traveling alone is absolutely unacceptable no longer applies; however, these societies continue to perpetuate ideas about female vulnerability. The prospect of negotiating an unfamiliar place can be scary for anyone, especially those who have little previous travel experience. Getting lost, facing language barriers, and dealing with cultural misunderstandings are fears that are common to many travelers. Women often admit their doubts about traveling alone and not having anyone to rely on but themselves. These fears are magnified by tourism industry guidelines for female travelers (e.g., dos and don'ts lists or security warnings), news and word-of-mouth stories (e.g., tourists being drugged, raped, or kidnapped), and concerns from family and friends. After announcing their intentions to travel alone, women are frequently subject to reactions that may range from surprise (e.g., "Are you sure that's safe?") to disapproval (e.g., "I do not think that's a good idea.") and even outrage (e.g., "How could you think of doing something like that?").[10] These reactions may be well intentioned, as they reflect a concern for the woman's well-being, but they also reproduce and reinforce the perception that women are vulnerable and thus solo independent travel is unwise.

Perceptions of acceptable behavior for women and toward women vary widely around the world. Consequently, solo independent female tourists may be judged by the sociocultural norms of the destination and subject to the reactions of local people. In some places, the idea of a woman traveling alone may still be unacceptable. When traveling to other culture regions, women often report feeling conspicuous, receiving unwanted attention, sensing hostility, experiencing some form of harassment, and feeling insecure or unsafe. In particular, women traveling without a male companion may be viewed as sexually available. For example, in some culturally conservative destinations, it may be unusual for a woman to appear in public unaccompanied, and those who do—including foreign tourists—will be thought of as prostitutes. In destinations that have received some female sex tourists, all foreign women may be perceived to be looking for that sort of relationship and approached accordingly.

All of these factors affect women's travel patterns and the ways they experience other places. They affect whether or not a woman decides to travel alone; if she does travel, they affect where she goes. She may choose to visit destinations that are closer to home, within her own country, or within similar culture regions where she is less likely to stand out. She may also consider her choice of destinations more carefully than a man, considering its reputation for safety and treatment of women. Likewise, these concerns affect where she stays at the destination, including the neighborhood in which an accommodation is located and the level of security it affords.[11]

These factors affect the places she visits at the destination and the type of activities she participates in. She may feel safest within a certain part of the destination, perhaps the front region where there are other people (especially other tourists) who make her feel less noticeable and less of a target. Some types of places are perceived to be less safe than others (e.g., large urban areas with high crime rates or remote, isolated forested areas) and may be avoided. Solo independent female tourists may also avoid certain "masculinized" places. For example, in some cultures, the café is a male-dominated space. If a woman enters that space, she is immediately the target of attention, and she may feel her presence is unwanted. In addition, time of day plays a role; a woman might feel comfortable in a particular place during the day, when it is well lit and populated by other women and children, but less comfortable at night when she might be more likely subjected to harassment. As a result, she may choose to stay in or close to her accommodation at night.

In addition, a woman may make other sorts of adjustments during a trip in an effort to minimize the risk of encountering problems while traveling alone. She may modify her patterns of behavior to be more in line with what is acceptable for women in that place, such as talking softly or lowering her eyes. She may change her patterns of dress to fit local norms (e.g., covering her hair in conservative Muslim countries, figure 14.5), to be more conservative (e.g., longer skirts or pants, longer sleeves, or higher necklines) so that she attracts less unwanted attention and/or incorrect assumptions, or simply to be more similar to local women and therefore less conspicuous. If she is not married, she may wear a band on her ring finger to send the message that, although she is not traveling with a man, she is not "available."

Given real and perceived security issues, a woman may feel that it is her responsibility to not put herself in unsafe places or dangerous situations. Consequently, she has to be constantly aware of her surroundings. This can increase the level of stress associated with travel, which can generate frustration because she is unable to relax and enjoy the experience. Moreover, she may feel that she has an incomplete experience of place because there are certain areas of the destination where she is not comfortable going, typically back regions where she would be conspicuous. Likewise, there may be certain activities that she would like to participate in but does not feel like she can.

These constraints play a real role in shaping women's travel patterns. Not all women will approach travel and experience destinations in the same ways. For some women, these issues may generate suppressed demand. Many women will acknowledge that places at home can also be unsafe, and the experiences faced by solo independent female travelers can also happen to single women at home. However, for women in their home environment, harassment usually occurs in isolated incidents over a long period of time, rather than being compounded during a week-long vacation. Moreover, women at home typically understand the situation and know how to best respond, whereas this is not always the case in foreign environments, which can be extremely unsettling and/or distressing. Yet, many women travel alone. It may be a matter of not giving in to fears or becoming a victim of social pressure. It may be that

Figure 14.5. This American tourist wore a headscarf while she was in Senegal, a predominantly Muslim West African country. *Source:* Pamela Rader

the rewards of travel outweigh the risks. Many female tourists find that solo independent travel is empowering and fulfilling, as they gain confidence in themselves and a sense of accomplishment for overcoming constraints.[12]

Of course, this is not the only category of tourist to experience perceptual and/or logistical constraints to travel. For example, members of various ethnic or religious groups, members of the LGBTQ community (box 14.2), and persons with disabilities will all have different considerations in the pre-trip stage as they make travel decisions and arrangements as well as different factors that affect their experiences.

Box 14.2. Experience: Gay Travel

As individuals, we all have certain things that we consider in our pre-trip planning and factors that affect our experiences in the places we visit. This is based on our personalities, past experiences, and worldviews—but it is also based on things like our age, gender, race, ethnicity, religion, or sexual orientation. Wesley has both lived and traveled extensively in Europe and Asia. In this box, he offers some thoughts on these experiences from his perspective as a gay man.

Travel is one of the best things a person can do for themselves no matter their race, gender, religion, or sexual orientation. It opens the mind like nothing else. It allows people to experience things and to see other points of view in a way that staying in one's own environment never could. It also forces people into situations that are sometimes scary and makes them extremely vulnerable. A friend once said, "being a foreigner is sometimes like putting on a purple bunny costume and walking down the street. You may be doing the most mundane of acts, but you are on display for everyone to see." Some people come across as undeniably gay no matter where they are, but because being gay is not something every homosexual person wears in the same way as skin color or gender, gay travelers often come across just as foreign. As a gay traveler, I think it is important to understand and judge the situations in which you find yourself to decide how much you want to reveal about yourself and how out you want to be.

How other people perceive homosexuality varies greatly from culture to culture. Every society has people that are averse to homosexuality. Some societies are more prone to this than others, whether it is based on the particular culture, religion, or simply a fear of the unfamiliar. In some countries, being gay will land you in jail. In others, same-sex couples are free to show affection openly in public. When I lived in South Korea I almost never talked about being gay except with people I knew well because I was concerned about losing my job. In Sweden, I don't hesitate to reveal that I am gay because I know that I am legally protected here. When I travel to other large European cities, like Copenhagen or Lisbon, I can walk down the street holding my husband's hand because I know that homosexuality is, for the most part, accepted in these cosmopolitan places. Still, traveling in another country can be nerve racking because it may be difficult to know what the legal rights and protections are, and you don't have family or friends in that place that can easily help you if you encounter problems.

Living in a culture that has such strong rights for LGBT people has encouraged me to be a much more confidant gay person and more willing to live an open life, even while traveling. For me, living a more authentic life has overtaken the fear of how others will perceive me. But I am still cautious. When I travel, I tend to stay in gay-friendly areas or frequent gay-friendly venues. Of course, I visit the same places and touristy locations as every other traveler when I visit a new country, but I choose these areas in particular to protect myself. I have used websites like misterbandb.com, which is specifically for gay people and sometimes also apps like Grindr to get insight from gay men in the area about fun and safe places to spend an evening out. However, if I go to a more rural area, there often is no such thing as a gay-friendly area. In this case, I use general tourism sites, like airbnb.com, although I may choose ambiguous words like "partner" to refer to my husband in the booking. This is not because I am ashamed of being gay but because it is hard to know what peoples' reactions might be.

I am very lucky in the sense that I have never had any extremely negative experiences with being gay while abroad. I have had amazing experiences and met beautifully open-minded people, even those who are from cultures where being gay is not accepted. Still, I think it behooves a person to be savvy when they are in a place with which they are not completely familiar. I would rather be more cautious and come away with a positive travel experience than risk the opposite. The world is a beautiful place with beautiful people; it can and should be enjoyed by everyone.

—*Wesley*

Consumption of Places

As we saw in the previous chapter, images are a tremendously important component of tourism representations that are intended to stimulate viewers' imaginations so that they will visualize themselves at the destination. This sort of daydream generates a demand for the actual experience, or consumption, of place. As a fundamentally place-based activity, places are the primary object of consumption in tourism, which is typically visual (i.e., sightseeing). In fact, tourism is often defined by the act of tourists traveling to other places to see new things.[13]

Sociologist John Urry has described this process of visual consumption as the tourist gaze. In particular, Urry distinguishes between two types of the tourist gaze: romantic and collective. The **romantic tourist gaze** is a private or personal experience, where the tourist can gaze in peace and feel as though he or she formed a connection with that place. This is typically undertaken in natural tourist sites (e.g., scenic vistas) and spiritual places (e.g., religious temples). Tourists who prefer the romantic gaze consider this the only way to experience such places; consequently, they may think the experience is "ruined" by the presence of others or the perceived inappropriate behavior of others given the character of the site (e.g., loud talking or laughing). In contrast, the **collective tourist gaze** depends on the presence of people. This occurs in public places that are at least partially characterized by the people found there. For example, the main square or plaza in a major metropolitan area may be a tourist attraction, not just for the architecture of the buildings that define the space, but also for the extent—and cacophony—of life there (box 14.3).[14]

Box 14.3. In-Depth: When the Romantic Gaze Becomes a Collective One

From media representations, we are all well familiar with iconic, awe-inspiring scenes such as Half Dome, Machu Picchu, Stonehenge, Victoria Falls, Angkor Wat, or the Great Wall of China. Some are purely spectacular natural environments; others combine tremendous human heritage with a dramatic setting. Still others are mystical in ways we cannot always explain. When presented with these images, we cannot help but imagine ourselves there, taking in the gaze. With few if any people present in these images, we can picture ourselves standing alone and appreciating the scene the way we want to and having a semi-spiritual experience of these exceptional places—in other words, the romantic gaze.

Yet, the experience of these places has become decidedly collective as millions of tourists visit every year. Countless cars, RVs, and tour buses create traffic congestion and smog and require extensive parking facilities located near the site or viewing area for convenience. Both tour groups and independent tourists may roam the site and accumulate in key spots. The concentration of tourists brings hawkers selling a range of products, as well as pickpockets and hustlers looking for opportunities. This mass of people affects the physical quality of the site by trampling paths, accumulating waste, increasing noise levels, and leaving graffiti or vandalizing property. This detracts not only from the view but also the sensory experience of the gaze and the relationship that the tourist has with the object of the gaze.

As places such as these become victims of their own popularity, the much-anticipated romantic gaze is fundamentally changed to a collective one. There are, of course, places sought specifically for the collective gaze. However, when a tourist is expecting a romantic gaze but gets a collective one instead, he or she is likely to be frustrated and dissatisfied. For the sites that have a distinct hold on our imagination, countless tourists will still visit, even though they know that the experience is not likely to live up to their expectations. Yet, other tourists will avoid popular sites in favor of others that may be considered secondary or tertiary attractions but continue to offer the more personal experience of the romantic gaze.

In a blog on Photo.net, Philip Greenspun suggests that we should revise the way we think about and photograph such sites.* The average tourist to any of the sites mentioned above will not experience it in solitude and stillness, as the media images of them tend to show. Other tourists are very much a part of the experience of these places today, yet we try to erase them from that experience by waiting for a "clear" photograph, free of people. If we are willing to accept that our experience of these sites will involve a collective gaze rather than a romantic one, we are more likely to be satisfied with our experience. In addition, we might find that there is more of interest in the collective gaze if we are willing to approach it with good nature and a little humor.

Discussion topic: Why do you think many tourists have such a strong desire for the romantic gaze?

Tourism on the web: Photo.net, "Tourists as Subjects," http://photo.net/travel/tourists-as-subjects/

* Philip Greenspun, "Tourists as Subjects," accessed July 19, 2016, http://photo.net/travel/tourists-as-subjects/

Whether tourists seek a romantic or a collective gaze, the places chosen must have something distinctive from other places, particularly the places from which tourists are coming. Some places are so different than anywhere else that they automatically attract the gaze. For example, many people feel that Venice is unique and, despite detractions (e.g., hostile local attitudes, high prices, overcrowding), it is ultimately a sight worth seeing. However, for many destinations, tourist offices must establish that their place has something worthy of being gazed upon. This may be a matter of selecting and promoting a particular characteristic or feature of that place. Asheville is a small city in North Carolina's Blue Ridge Mountains. Although there are many opportunities for outdoor recreation in the vicinity, the city has one distinctive attraction: the Biltmore Estate. This 8,000-acre estate and late-nineteenth-century Châteauesque-style mansion is on the U.S. National Register of Historic Places and frequently appears in articles such as *Forbes'* list of America's most beautiful mansions.[15] In other cases, the object of the gaze must be created. In one of the most notable examples, the aging London Bridge was sold to a real estate developer who reconstructed it in Lake Havasu City, Arizona, for the purpose of bringing tourists to the little-known town.

While efforts to highlight a place (or part of a place) as "worthy" of the tourist gaze often takes place in the form of media representations, it can also be done onsite. Tour guides or self-guided-tour information identifies the sights to be seen. Most blatantly, signs literally identify the places or objects that should be gazed upon and direct tourists to them. Of course, this signposting is necessary, to some extent, to help tourists reach the sights to be consumed. Moreover, this may help destinations concentrate tourists in specific areas (e.g., the preferred sites identified in chapter 11) and steer them away from other areas that might not match up with their expectations.

However, someone must decide what is significant or interesting, which may or may not be what tourists would actually choose to see.

In addition to the criticisms of tourism discussed above, Edward Relph argues that this practice of identifying and highlighting those things worth seeing for tourists creates an inauthentic sense of place. Tourists who accept this professionally prepackaged portrayal may think they have a sense of that place. Yet, this is not a deep association with and emotional attachment to a place but a partial, selective, and superficial impression of a place that is not even based on their own explorations, findings, and assessments.[16]

For most tourists, it is not just about seeing the sights; it is also about recording them. Scholars Mike Robinson and David Picard argue, "To be a tourist, it would seem, involves taking photographs. Whilst photography is clearly not the exclusive preserve of tourists, it is nonetheless one of the markers of *being* a tourist."[17] Tourists have long sought to "capture" the scene and bring it home with them as evidence of having been there and a tool to remember the experience. In the earliest eras of tourism, tourists would sketch the places visited or purchase paintings and replicas of famous sites. This practice became even more firmly embedded in tourism with the development of small, portable, easy-to-use personal cameras. While the camera became a symbol of "the tourist"—with all its negative connotations—most tourists were nonetheless willing to endure potential derision to be able to record the places they visit as well as themselves in those places. With the ubiquity of mobile phones, today's tourists can be more inconspicuous (so long as they are not traveling with a selfie stick). Tourist photographs and video recordings comprise another type of representation for the destination, as they are shown to family, friends, and potentially limitless audiences through social media.

This desire to record can ultimately shape the ways tourists experience the destination. The itineraries of both package and self-guided tours are often structured around stops at locations that have been predetermined to offer the best photographic opportunities. Thus, the first and perhaps the only thing tourists do at these locations is take a picture. These tourists may not even be aware of what they are taking pictures of or why. At well-known destinations, tourists commonly look for the sights they have seen countless times before in media representations, places like the Eiffel Tower, the Sydney Opera House, or the Christ the Redeemer statue, so they can capture it for themselves.

Tourists are often so focused on taking pictures that they lose the opportunity to explore the character of places for themselves and miss other sites and scenes that are equally or more interesting. While such tourists have visual evidence of the places they visit, they do not really see or experience them through their other senses. They are, in a sense, merely "collecting" places (figure 14.6). In the worst-case scenario, tourists die trying to get that perfect shot or selfie. On June 29, 2016, two tourists died in Peru in unrelated incidents as they attempted to get photographs of themselves. In the first incident, a South Korean tourist looking for the best place to take a selfie slipped and fell under the Gocta waterfall in the Amazon region. In the second, a German tourist ignored a security cordon to position himself on a ledge at Machu Picchu before falling to his death.[18]

Figure 14.6. "Selfies" are a part of modern tourism. *Source:* Velvet Nelson

Conclusion

Tourist experiences have been considered the realm of related fields in tourism studies, such as psychology and sociology; geography has had generally little to contribute. As a fundamentally place-based activity, there is much potential for cross-fertilization between the geography of tourism and the geography of place. Just as representations of place are an extraordinarily important part of tourism, so is the experience of places. Yet, we must always remember, as John Urry wrote in *The Tourist Gaze*, "There is no universal experience that is true for all tourists at all times."[19]

Key Terms

- collective tourist gaze
- romantic tourist gaze
- sense of place

Notes

1. Jaakko Suvantola, *Tourist's Experience of Place* (Aldershot: Ashgate, 2002).
2. Edward Relph, *Place and Placelessness* (London: Pion, 1976), 93.
3. Relph, *Place and Placelessness*, 95.
4. Relph, *Place and Placelessness*; Edward Relph, "Sense of Place," in *Ten Geographic Ideas That Changed the World*, ed. Susan Hanson (New Brunswick, NJ: Rutgers University Press, 1997).
5. Yi-Fu Tuan, "Place: An Experiential Perspective," *Geographical Review* 65 (1975): 164.
6. Relph, *Place and Placelessness*, 142; Relph, "Sense of Place," 208.
7. Jelmer H.G. Jeuring and Karin B.M. Peters, "The Influence of the Weather on Tourist Experiences: Analysing Travel Blog Narratives," *Journal of Vacation Marketing* 19, no. 3 (2013): 209.
8. Erik Cohen, "The Tourist Guide: The Origins, Structure, and Dynamics of a Role," *Annals of Tourism Research* 12 (1985): 5–8.
9. John Ap and Kevin K.F. Wong. "Case Study on Tour Guiding: Professionalism, Issues, and Problems," *Tourism Management* 22 (2001): 551.
10. Erica Wilson and Donna E. Little, "The Solo Female Travel Experience: Exploring the 'Geography of Women's Fear,'" *Current Issues in Tourism* 11, no. 2 (2008): 174.
11. Erica Wilson and Donna E. Little, "A 'Relative Escape'? The Impact of Constraints on Women Who Travel Solo," *Tourism Review International* 9 (2005): 165–6.
12. Fiona Jordan and Heather Gibson, "'We're Not Stupid . . . But We'll Not Stay Home Either': Experiences of Solo Women Travelers," *Tourism Review International* 9 (2005): 205.
13. James Moir, "Seeing the Sites: Tourism as Perceptual Experience," in *Tourism and Visual Culture, vol. 1, Theories and Concepts*, ed. Peter M. Burns, Cathy Palmer, and Jo-Anne Lester (Oxfordshire: CABI, 2010), 165.
14. John Urry, *The Tourist Gaze* (London: Sage, 1990); John Urry, *Consuming Places* (London: Routledge, 1995), 131.
15. Bethany Lytle, "America's Most Beautiful Mansions," *Forbes*, July 18, 2012, accessed July 19, 2016, http://www.forbes.com/sites/forbeslifestyle/2012/07/18/americas-most-beautiful-mansions/#1fb3ecff1fcd
16. Relph, *Place and Placelessness*.
17. Mike Robinson and David Picard, "Moments, Magic, and Memories: Photographic Tourists, Tourist Photographs, and Making Worlds," in *The Framed World: Tourism, Tourists, and Photography*, ed. Mike Robinson and David Picard (Farnham: Ashgate, 2009), 1.
18. Mark Piggott, "Deadly Selfie Craze: South Korean Dies in Peru Just Hours After Death of German at Machu Picchu," *International Business Times*, July 6, 2016, accessed July 19, 2016, http://www.ibtimes.co.uk/selfie-dangers-2-tourists-die-same-day-taking-risky-photographs-peru-beauty-spots-1569096
19. Urry, *The Tourist Gaze*, 1.

Glossary

accessibility. The relative ease with which one location may be reached from another

acculturation. The process of exchange that takes place when two groups of people come into contact over time

affect. To act on or produce a change in something

back region. The part of a destination that is not intended for, or is closed to, tourists

beautiful. An aesthetic landscape concept dating back to the eighteenth century, describing a landscape that is soft, smooth, and harmonious in appearance, the experience of which is reassuring and pleasurable

biogeography. The study of living things

circular itinerary. A trip in which tourists travel from home to multiple destinations before returning home

climate change adaptation. The technological, economic, and sociocultural changes that are intended to minimize the risks and capitalize on the opportunities created by climate change

climate change mitigation. The technological, economic, and sociocultural changes that can lead to reductions in greenhouse gas emissions

climatology. The study of climate

code of conduct. A set of voluntary principles intended to inform patterns of behavior among tourism stakeholders and tourists to minimize the negative environmental effects of tourism

collective tourist gaze. The visual consumption of public places that are characterized by the presence of other people

commodification. The transformation of something of intrinsic value into a product that can be packaged and sold for consumption

complementarity. The relationship between people who have a desire for certain travel experiences and the place that has the ability to satisfy that desire

cultural geography. A broad topical branch in human geography that studies various issues pertaining to how societies make sense of, give meaning to, interact with, and shape space and place

deferred demand. Those people who wish to travel but do not because of a problem or barrier at the desired destination or in the tourism infrastructure

demonstration effect. Changes in attitudes, values, or patterns of behavior experienced by local people as a result of observing tourists

direct economic effect. The introduction of tourist dollars to the local economy

discretionary income. The money that is left over after taxes and all other necessary expenses have been taken care of

distance decay. Exponential decrease in demand for a product or service as the distance traveled to obtain that product or service increases

domestic tourism. Tourists traveling within their own country

drifter. A type of tourist that seeks out new tourism destinations, utilizes local infrastructure, and immerses himself or herself in the local culture

economic development. A process of change that creates the conditions for improvements in productivity and income of the population

economic geography. The study of the spatial patterns of economic activities, including locations, distributions, interactions, and outcomes

ecotourism (the International Ecotourism Society definition). Responsible travel to natural areas that conserves the environment and improves the well-being of local people

effect. Something that is produced by an agency or cause; a result or a consequence

effective demand. Those people who wish to and have the opportunity to travel

electronic word-of-mouth. The sharing of information using Web 2.0 technologies

enclave tourism. Geographically isolated and spatially concentrated tourism facilities and activities

environmental carrying capacity. The extent of tourism that can take place at a site before its environment experiences negative effects

environmental geography. A topical branch of geography that lies at the intersection of physical geography and human geography and is concerned with the ways in which the environment affects people and people affect the environment

experience stage. The primary stage of the tourism process, in which tourists participate in a variety of activities at a destination

experiential marketing. Marketing that uses multiple senses in such a way that will prompt the audience to imagine the experience and become emotionally invested in it

explorer. A type of tourist that travels for more than pleasure or diversion, utilizes a combination of tourist and local infrastructure, and seeks interaction with local people

front region. The part of a destination that has been entirely constructed for the purpose of tourism

geomorphology. The study of landforms

geotourism (National Geographic Society definition). Tourism that sustains or enhances the geographical character of a place, including its environment, culture, aesthetics, heritage, and the well-being of its residents

globalization. The increasing interconnectedness of the world

historical geography. The study of the geography and geographic conditions of past periods and the processes of change that have taken place over time to better understand the geography of the present

hub-and-spoke itinerary. A trip in which tourists travel from home to a destination and use that destination as a base from which to visit other destinations

human geography. One of the two main subdivisions of geography, which focuses on the study of the patterns of human occupation of the earth

hydrology. The study of water

inbound tourism. Tourists traveling to a place of destination

indirect economic effect. The second round of spending, in which recipients of tourist dollars pay the expenses of and reinvest in their tourism business

individual mass tourist. A type of tourist that travels for pleasure and seeks experiences different from those that may be obtained at home without straying too far from his or her comfort zone

induced economic effect. An additional round of spending after the recipients of tourist dollars pay the government, employees, suppliers, etc.; money spent by these new recipients for their own purposes

interchange. A node within a transportation network

international tourism. Tourists traveling to another country

last chance tourism. A recent trend in tourism in which tourists seek environments that are experiencing fundamental changes and might ultimately "disappear"

leakages. The portion of the income from tourism that does not get reinvested in the local economy; occurs with each round of spending

leisure time. The free time left over after necessary activities have been completed, in which an individual may do what he or she chooses

lingua franca. A language used for the purpose of communication between people speaking different languages

linkages. The connections formed between tourism and other local economic sectors that can support tourism and help provide the goods and services demanded by tourists

mass tourism. The production of standardized experiences made available to large numbers of tourists at a low cost

meteorology. The study of weather

movement stage. The stage of the tourism process in which tourists use some form of transportation to reach the destination and to return home; may be a means to an end or a part of the experience stage

multiplier effect. A ratio of the additional income generated by the indirect and induced economic effects from the re-spending of tourist dollars in the local economy

niche tourism. The production of specialized experiences for relatively small markets based on a particular resource at the destination or a specific tourism product

no demand. Those people who do not travel and do not wish to travel

organized mass tourist. A type of tourist that travels purely for diversion, in which place is less important than experience, and is entirely dependent on the tourism infrastructure

outbound tourism. Tourists traveling from their home environment

perceptual carrying capacity. The extent of tourism that can take place at a site before tourist dissatisfaction occurs

physical carrying capacity. The limits of a particular space, such as the number of tourists a site can contain

physical geography. One of the two main subdivisions of geography, which focuses on the study of the earth's physical systems

place. A unit of the earth's surface that has meaning based on the physical and human features of that location

placelessness. A loss of identity where one place looks and feels like other places, often as a result of the superficial, stereotypical images circulated by the media

place promotion. The deliberate use of marketing tools to communicate both specific and selective ideas and images about a particular place to a desired audience for the purpose of shaping perceptions of that place and ultimately influencing decisions

place representation. The way places are summarized and portrayed to an audience that then creates ideas and images about those places

place reputation. The composite of ideas and impressions of a place held by external audiences

place reputation management. Efforts to adjust place reputation so that it is closer to how stakeholders would like the place to be perceived

point-to-point itinerary. A trip in which tourists travel from home to a destination and back

political geography. The study of the ways states relate to each other in a globalized world

post-trip stage. The final stage in the tourism process after the tourists return home, in which they relive their trip through memories, pictures, and souvenirs

potential demand. Those people who wish to travel and will do so when their circumstances change

preferred sites. Planned locations that have sufficient tourist facilities to spatially concentrate visitors, thereby limiting the environmental effects of tourism to a particular area

pre-trip stage. The first stage in the tourism process, in which potential tourists evaluate their travel options, make decisions, and complete all arrangements for a trip

price discrimination. A system of pricing policies in which different categories of visitors are charged different fees for the same activities or services

pro-poor tourism (Pro-Poor Tourism Partnership definition). Tourism that results in increased net benefits for poor people and ensures that tourism growth contributes to poverty reduction

protected area (Convention on Biological Diversity definition). A geographically defined area that is designated or regulated and managed to achieve specific conservation objectives

pull factor. Something in the destination environment that attracts people to visit that place over another

push factor. Something in the home environment that impels people to leave temporarily and travel somewhere else

region. A unit of the earth's surface that is distinguished from other areas by certain characteristics

regional geography. An approach in geography that studies the varied geographic characteristics of a region

relative location. The position of a place in relation to other places

romantic tourist gaze. A private, personal experience in which tourists feel they form a connection with a place through the visual consumption of that place

rural geography. The study of contemporary rural landscapes, societies, and economies

scale. The size of the area studied

sense of place. The association with and emotional attachment to places

social carrying capacity. The extent of tourism that can take place at a site before the local community becomes dissatisfied

social geography. The topical branch of geography concerned with the relationships between society and space, such as space as a setting for social interaction or the ways in which spaces are shaped by these interactions

space. Locations on the earth's surface

spatial distribution. The organization of various phenomena on the earth's surface

spatial zoning. A land management strategy that designates permissible uses of an area based on its resources and/or character—that is, what tourism activities may be undertaken where

sublime. An aesthetic landscape concept dating back to the eighteenth century, describing a landscape that is rugged, vast, or dark, the experience of which may be frightening and thrilling

suppressed demand. Those people who wish to travel but do not

sustainable tourism (United Nations World Tourism Organization definition). Tourism that takes full account of its current and future economic, social, and environmental impacts, addressing the needs of visitors, the industry, the environment, and host communities

terminal. A node where transport flows begin and end

topical geography. An approach in geography that studies a particular geographic topic in various place or regional contexts

tourism (United Nations World Tourism Organization definition). The activities of persons traveling to and staying in places outside of their usual environment for not more than one consecutive year for leisure, business, and other purposes

tourism attractions. Aspects of places that are of interest to tourists and can include things to be seen, activities to be done, or experiences to be had

tourism carrying capacity. Refers to the number of tourists a destination or attraction can support and sustain

tourism demand. The total number of persons who travel, or wish to travel, to use tourist facilities and services at places away from their places of work and residence

tourism itinerary. The planned route or journey for a trip

tourism products. The increasingly specialized types of experiences provided in the supply of tourism

tourism resource. A component of the destination's physical or cultural environment that has the potential to facilitate tourism or provide the basis for a tourism attraction

tourism resource audit. A tool that can be used by destination stakeholders to systematically identify, classify, and assess all of those features of a place that will impact the supply of tourism

tourism stakeholders. The various individuals and/or organizations that have an interest in tourism

tourism supply. The aggregate of all businesses that directly provide goods or services to facilitate business, pleasure, and leisure activities away from the home environment

tourist area life cycle. A model proposed to explain the process of development and evolution of tourism destinations over six stages, including exploration, involvement, development, consolidation, stagnation, and an undetermined post-stagnation stage

tourist dollars. The money that tourists bring with them and spend at the destination on lodging, food, souvenirs, excursions, and other activities or services

tourist-generating regions. The source areas or origins for tourists

tourist inversions. The theory that the experience that a tourist seeks in his or her temporary escape is one of contrasts and involves a shift in attitudes or patterns of behavior away from the norm to a temporary opposite

tourist-receiving regions. The destination areas for tourists

tourist typology. An organizational framework to identify categories of tourists based on motivations, behavior, demographic characteristics, or other variables

transport geography. The topical branch of geography concerned with the movement of goods and people from one place to another, including the spatial patterns of this movement and the geographic factors that allow or constrain it

transportation mode. The means of movement or type of transportation; generally air, surface, or water

transportation network. The spatial structure and organization of the infrastructure that supports, and to some extent determines, patterns of movement

transportation node. An access point on a transportation network

travel account. The difference between the income that the destination country receives from tourism and the expenditures of that country's citizens when they travel abroad

travel 2.0. A new interactive approach to travel technologies in which tourists are both consumers and producers of travel information via the Internet.

urban geography. The study of the relationships between or patterns within cities and metropolitan areas

word-of-mouth. The passing of information, typically through person-to-person communication, that shapes ideas about places and influences travel decisions

Bibliography

Agapito, Dora, Patrícia Oom do Valle, and Júlio da Costa Mendes. "Sensory Marketing and Tourist Experiences." *Spatial and Organizational Dynamics* 10 (2012): 7–19.

"Airport Attack in Istanbul Is the Latest in a Year of Terror in Turkey." *The New York Times*, June 30, 2016. Accessed August 2, 2016. http://www.nytimes.com/interactive/2016/06/28/world/middleeast/turkey-terror-attacks-bombings.html?_r=0

Albalate, Daniel, and Germà Bel. "Tourism and Urban Public Transport: Holding Demand Pressure under Supply Constraints." *Tourism Management* 31 (2010): 425–33.

Alderman, Derek H. "Surrogation and the Politics of Remembering Slavery in Savannah, Georgia (USA)." *Journal of Historical Geography* 36, no. 1 (2010): 90–101.

Alderman, Derek H. and Rachel M. Campbell. "Symbolic Excavation and the Artifact Politics of Remembering Slavery in the American South: Observations from Walterboro, South Carolina." *Southeastern Geographer* 48, no. 3 (2008): 338–5.

Alderman, Derek H. and Joshua Inwood. "Toward a Pedagogy of Jim Crow: A Geographic Reading of The Green Book in Teaching Ethnic Geography in the 21st Century." In *Teaching Ethnic Geography in the 21st Century*, edited by. Lawrence E. Estaville, Edris J. Montalvo, and Fenda A. Akiwumi, 68–78. Washington, DC: National Council for Geographic Education, 2014.

Almeida, António, and Antónia Correia, "Tourism Development in Madeira: An Analysis Based on the Life Cycle Approach." *Tourism Economics* 16, no. 2 (2010): 427–41.

American Association of Geographers. "Highlights from the 2016 Annual Meeting in San Francisco." Accessed June 17, 2016. http://www.aag.org/cs/annualmeeting/annual_meeting_archives/2016_san_francisco/2016_san_francisco_highlights

Amor, Bani. "Dispatches." Accessed July 12, 2016. https://baniamor.com/dispatches/

Anderson, Wineaster. "Enclave Tourism and Its Socio-Economic Impact in Emerging Destinations." *Anatolia—An International Journal of Tourism and Hospitality Research* 22, no. 3 (2011): 361–77.

Andersson, Petra, Sara Crone, Jesper Stage, and Jorn Stage. "Potential Monopoly Rents from International Wildlife Tourism: An Example from Uganda's Gorilla Tourism." *Eastern Africa Social Science Research Review* 21, no. 1 (2005): 1–18.

Andrews, Hazel. "Feeling at Home: Embodying Britishness in a Spanish Charter Tourists Resort." *Tourist Studies* 5, no. 3 (2005): 247–66.

Anholt, Simon. "Competitive Identity." In *Destination Brands: Managing Place Reputation.* 3rd ed., edited by. Nigel Morgan, Annette Pritchard, and Roger Pride, 21–31. Florence, KY: Routledge, 2011.

———. *Places: Identity, Image and Reputation*. Houndmills: Palgrave Macmillan, 2010.

Ansari, Azadeh. "Bahamas Tells Its Citizens Traveling to U.S.: Be Careful." *CNN*, July 10, 2016. Accessed August 2, 2016. http://www.cnn.com/2016/07/09/travel/bahamas-us-travel-advisory/

Ap, John, and Kevin K.F. Wong. "Case Study on Tour Guiding: Professionalism, Issues, and Problems." *Tourism Management* 22 (2001): 551–63.

"Arenal Volcano Costa Rica." Accessed July 28, 2016. http://www.arenal.net

Armstead, Myra B. Young. "Revisiting Hotels and Other Lodgings: American Tourist Spaces through the Lens of Black Pleasure-Travelers, 1880–1950." *The Journal of Decorative and Propaganda Arts* 25 (2005): 136–159.

Ashley, Caroline, Charlotte Boyd, and Harold Goodwin. "Pro-Poor Tourism: Putting Poverty at the Heart of the Tourism Agenda." *Natural Resource Perspectives* 51 (2000): 1–6.

Ashley, Caroline, Dilys Roe, and Harold Goodwin. "Pro-Poor Strategies: Making Tourism Work for the Poor." *Pro-Poor Tourism Report* 1 (2001). Available at https://www.odi.org/sites/odi.org.uk/files/odi-assets/publications-opinion-files/3246.pdf

Aspinall, Algernon E. *The Pocket Guide to the West Indies*. 2nd ed. London: Duckworth & Co., 1910.

Awwad, Ramadan A., T.N. Olsthoorn, Y. Zhou, Stefan Uhlenbrook, and Ebel Smidt. "Optimum Pumping-Injection System for Saline Groundwater Desalination in Sharm El Sheikh." *WaterMill Working Paper Series* 11 (2008). Accessed October 26, 2011. http://www.unesco-ihe.org/WaterMill-Working-Paper-Series/Working-Paper-Series

Bagchi-Sen, Sharmistha and Helen Lawton Smith. "Introduction: The Past, Present, and Future of Economic Geography." In *Economic Geography: Past, Present, and Future*, edited by. Sharmistha Bagchi-Sen and Helen Lawton Smith, 1–8. London: Routledge, 2006.

Barbados Tourism Authority. "Perfect Weather." Accessed July 14, 2016. http://www.visitbarbados.org/perfect-weather

Basu, Paul. "Route Metaphors of 'Roots-Tourism' in the Scottish Highland Diaspora." In *Reframing Pilgrimage: Cultures in Motion*, edited by. Simon Coleman and John Eade, 150–74. London: Routledge, 2004.

Baum, Tom. "Images of Tourism Past and Present." *International Journal of Contemporary Hospitality Management* 8, no. 4 (1996): 25–30.

Beckerson, John, and John K. Walton. "Selling Air: Marketing the Intangible at British Resorts." In *Histories of Tourism: Representation, Identity, and Conflict*, edited by. John Walton, 55–68. Clevedon, UK: Channel View Publications, 2005.

Bhattacharyya, Deborah P. "Mediating India: An Analysis of a Guidebook." *Annals of Tourism Research* 24, no. 2 (1997): 371–89.

Blacksell, Mark. *Political Geography*. London: Routledge, 2006.

Boniface, Brian, and Chris Cooper. *Worldwide Destinations: The Geography of Travel and Tourism*. 4th ed. Amsterdam: Elsevier Butterworth Heinemann, 2005.

Boorstin, Daniel J. *The Image: A Guide to Pseudo-Events in America*. New York: Vintage Books, 1961; 50th Anniversary Edition, 2012.

Borelli, Simone, and Stefania Minestrini. "WWF Mediterranean Programme." Accessed July 14, 2016. http://www.monachus-guardian.org/library/medpro01.pdf

Bosangit, Carmela, Juline Dulnuan, and Miguela Mena, "Using Travel Blogs to Examine Postconsumption Behavior of Tourists." *Journal of Vacation Marketing* 18, no. 3 (2012): 207–19.

Bricker, Kelly S., Rosemary Black, and Stuart Cottrell (editors). *Sustainable Tourism & the Millennium Development Goals: Effecting Positive Change*. Burlington: Jones & Bartlett Learning, 2013.

Buckley, Ralf. "Environmental Inputs and Outputs in Ecotourism: Geotourism with a Positive Triple Bottom Line?" *Journal of Ecotourism* 2, no. 1 (2003): 76–82.

Bullen, Frank T. *Back to Sunny Seas*. London: Smith, Elder & Co., 1905.

Burns, Peter and Lyn Bibbings. "Climate Change and Tourism." In *The Routledge Handbook of Tourism and the Environment*, edited by. Andrew Holden and David Fennell, 406–20. London: Routledge, 2013.

Butler, David L., "Whitewashing Plantations." *International Journal of Hospitality & Tourism Administration* 2, no. 3–4 (2001): 163–75.

Butler, David L., Perry L. Carter, and Owen J. Dwyer, "Imagining Plantations: Slavery, Dominant Narratives, and the Foreign Born." *Southeastern Geographer* 48, no. 3 (2008): 288–302.

Butler, R.W. "The Concept of a Tourist Area Cycle Evolution: Implications for Management of Resources." *Canadian Geographer* 24, no. 1 (1980): 5–12.

Butler, Richard. "The Resort Cycle Two Decades On." In *Tourism in the 21st Century: Lessons from Experience*, edited by. Bill Faulkner, Gianna Moscardo, and Eric Laws, 284–99. London: Continuum, 2000.

Butler, Richard, and Wantanee Suntikul. "Tourism and War: An Ill Wind?" In *Tourism and War*, edited by. Richard Butler and Wantanee Suntikul, 1–11. London: Routledge, 2013.

Buzinde, Christine N., and Carla Almeida Santos, "Interpreting Slavery Tourism." *Annals of Tourism Research* 36, no. 3 (2009): 439–58.

Caribbean Tourism Organization. "Country Statistics and Analysis: Anguilla, Antigua & Barbuda, Aruba, The Bahamas, Barbados, Belize, Bermuda, Bonaire." Accessed June 9, 2016. http://www.onecaribbean.org/content/files/Strep1AnguillaToBonaire2010.pdf

———. *Caribbean Vacation Planner*. Coral Gables, FL: Gold Book, 2002.

Carter, Perry L. "Coloured Places and Pigmented Holidays: Racialized Leisure Travel," *Tourism Geographies* 10, no. 3 (2008): 265–84.

Central Intelligence Agency. "The World Factbook: Croatia." Accessed June 15, 2016. https://www.cia.gov/library/publications/the-world-factbook/geos/hr.html

———. "The World Factbook: Slovenia." Accessed June 15, 2016. https://www.cia.gov/library/publications/the-world-factbook/geos/si.html

———. "The World Factbook: Zimbabwe." Accessed June 23, 2016. https://www.cia.gov/library/publications/the-world-factbook/geos/zi.html

Central West Virginia Regional Airport Authority. "Yeager Airport History." Accessed July 28, 2016. http://www2.yeagerairport.com/history/

Chang, T.C. and Shirlena Huang. "Urban Tourism: Between the Global and the Local." In *A Companion to Tourism*, edited by. Alan A. Lew, C. Michael Hall, and Allan M. Williams, 223–34. Malden, MA: Blackwell, 2004.

Chhabra, Deepak, Robert Healy, and Erin Sills. "Staged Authenticity and Heritage Tourism." *Annals of Tourism Research* 30, no. 3 (2003): 702–19.

Christopherson, Robert W. *Geosystems: An Introduction to Physical Geography*. 7th ed. Upper Saddle River, NJ: Pearson Prentice Hall, 2009.

Christou, Evangelos. "Introduction to Part 2." In *Social Media in Travel, Tourism and Hospitality: Theory, Practice and Cases*, edited by. Marianna Sigala, Evangelos Christou and Ulrike Gretzel, 69–71. Ashgate: Surrey, 2012.

Chung, Jin Young, and Taehee Whang. "The Impact of Low Cost Carriers on Korean Island Tourism." *Journal of Transport Geography* 19 (2011): 1335–40.

Churchill Downs Incorporated. "What to Expect at the Kentucky Derby." Accessed June 17, 2016. https://www.kentuckyderby.com/visit/what-to-expect

City of Leavenworth, Washington. "History." Accessed July 19, 2016. http://cityofleavenworth.com/about-leavenworth/leavenworth-history/

Claval, Paul. "Regional Geography: Past and Present (A Review of Ideas, Approaches and Goals)." *Geographia Polonica* 80, no. 1 (2007): 25–42.

Cleave, Evan, and Arku Godwin. "Place Branding and Economic Development at the Local Level in Ontario, Canada." *Geojournal* 80, no. 3 (2015): 323–38.

"Climate Change Could Bring Tourists to UK— Report." *The Guardian*, July 28, 2006. Accessed July 27, 2016. https://www.theguardian.com/travel/2006/jul/28/travelnews.uknews.climatechange

Cohen, Erik. "Toward a Sociology of International Tourism." *Social Research* 39, no. 1 (1972): 164–82.

———. "Authenticity and Commoditization in Tourism." *Annals of Tourism Research* 15 (1988): 371–86.

———. "The Tourist Guide: The Origins, Structure, and Dynamics of a Role." *Annals of Tourism Research* 12 (1985): 5–29.

Cohen, Scott A., and James E.S. Higham. "Contradictions in Climate Concern: Performances at Home and Away." In *Tourism, Climate Change and Sustainability*, edited by. Maharaj Vijay Reddy and Keith Wilkes, 257–70. London: Routledge, 2013.

Cole, Sam. "Space Tourism: Prospects, Positioning, and Planning." *Journal of Tourism Futures* 1, no. 2 (2015): 131–40.

Columbus Zoo and Aquarium. "The Wilds—History." Accessed July 14, 2016. https://thewilds.columbuszoo.org/home/about/about-the-wilds/history

Cook, Matthew R. "Counter-Narratives of Slavery in the Deep South: The Politics of Empathy Along and Beyond River Road." *Journal of Heritage Tourism* 11, no. 3 (2016): 290–308.

Cresswell, Tim. *Place: A Short Introduction*. Malden, MA: Blackwell, 2004.

Croatian National Tourist Board. "Regions." Accessed June 15, 2016. http://www.croatia.hr/en-GB/Destinations/Regions

Davie, Tim. *Fundamentals of Hydrology*. 2nd ed. London: Routledge, 2002.

De Freitas, C.R. "Tourism Climatology: Evaluating Environmental Information for Decision Making and Business Planning in the Recreation and Tourism Sector." *International Journal of Biometeorology* 48 (2003): 45–54.

deRios, Marlene Dobkin. "Drug Tourism in the Amazon." *Anthropology of Consciousness* 5, no. 1 (1994): 16–19.

Detroit Metro Convention & Visitors Bureau. "Metro Detroit—Tourism and Travel." Accessed July 15, 2016. http://visitdetroit.com

Diab, Atef M. "Bacteriological Studies on the Potability, Efficacy, and EIA of Desalination Operations at Sharm El-Sheikh Region, Egypt." *Egyptian Journal of Biology* 3 (2001): 59–65.

Dominica Hotel and Tourism Association. *Destination Dominica*. North Miami, FL: Ulrich Communications Corporation, 2003.

Driver, Felix. "Distance and Disturbance: Travel, Exploration and Knowledge in the Nineteenth Century." *Transactions of the Royal Historical Society* 14 (2004): 73–92.

Duncan, James, and Derek Gregory. "Introduction." In *Writes of Passage: Reading Travel Writing*, edited by. James Duncan and Derek Gregory, 1–13. London: Routledge, 1999.

Eichstedt, Jennifer L., and Stephen Small. *Representations of Slavery: Race and Ideology in Southern Plantation Museums*. Washington, DC: Smithsonian Institution Press, 2002.

Eslrud, Torun. "Risk Creation in Traveling: Backpacker Adventure Narration." *Annals of Tourism Research* 28, no. 3 (2001): 597–617.

Eugenio-Martin, Juan L., and Federico Inchuasti-Sintes. "Low-Cost Travel and Tourism Expenditures." *Annals of Tourism Research* 57 (2016): 140–59.

Fair Trade Tourism. "About Us." Accessed July 14, 2016. http://www.fairtrade.travel/corporate

Feifer, Maxine. *Tourism in History: From Imperial Rome to the Present*. New York: Stein and Day, 1986.

Fennell, David A. *Ecotourism*. 4th ed. London: Routledge, 2015.

Filier, Raffaele, Salma Alguezaui, and Frazer McLeay. "Why Do Travelers Trust TripAdvisor? Antecedents of Trust towards Consumer-Generated Media and Its Influence on Recommendation Adoption and Word of Mouth." *Tourism Management* 51 (2015): 174–85.

Florida, Richard. *The Rise of the Creative Class, Revisited.* New York: Basic Books, 2012.

Ford, Caroline. "A Summer Fling: The Rise and Fall of Aquariums and Fun Parks on Sydney's Ocean Coast 1885–1920." *Journal of Tourism History* 1, no. 2 (2009): 95–112.

Frenkel, Stephen, and Judy Walton. "Bavarian Leavenworth and the Symbolic Economy of a Theme Town." *The Geographical Review* 90, no. 4 (2000): 559–84.

Froude, James A. *The English in the West Indies or the Bow of Ulysses.* London: Longmans, Green and Co., 1909.

Gassan, Richard H. *The Birth of American Tourism: New York, the Hudson Valley, and American Culture, 1790–1830.* Amherst: University of Massachusetts Press, 2008.

Gay, *Kathlyn African-American Holidays, Festivals, and Celebrations: The History, Customs, and Symbols Associated with Both Traditional and Contemporary Religious and Secular Events Observed by Americans of African Descent.* Detroit: Omnigraphics, 2007.

Gibson, Chris. "Locating Geographies of Tourism." *Progress in Human Geography* 32, no. 3 (2008): 407–22.

Godfrey, Kerry, and Jackie Clarke. *The Tourism Development Handbook: A Practical Approach to Planning and Marketing.* London: Cassell, 2000.

Goeldner, Charles R., and J.R. Brent Ritchie. *Tourism: Principles, Practices, Philosophies.* 9th ed. Hoboken, NJ: Wiley, 2006.

Goh, Carey. "Exploring Impact of Climate on Tourism Demand." *Annals of Tourism Research* 39, no. 4 (2012): 1859–83.

Gold, John R., and Margaret M. Gold. "Introduction." In *Olympic Cities: City Agendas, Planning and the World's Games, 1896–2016.* 2nd ed., edited by. John R. Gold and Margaret M. Gold, 1–13. London: Routledge, 2011.

Gómez, Martín, Ma. Belén. "Weather, Climate, and Tourism: A Geographical Perspective." *Annals of Tourism Research* 32, no. 3 (2005): 571–91.

Graburn, Nelson. "The Anthropology of Tourism." *Annals of Tourism Research* 10 (1983): 9–33.

Grand Duchy of Luxembourg. "Blog—Visit Luxembourg." Accessed July 18, 2016. http://www.visitluxembourg.com/en/blog

Green, Victor H. *The Negro Motorist Green Book.* New York: Victor H. Green & Co., Publishers, 1949.

Greenspun, Philip. "Tourists as Subjects." Accessed July 19, 2016. http://photo.net/travel/tourists-as-subjects/.

Gregory, Derek. "Scripting Egypt: Orientalism and the Cultures of Travel." In *Writes of Passage: Reading Travel Writing,* ed. James Duncan and Derek Gregory, 114–50. London: Routledge, 1999.

Gregory, Derek, Ron Johnston, and Geraldine Pratt. *Dictionary of Human Geography.* 5th ed. Hoboken, NJ: Wiley-Blackwell, 2009.

Gretzel, Ulrike. "Introduction to Part 3." In *Social Media in Travel, Tourism and Hospitality: Theory, Practice and Cases,* edited by. Marianna Sigala, Evangelos Christou, and Ulrike Gretzel, 167–69. Ashgate: Surrey, 2012.

Grey, Lisa. "A Free Day in Houston, Texas." *Business Week,* April 22, 2010. Accessed May 6, 2014. http://www.visithoustontexas.com/about-houston/in-the-news/

Gross, Sven, and Louisa Klemmer. *Introduction to Tourism Transport.* Oxfordshire: CABI, 2014,

Gunn, Clare A., and Turgut Var. *Tourism Planning: Basics, Concepts, Cases.* 4th ed. New York: Routledge, 2002.

Halewood, Chris, and Kevin Hannam. "Viking Heritage Tourism: Authenticity and Commodification." *Annals of Tourism Research* 28, no. 3 (2001): 565–80.

Hall, Derek R. "Conceptualising Tourism Transport: Inequality and Externality Issues." *Journal of Transport Geography* 7 (1999): 181–88.

Hall, Michael C., and Alan Lew. *Understanding and Managing Tourism Impacts: An Integrated Approach*. New York: Routledge, 2009.

Hanna, Stephen P. "Placing the Enslaved at Oak Alley Plantation: Narratives, Spatial Contexts, and the Limits of Surrogation." *Journal of Heritage Tourism* 11, no. 3 (2016): 219–34.

Hanson, Susan. "Thinking Back, Thinking Ahead: Some Questions for Economic Geographers." In *Economic Geography: Past, Present, and Future*, ed. Sharmistha Bagchi-Sen and Helen Lawton Smith, 25–33. London: Routledge, 2006.

Hays, Stephanie, Stephen John Page, and Dimitrios Buhalis, "Social Media as a Destination Marketing Tool: Its Use by National Tourism Organizations." *Current Issues in Tourism* 16, no. 3 (2013): 211–39.

Higgins, Charlotte. "Art in the Countryside: Why More and More UK Creatives Are Leaving the City." *The Guardian*, August 26, 2013. Accessed August 4, 2016. https://www.theguardian.com/artanddesign/2013/aug/26/art-countryside-uk-creatives

Holden, Andrew. *Environment and Tourism*. 2nd ed. London: Routledge, 2008.

Horner, Susan, and John Swarbrooke. *International Cases in Tourism Management*. Burlington, MA: Elsevier Butterworth-Heinemann, 2004.

Hose, Thomas A. "Selling the Story of Britain's Stone." *Environmental Interpretation* 10 (1995): 16–17.

Hosney Fahmy, Faten, Ninet Mohamed Ahmed, and Hanaa Mohamed Farghally. "Optimization of Renewable Energy Power System for Small Scale Brackish Reverse Osmosis Desalination Unit and a Tourism Motel in Egypt." *Smart Grid and Renewable Energy* 3 (2012): 43–50.

Hudson, Simon, and J.R. Brent Ritchie. "Branding a Memorable Destination Experience. The Case of 'Brand Canada." *International Journal of Tourism Research* 11 (2009): 217–28.

Hunt, Katrina Brown. "America's Best Cities for Foodies." *Travel + Leisure*. Accessed May 26, 2015. http://www.travelandleisure.com/slideshows/americas-best-cities-for-foodies

Hyundai Motor Manufacturing Alabama. "About HMMA." Accessed July 12, 2016. http://www.hmmausa.com/our-company/about-hmma/

International Civil Aviation Organization. "Air Transport Policy and Regulation." Accessed July 22, 2016. http://www.icao.int/sustainability/Pages/Low-Cost-Carriers.aspx

The International Ecotourism Society. "What Is Ecotourism?" Accessed July 14, 2016. http://www.ecotourism.org/what-is-ecotourism

International Olympic Committee. *London 2012 Summer Olympics*. Accessed June 17, 2016. https://www.olympic.org/london-2012

———. *Olympic Charter*. Accessed July 20, 2016. https://stillmed.olympic.org/Documents/olympic_charter_en.pdf

Ioannides, Dimitri. "The Economic Geography of the Tourist Industry: Ten Years of Progress in Research and an Agenda for the Future." *Tourism Geographies* 8 (2006): 76–86.

———. "Strengthening the Ties between Tourism and Economic Geography: A Theoretical Agenda." *Professional Geographer* 47 (1995): 49–60.

Ioannides, Dimitri, and Keith G. Debbage. "Introduction: Exploring the Economic Geography and Tourism Nexus." In *The Economic Geography of the Tourist Industry*, edited by. Dimitri Ioannides and Keith G. Debbage, 1–13. London: Routledge, 1998.

Isacsson, Annica, Leena Alakoski, and Asta Bäck. "Using Multiple Senses in Tourism Marketing: The Helsinki Expert, Eckerö Line and Linnanmäki Amusement Park Cases." *Turismos: An International Multidisciplinary Journal of Tourism* 4, no. 3 (2009): 167–84.

Islas Travel Guides. "Welcome to Magaluf." Accessed July 14, 2016. http://www.majorca-mallorca.co.uk/magaluf.htm

Ivanovic, Milena. *Cultural Tourism*. Cape Town: Juta, 2008.

Jaackson, Reiner. "Beyond the Tourist Bubble? Cruiseship Passengers in Port." *Annals of Tourism Research* 31, no. 1 (2004): 44–60.

Jasne-Verbeke, Myriam, and Wanda George. "Reflections on the Great War Centenary: From Warscapes to Memoryscapes in 100 Years." In *Tourism and War*, edited by. Richard Butler and Wantanee Suntikul, 273–87. London: Routledge, 2013.

Jeuring, Jelmer H.G., and Karin B.M. Peters. "The Influence of the Weather on Tourist Experiences: Analysing Travel Blog Narratives." *Journal of Vacation Marketing* 19, no. 3 (2013): 209–19.

Jones, Calvin, and ShiNa Li. "The Economic Importance of Meetings and Conferences: A Satellite Account Approach." *Annals of Tourism Research* 52 (2015): 117–33.

Jordan, Fiona, and Heather Gibson. "'We're Not Stupid . . . But We'll Not Stay Home Either': Experiences of Solo Women Travelers." *Tourism Review International* 9 (2005): 195–211.

Kaplan, David H., Steven R. Holloway, and James O. Wheeler. *Urban Geography*. 3rd ed. Hoboken, NJ: Wiley, 2014.

Keeling, David J. "Transportation Geography: New Directions on Well-Worn Trails." *Progress in Human Geography* 31 (2007): 217–25.

Kershaw, Steve. *Oceanography: An Earth Science Perspective*. Cheltenham, UK: Stanley Thornes, 2000.

Kevan, Simon. "Quests for Cures: A History of Tourism for Climate and Health." *International Journal of Biometeorology* 37 (1993): 113–24.

Khair, Tabish, Martin Leer, Justin D. Edwards, and Hanna Ziadeh. *Other Routes: 1500 Years of African and Asian Travel Writing*. Bloomington: Indiana University Press, 2005.

Kingsley, Charles. *At Last: A Christmas in the West Indies*. New York: Harper & Brothers Publishers, 1871.

Kladou, Stella, and Eleni Mavragani. "Assessing Destination Image: An Online Marketing Approach and the Case of TripAdvisor." *Journal of Destination Marketing & Management* 4 (2015): 187–93.

Konecnik Maja, and Frank Go. "Tourism Destination Brand Identity: The Case of Slovenia." *Brand Management* 15, no. 3 (2008): 177–89.

Koshar, Rudy. "'What Ought to Be Seen': Tourists' Guidebooks and National Identities in Modern Germany and Europe." *Journal of Contemporary History* 33 (1998): 323–40.

Krabi Tourism. "Maya Bay, Krabi—Thailand." Accessed July 11, 2016. http://www.krabitourism.com/phiphi/maya-bay.htm

Kumbh Mela. "Official Website of Kumbh Mela 2013." Accessed June 17, 2016. http://kumbhmelaallahabad.gov.in/english/index.html

Laarman, Jan G., and Hans M. Gregersen. "Pricing Policy in Nature-Based Tourism." *Tourism Management* 17, no. 4 (1996): 247–54.

Laderman, Scott. "Shaping Memory of the Past: Discourse in Travel Guidebooks for Vietnam." *Mass Communication & Society* 5, no. 1 (2002): 87–110.

Lanegran, David A., and Salvatore J. Natoli. *Guidelines for Geographic Education in the Elementary and Secondary Schools*. Washington, DC: Association of American Geographers, 1984.

Las Vegas Convention and Visitors Authority. "2015 Las Vegas Year-to-Date Executive Summary." Accessed July 27, 2016. http://www.lvcva.com/includes/content/images/media/docs/ES-YTD-2015.pdf

Leavenworth Nutcracker Museum. "NM About Us." Accessed July 19, 2016. http://www.nutcrackermuseum.com/about.htm

Lemelin, Raynald Harvey, Emma Stewart, and Jackie Dawson. "An Introduction to Last Chance Tourism." In *Last Chance Tourism: Adapting Tourism Opportunities in a Changing World*, edited by. Raynald Harvey Lemelin, Jackie Dawson, and Emma J. Stewart, 3–9. London, Routledge, 2012.

Leung, Daniel, Rob Law, Hubert van Hoof, and Dimitrios Buhalis. "Social Media in Tourism and Hospitality: A Literature Review." *Journal of Travel & Tourism Marketing* 30 (2013): 3–22.

Light, Duncan. "'Facing the Future': Tourism and Identity-Building in Post-Socialist Romania." *Political Geography* 20 (2001): 1053–1074.

Löfgren, Orvar. *On Holiday: A History of Vacationing.* Berkeley: University of California Press, 1999.

Lomine, Loykie. "Tourism in Augustan Society (44 BC–AD 69)." In *Histories of Tourism: Representation, Identity, and Conflict*, edited by. John Walton, 69–87. Clevedon, UK: Channel View Publications, 2005.

London Connection. "Think You've Seen It All…London Will Surprise You!" Accessed August 4, 2016. https://londonconnection.com/think-youve-seen-it-all-london-will-surprise-you/

Longjit, Chootima, and Douglas G. Pearce. "Managing a Mature Coastal Destination: Pattaya, Thailand." *Journal of Destination Marketing & Management* 2 (2013): 165–75.

Lowen, Mark. "Turkey Tourism: An Industry in Crisis." *BBC News*, June 17, 2016. Accessed August 2, 2016. http://www.bbc.com/news/world-europe-36549880

Lozanski, Kristin. "Desire for Danger, Aversion to Harm: Violence in Travel to 'Other' Places." In *Tourism and Violence*, edited by. Hazel Andrews, 33–47. London: Routledge, 2016.

Lucas, Eric. "High Energy Houston: Alaska Airlines' Newest City." *Alaska Airlines*, September 2009. 23–35, 120–3.

Lumsdon, Les, and Stephen J. Page. "Progress in Transport and Tourism Research: Reformulating the Transport-Tourism Interface and Future Research Agendas." In *Tourism and Transport: Issues and Agenda for the New Millennium*, edited by. Les Lumsdon and Stephen J. Page, 1–28. Amsterdam: Elsevier, 2004.

Lyttle, Bethany. "America's Most Beautiful Mansions." *Forbes*, July 18, 2012. Accessed July 19, 2016. http://www.forbes.com/sites/forbeslifestyle/2012/07/18/americas-most-beautiful-mansions/#1fb3ecff1fcd

MacCannell, Dean. "Staged Authenticity: Arrangements of Social Space in Tourist Settings." *American Journal of Sociology* 79, no. 3 (1973): 589–603.

———. *The Tourist: A New Theory of the Leisure Class.* New York: Schocken Books, 1976. Reprinted with foreword by Lucy R. Lippard. Berkeley: University of California Press, 1999.

MacKay, Kelly J., and Daniel R. Fesenmaier. "Pictorial Element of Destination in Image Formation." *Annals of Tourism Research* 24, no. 3 (1997): 537–65.

MacLeod, Nicola. "Cultural Tourism: Aspects of Authenticity and Commodification." In *Cultural Tourism in a Changing World: Politics, Participation, and (Re)presentation*, edited by. Melanie K. Smith and Mike Robinson, 177–90. Clevedon, UK: Channel View Publications, 2006.

Mahrouse, Gada. "War-Zone Tourism: Thinking Beyond Voyeurism and Danger." *ACME: An International Journal for Critical Geographies* 15 no. 2 (2016): 330–45.

Mair, Heather. "Trust and Participatory Tourism Planning." In *Trust, Tourism Development and Planning*, edited by. Robin Nunkoo and Stephen L.J. Smith, 46–63. London: Routledge, 2015.

"Majority of Brazilians Think Rio Olympics Will Cause More Harm Than Good." *Associated Press*, July 19, 2016. Accessed July 20, 2016. http://olympics.cbc.ca/news/article/majority-brazilians-think-rio-olympics-will-cause-more-harm-than-good.html

Mak, Athena H.N., Margaret Lumbers, and Anita Eves. "Globalisation and Food Consumption in Tourism." *Annals of Tourism Research* 39, no. 1 (2012): 171–96.

Martin, Geoffrey J. *All Possible Worlds: A History of Geographical Ideas.* New York: Oxford University Press, 2005.

Maruyama, Naho, and Amanda Stronza. "Roots Tourism of Chinese Americans." *Ethnology* 49, no. 1 (2010): 23–44.

Mason, Peter. *Tourism Impacts, Planning and Management.* 3rd ed. London: Routledge, 2016.

Mathieson, Alister, and Geoffrey Wall. *Tourism: Economic, Physical, and Social Impacts*. London: Longman, 1982.

Mbaiwa, Joseph E. "Enclave Tourism and Its Socio-Economic Impacts in the Okavango Delta, Botswana." *Tourism Management* 26 (2005): 157–72.

McElroy, Jerome L., and Courtney E. Parry. "The Characteristics of Small Island Tourist Economies." *Tourism and Hospitality Research* 10, no. 4 (2010): 315–28.

McGill, Kevin. "Jindal: BP Funding Millions for Oil Spill Recovery." *Associated Press*, November 1, 2010. Accessed July 14, 2016. https://www.yahoo.com/news/jindal-bp-funding-millions-oil-spill-recovery.html?ref=gs

McKercher, Bob, and Alan A. Lew. "Tourist Flows and the Spatial Distribution of Tourists." In *A Companion to Tourism*, edited by. Alan A. Lew, C. Michael Hall, and Allan M. Williams, 36–48. Malden, MA: Blackwell, 2004.

McKnight, Tom, and Darrel Hess. *Physical Geography: A Landscape Appreciation*. Upper Saddle River, NJ: Prentice Hall, 2000.

McNamara, Karen Elizabeth, and Bruce Prideaux. "A Typology of Solo Independent Women Travellers." *International Journal of Tourism Research* 12 (2010): 253–64.

McVeigh, Tracy. "Magaluf's Days of Drinking and Casual Sex Are Numbered—Or So Mallorca Hopes." *The Observer*, April 18, 2015. Accessed June 17, 2016. http://www.theguardian.com/travel/2015/apr/18/vodka-sex-magaluf-tourists-spain-mallorca-shagaluf

Meethan, Kevin. *Tourism in Global Society: Place, Culture, Consumption*. Houndmills, UK: Palgrave, 2001.

Mmopelwa, G., D.L. Kgathi, and L. Molefhe. "Tourists' Perceptions and Their Willingness to Pay for Park Fees: A Case Study of Self-Drive Tourists and Clients for Mobile Tour Operators in Moremi Game Reserve, Botswana." *Tourism Management* 28 (2007): 1044–56.

Mol, Arthur P.J., "Sustainability as Global Attractor: The Greening of the 2008 Beijing Olympics." *Global Networks* 10, no. 4 (2010): 510–28.

Moir, James. "Seeing the Sites: Tourism as Perceptual Experience." In *Tourism and Visual Culture*, vol. 1, *Theories and Concepts*, edited by. Peter M. Burns, Cathy Palmer, and Jo-Anne Lester, 165–69. Oxfordshire, UK: CABI, 2010.

Moon, Seung-Il, Sean Il-Kwon, In-Ho Choi, and Ju-Seok Park. "Research on the Introduction of Tourist Police System." *Proceedings, the 2nd International Conference on Information Science and Technology* 23 (2013): 353–6.

Morgan, Nigel, Annette Pritchard, and Roger Pride. "Tourism Places, Brands, and Reputation Management." In *Destination Brands: Managing Place Reputation*, 3rd ed., edited by. Nigel Morgan, Annette Pritchard, and Roger Pride, 3–20. Florence, KY: Routledge, 2011.

Moscardo, Gianna, Philip Pearce, Alastair Morrison, David Green, and Joseph T. O'Leary. "Developing a Typology for Understanding Visiting Friends and Relatives Markets." *Journal of Travel Research* 38, no. 3 (2000): 251–59.

Mulongoy, Kalemani Jo, and Stuart Chape. "Protected Areas and Biodiversity: An Overview of Key Issues." United Nations Environment Programme. Accessed July 14, 2016. https://portals.iucn.org/library/sites/library/files/documents/2004-011.pdf

Munar, Ana María, and Jens Kr. Steen Jacobsen. "Motivations for Sharing Tourism Experiences through Social Media." *Tourism Management* 43 (2014): 46–54.

———. "Trust and Involvement in Tourism Social Media and Web-Based Travel Information Sources." *Scandinavian Journal of Hospitality and Tourism* 13, no. 1 (2013): 1–19.

Mundt, Jörn W. *Tourism and Sustainable Development: Reconsidering a Concept of Vague Policies*. Berlin: Erich Schmidt Verlag, 2011.

Myrtle Beach Area Chamber of Commerce. "2014 Annual Marketing Report." Accessed June 29, 2016. http://www.myrtlebeachareamarketing.com/docs/preso/2014AnnualMarketingReport.pdf

Naidoo, Perunjodi, and Richard Sharpley. "Local Perceptions of the Relative Contributions of Enclave Tourism and Agritourism to Community Well-Being: The Case of Mauritius." *Journal of Destination Marketing & Management* 5 (2016): 16–25.

National Geographic Education Foundation. "Survey Results: U.S. Young Adults Are Lagging." Accessed July 14, 2016. http://www.nationalgeographic.com/geosurvey/highlights.html

National Geographic Society. "The Geotourism Charter." Accessed July 14, 2016. http://travel.nationalgeographic.com/travel/sustainable/pdf/geotourism_charter_template.pdf

———. "Nature." *Mesa Verde*. Accessed May 17, 2016. https://www.nps.gov/meve/learn/nature/index.htm

Nelson, Velvet. "Experiential Branding of Grenada's Spice Island Brand." In *Travel, Tourism, and Identity: Culture & Civilization, Volume 7*, edited by. Gabriel Ricci, 115–26. New Brunswick, NJ: Transaction Publishers.

———. "Investigating Energy Issues in Dominica's Accommodations." *Tourism and Hospitality Research* 10 (2010): 345–58.

———. "Place Reputation: Representing Houston, Texas as a Creative Destination through Culinary Culture." *Tourism Geographies* 17, no. 2 (2015): 192–207.

———. " 'R.I.P. Nature Island': The Threat of a Proposed Oil Refinery on Dominica's Identity." *Social & Cultural Geography* 11, no. 8 (2010): 903–19.

———."Tourist Identities in Narratives of Unexpected Adventure in Madeira." *International Journal of Tourism Research* 17, no. 6 (2015): 537–44.

———. "Traces of the Past: The Cycle of Expectation in Caribbean Tourism Representations." *Journal of Tourism and Cultural Change* 5 (2007): 1–16.

New Orleans Tourism Marketing Corporation. *"2016 New Orleans Festivals."* Accessed September 12, 2016. http://www.neworleansonline.com/neworleans/festivals/festivals.html.

Novelli, Marina, and Alexander Hellwig. "The UN Millennium Development Goals, Tourism and Development: The Tour Operators' Perspective." *Current Issues in Tourism* 14, no. 3 (2011): 205–20.

Noy, Chaim. "This Trip Really Changed Me: Backpackers' Narratives of Self-Change." *Annals of Tourism Research* 31, no. 1 (2004): 78–102.

Nunkoo, Robin, and Haywantee Ramkissoon. "Stakeholders' Views of Enclave Tourism: A Grounded Theory Approach." *Journal of Hospitality & Tourism Research* 40, no. 5 (2016): 557–88.

Nusair, Khaldoon, Mehmet Erdem, Fevzi Okumus, and Anil Bilgihan. "Users' Attitudes toward Online Social Networks in Travel." In *Social Media in Travel, Tourism and Hospitality: Theory, Practice and Cases*, edited by. Marianna Sigala, Evangelos Christou, and Ulrike Gretzel, 207–24. Ashgate: Surrey, 2012.

O'Connell, John F., and George Williams. "Passengers' Perceptions of Low Cost Airlines and Full Service Carriers: A Case Study Involving Ryanair, Aer Lingus, Air Asia and Malaysia Airlines." *Journal of Air Transport Management* 11 (2005): 259–72.

Oktoberfest. "The Oktoberfest 2015 Roundup." Accessed June 17, 2016. http://www.oktoberfest.de/en/article/Oktoberfest+2016/About+the+Oktoberfest/The+Oktoberfest+2015+roundup/4408/

Ohio Sauerkraut Festival. "Festival History." Accessed July 14, 2016. http://www.sauerkraut-festival.com

Oppermann, Martin. "Sex Tourism." *Annals of Tourism Research* 26, no. 2 (1999): 251–66.

Pacific Asia Travel Association. "About PATA." Accessed July 14, 2016. https://www.pata.org/about-pata/

Page, Stephen. *Transport and Tourism: Global Perspectives*. 2nd ed. Harlow, UK: Pearson Prentice Hall, 2005.

Page, Stephen, and Joanne Connell. "Transport and Tourism." In *The Wiley Blackwell Companion to Tourism*, edited by. Alan A. Lew, C. Michael Hall, and Allan M. Williams, 155–67. Malden, MA: Wiley Blackwell, 2014.

Pain, Rachel, Michael Barke, Duncan Fuller, Jamie Gough, Robert MacFarlane, and Graham Mowl. *Introducing Social Geographies*. London: Arnold, 2001.

Parra-López, Eduardo, Desiderio Gutiérrez-Taño, Ricardo J. Díaz-Armas, and Jacques Bulchand-Gidumal. "Travellers 2.0: Motivation, Opportunity and Ability to Use Social Media." In *Social Media in Travel, Tourism and Hospitality: Theory, Practice and Cases*, edited by. Marianna Sigala, Evangelos Christou, and Ulrike Gretzel, 171–87. Ashgate: Surrey, 2012.

Patnaude, Art, and Nicolas Parasie. "Next Big Travel Destination: Iran?" *The Wall Street Journal*, May 17, 2016. Accessed August 2, 2016. http://www.wsj.com/articles/next-big-travel-destination-iran-1463490999

Peters, F.E. *The Hajj: The Muslim Pilgrimage to Mecca and the Holy Places*. Princeton: Princeton University Press, 1994.

Piggott, Mark. "Deadly Selfie Craze: South Korean Dies in Peru Just Hours after Death of German at Machu Picchu." *International Business Times*, July 6, 2016. Accessed July 19, 2016. http://www.ibtimes.co.uk/selfie-dangers-2-tourists-die-same-day-taking-risky-photographs-peru-beauty-spots-1569096

Pizam, Abraham, Peter E. Tarlow, and Jonathan Bloom. "Making Tourists Feel Safe: Whose Responsibility Is It?" *Journal of Travel Research* 36, no. 1 (1997): 23–8.

Pratt, Mary Louise. *Imperial Eyes: Travel Writing and Transculturation*. 2nd ed. London: Routledge, 2008.

Pratt, Stephen, and Anyu Liu, "Does Tourism Really Lead to Peace? A Global View." *International Journal of Tourism Research* 18 (2016): 82–90.

Prideaux, Bruce. "The Role of the Transport System in Destination Development." *Tourism Management* 21 (2000): 53–63.

Project REVITAS. "Project Description." Accessed June 15, 2016. http://revitas.org/en/project/project-description/

Rayan, Magdy Abou, Berge Djebedjian, and Ibrahim Khaled. "Water Supply and Demand and a Desalination Option for Sinai, Egypt." *Desalination* 136 (2001): 73–81.

Reddy, Maharaj Vijay, Mirela Nica, and Keith Wilkes. "Space Tourism: Research Recommendations for the Future of the Industry and Perspectives of Potential Participants." *Tourism Management* 33 (2012): 1093–1102.

Reid, Alastair. "Reflections: Waiting for Columbus." *New Yorker*, February 24, 1992, 57–75.

Reilly, Jennifer, Peter Williams, and Wolfgang Haider. "Moving towards More Eco-Efficient Tourist Transportation to a Resort Destination: The Case of Whistler, British Columbia." *Research in Transportation Economics* 26 (2010): 66–73.

Relph, Edward. *Place and Placelessness*. London: Pion, 1976.

———. "Sense of Place." In *Ten Geographic Ideas That Changed the World*, edited by. Susan Hanson, 205–26. New Brunswick, NJ: Rutgers University Press, 1997.

Richards, Greg. "Creativity and Tourism: The State of the Art." *Annals of Tourism Research* 38 (2011): 1225–53.

Rivera, Lauren A. "Managing 'Spoiled' National Identity: War, Tourism, and Memory in Croatia." *American Sociological Review* 73 (2008): 613–34.

Roach, John. "Young Americans Geographically Illiterate, Survey Suggests." *National Geographic News*, May 2, 2006. Accessed July 14, 2016. http://news.nationalgeographic.com/news/2006/05/0502_060502_geography.html

Robinson, Mike, and David Picard. "Moments, Magic, and Memories: Photographic Tourists, Tourist Photographs, and Making Worlds." In *The Framed World: Tourism, Tourists, and Photography*, edited by. Mike Robinson and David Picard, 1–38. Farnham, UK: Ashgate, 2009.

Rodrigue, Jean-Paul, Claude Comtois, and Brian Slack. *The Geography of Transport Systems*. 2nd ed. New York: Routledge, 2009. Accessed February 10, 2011. http://people.hofstra.edu/geotrans

Rogers, Peter, Kazi F. Jalal, and John A. Boyd. *An Introduction to Sustainable Development*. London: Earthscan, 2008.

Ross, Winston. "Holland's New Marijuana Laws Are Changing Old Amsterdam." *Newsweek*, February 22, 2015. Accessed June 16, 2016. http://www.newsweek.com/marijuana-and-old-amsterdam-308218

Sachs, Jeffrey D. *The Age of Sustainable Development*. New York: Columbia University Press, 2015.

Saha, Shrabani, and Ghialy Yap. "The Moderation Effects of Political Instability and Terrorism on Tourism Development: A Cross-Country Panel Analysis." *Journal of Travel Research* 53 no. 4 (2014): 509–21.

Santich, Barbara. "The Study of Gastronomy and Its Relevance to Hospitality Education and Training." *Hospitality Management* 23 (2004): 15–24.

Saarinen, Jarkko, Christian M. Rogerson, and Haretsebe Manwa (editors). *Tourism and the Millennium Development Goals: Tourism, Local Communities and Development.* London: Routledge, 2013.

SBB. "Visit Switzerland." Accessed July 22, 2016. https://www.sbb.ch/en/leisure-holidays/holidays--short-breaks-in-switzerland/swisstravelsystem.html

Scheyvens, Regina. *Tourism and Poverty*. New York: Routledge, 2011.

Sciochet, Catherine E. "Is Rio Ready? Other Olympic Host Cities Faced Problems, Too." *CNN*, July 20, 2016. Accessed July 20, 2016. http://www.cnn.com/2016/07/20/world/olympics-problems/

Scott, Daniel. "Climate Change Implications for Tourism." In *The Wiley Blackwell Companion to Tourism*, edited by. Alan A. Lew, C. Michael Hall, and Allan M. Williams, 466–78. Malden, MA: Wiley Blackwell, 2014.

Scott, Daniel, and Christopher Lemieux. "The Vulnerability of Tourism to Climate Change." In *The Routledge Handbook of Tourism and the Environment*, edited by. Andrew Holden and David Fennell, 241–58. London: Routledge, 2013.

Serenity Holidays Ltd. "Mandina and Makasutu: Our Story." Accessed July 14, 2016. http://www.mandinalodges.com/our-story

Shaffer, Marguerite. *See America First: Tourism and National Identity, 1880–1940*. Washington, DC: Smithsonian Institution Press, 2001.

Sharpley, Richard. "Tourism and the Countryside." In *A Companion to Tourism*, ed. Alan A. Lew, C. Michael Hall, and Allan M. Williams, 374–81. Malden, MA: Blackwell, 2004.

Shattuck, Harry. "Midtown Houston." *Executive Travel*, October 2009, 69–74.

Shaw, Gareth, and Allan M. Williams. *Critical Issues in Tourism: A Geographical Perspective*. 2nd ed. Malden, MA: Blackwell, 2002.

Shehata, M., M. Mahgoub, and R. Hinklemann. "High Resolution 3D Model of Desalination Brine Spreading: Test Cases and Field Case El-Gouna, Egypt." *E-Proceedings of the 36th IAHR World Congress*, June 28–July 3 (2015), 1–10.

Simmons, Jack. "Railways, Hotels, and Tourism in Great Britain, 1839–1914." *Journal of Contemporary History* 19 (1984): 201–22.

Simpson, Murray C., Stefan Gossling, Daniel Scott, C. Michael Hall, and Elizabeth Gladin. *Climate Change Adaptation and Mitigation in the Tourism Sector: Frameworks, Tools and Practices*. Paris: UNEP, University of Oxford, UNWTO and WMO, 2008.

Slovenian Tourist Board. "Regions." Accessed June 15, 2016. http://www.slovenia.info/en/Regions.htm?_ctg_regije=0&lng=2

Smith, Stephen L.J. "Defining Tourism: A Supply Side View." *Annals of Tourism Research* 15, no. 2 (1988): 179–90.

Sönmez, Sevil F. "Tourism, Terrorism, and Political Instability." *Annals of Tourism Research* 25, no. 2 (1998): 416–56.

Spindler, Edmund A. "The History of Sustainability: The Origins and Effects of a Popular Concept." In *Sustainability in Tourism: A Multidisciplinary Approach*, edited by. Ian Jenkins and Roland Schröder, 9–31. Wiesbaden: Spinger Gabler, 2013.

Stankov, Uglješa, Lazar Lazić, and Vanja Dragiević. "The Extent and Use of Basic Facebook User-Generated Content by the National Tourism Organizations in Europe." *European Journal of Tourism Research* 3, no. 2 (2010): 105–13.

Stephens, Ronald J. *Idlewild: The Black Eden of Michigan.* Charleston: Arcadia, 2001.

Steward, Jill. "'How and Where to Go': The Role of Travel Journalism in Britain and the Evolution of Foreign Travel, 1840–1914." In *Histories of Tourism: Representation, Identity, and Conflict*, edited by. John Walton, 39–54. Clevedon, UK: Channel View Publications, 2005.

Stone, Philip R. "A Dark Tourism Spectrum: Towards a Typology of Death and Macabre Related Tourist Sites, Attractions and Exhibition." *Tourism* 54, no. 2 (2006): 145–60.

Stone, Philip R., and Richard Sharpley. "Consuming Dark Tourism: A Thanatological Perspective." *Annals of Tourism Research* 35, no. 2 (2008): 574–95.

Strasdas, Wolfgang. "Ecotourism and the Challenge of Climate Change: Vulnerability, Responsibility, and Mitigation Strategies." In *Sustainable Tourism & the Millennium Development Goals: Effecting Positive Change*, edited by. Kelly S. Bricker, Rosemary Black, and Stuart Cottrell, 209–30. Burlington: Jones & Bartlett Learning, 2013.

Suvantola, Jaakko. *Tourist's Experience of Place.* Aldershot, UK: Ashgate, 2002.

Swarbrooke, John. *The Development and Management of Visitor Attractions.* 2nd ed. Burlington, MA: Butterworth-Heinemann, 2002.

Swiss Travel System. "Glacier Express." Accessed July 22, 2016. http://www.swisstravelsystem.com/en/highlights-en/train-bus-boat/railway/glacier-express-st-moritz-zermatt.html

Tarlow, Peter E. "Tourism Police Help Create Destination Image." *Tourism Review*, August 25, 2014. Accessed August 1, 2016. http://www.tourism-review.com/travel-tourism-magazine-tourism-police-create-the-image-of-the-destination-article2450

"3,300 British Tourists Evacuated from Tunisia as Riot Chaos Deepens after President Flees." *Daily Mail*, January 14, 2011. Accessed August 2, 2016. http://www.dailymail.co.uk/news/article-1347112/Tunisia-riots-3-300-British-tourists-evacuated-travel-companies-23-dic.html

Thompson, Paul. "Blarney Stone 'Most Unhygienic Tourist Attraction in the World.'" *Daily Mail*, June 16, 2009. Accessed July 14, 2016. http://www.dailymail.co.uk/news/article-1193477/Blarney-Stone-unhygienic-tourist-attraction-world.html

Timothy, Dallen J. "Tourism, War, and Political Instability: Territorial and Religious Perspectives." In *Tourism and War*, edited by. Richard Butler and Wantanee Suntikul, 12–25. London: Routledge, 2013.

Torres, Rebecca. "Linkages between Tourism and Agriculture in Mexico." *Annals of Tourism Research* 30, no. 3 (2003): 546–66.

———. "Toward a Better Understanding of Tourism and Agriculture Linkages in the Yucatan: Tourist Food Consumption and Preferences." *Tourism Geographies* 4, no. 3 (2002): 282–306.

Torres, Rebecca and Janet Henshall Momsen. "Challenges and Potential for Linking Tourism and Agriculture to Achieve Pro-Poor Tourism Objectives." *Progress in Development Studies* 4, no. 4 (2004): 294–318.

Torres, Rebecca Maria and Janet D. Momsen. "Gringolandia: The Construction of a New Tourist Space in Mexico." *Annals of the Association of American Geographers* 95, no. 2 (2005): 314–35.

Tourism BC. "Vancouver Things to Do." Accessed February 1, 2012. http://www.hellobc.com/vancouver/things-to-do.aspx

Towner, John. "The Grand Tour: A Key Phase in the History of Tourism." *Annals of Tourism Research* 12 (1985): 297–333.

———. "What is Tourism's History?" *Tourism Management* 16, no. 5 (1995): 339–43.

TravelBlog. "About TravelBlog." Accessed July 8, 2016. http://www.travelblog.org/about.html

Trollope, Anthony. *The West Indies and the Spanish Main.* 4th ed. London: Dawsons of Pall Mall, 1968.

Tuan, Yi-Fu. "Place: An Experiential Perspective." *Geographical Review* 65 (1975): 151–65.

Uncyclopedia. "Tourist—The Stereotype." Accessed July 14, 2016. http://uncyclopedia.wikia.com/wiki/Tourist_-_the_stereotype

United Nations. "SDGs: Sustainable Development Knowledge Platform." Accessed July 1, 2016. https://sustainabledevelopment.un.org/sdgs

———. "United Nations Millennium Development Goals." Accessed July 1, 2016. http://www.un.org/millenniumgoals/

United Nations Educational, Scientific, and Cultural Organization. "The Criteria for Selection." Accessed September 10, 2011. http://whc.unesco.org/en/criteria

———."Cornwall and West Devon Mining Landscape." Accessed June 29, 2016. http://whc.unesco.org/en/list/1215

———. "UNESCO Global Geoparks." *Earth Sciences.* Accessed May 17, 2016.
http://www.unesco.org/new/en/natural-sciences/environment/earth-sciences/unesco-global-geoparks/

United Nations International Children's Emergency Fund. "Orphanage Voluntourism in Nepal: What You Should Know." Accessed June 16, 2016. http://unicef.org.np/media-centre/reports-and-publications/2015/05/10/orphanage-voluntourism-in-nepal-what-you-should-know

United Nations Office on Drugs and Crime. "United Nations Organizations Cooperate to Stamp Out Human Trafficking and Sex Tourism." April 2012. Accessed July 14, 2016. http://www.unodc.org/unodc/en/frontpage/2012/April/united-nations-organizations-cooperate-to-stamp-out-human-trafficking-and-sex-tourism.html

United Nations Office of the High Representative for the Least Developed Countries, Landlocked Developing Countries, and the Small Island Developing States. "Small Island Developing States." Accessed July 14, 2016. http://unohrlls.org/about-sids/

United Nations World Tourism Organization. "FAQ—Climate Change and Tourism." Accessed July 14, 2016. http://sdt.unwto.org/content/faq-climate-change-and-tourism

———. "International Tourist Arrivals Up 4% Reach a Record 1.2 billion in 2015," January 18, 2016, Accessed June 8, 2016, http://media.unwto.org/press-release/2016-01-18/international-tourist-arrivals-4-reach-record-12-billion-2015.

———. "Sustainable Development of Tourism." Accessed July 5, 2016. http://sdt.unwto.org/content/about-us-5

———. "Tourism & the Millennium Development Goals (MDGs)." Accessed July 5, 2016. http://icr.unwto.org/en/content/tourism-millennium-development-goals-mdgs

———. "Tourism and the Sustainable Development Goals." Accessed July 5, 2016. http://www.e-unwto.org/doi/pdf/10.18111/9789284417254

———. *Tourism Highlights 2015 Edition* 8, no. 1 (2015), Accessed June 8, 2016, http://www.e-unwto.org/doi/pdf/10.18111/9789284416899.

Uriely, Natan, and Yniv Belhassen. "Drugs and Tourists' Experiences." *Journal of Travel Research* 43 (2005): 238–46.

United States Census Bureau. "2010 Census Urban and Rural Classification and Urban Area Criteria." Accessed August 4, 2016. http://www.census.gov/geo/www/ua/2010urbanrural-class.html

United States Department of State. "Alerts and Warnings." Accessed August 2, 2016. https://travel.state.gov/content/passports/en/alertswarnings.html

Urry, John. *Consuming Places*. London: Routledge, 1995.

———. *The Tourist Gaze*. London: Sage, 1990.

Vacanti Brondo, Keri. "The Spectacle of Saving: Conservation Voluntourism and the New Neoliberal Economy on Utila, Honduras." *Journal of Sustainable Tourism* 23, no. 10 (2015): 1405–25.

The Venetian Las Vegas. "Human Resources." Accessed March 31, 2011. http://www.venetian.com/Company-Information/Human-Resources

Visit Baltimore. "Baltimore Harbor." Accessed July 14, 2016. http://baltimore.org/neighborhoods-maps-transportation/inner-harbor

Visit Scotland. "Climate & Weather in Scotland." Accessed July 27, 2016. https://www.visitscotland.com/about/practical-information/weather/

Vodeb, Ksenija. "Cross-Border Regions as Potential Tourist Destinations along the Slovene Croatian Frontier." *Tourism and Hospitality Management* 16 (2010): 219–28.

Waitt, Gordon, and Chris Gibson. "Tourism and Creative Economies." In *The Wiley Blackwell Companion to Tourism*, edited by Alan A. Lew, C. Michael Hall, and Allan M. Williams, 230–9. Malden, MA: Wiley Blackwell, 2014.

Walton, John K. "Prospects in Tourism History: Evolution, State of Play, and Future Development." *Tourism Management* 30 (2009): 783–93.

Ward, Julian. *Xu Xiake (1587–1641): The Art of Travel Writing*. London: Routledge, 2001.

Watts, Jonathan, and Bruce Douglas. "Rio Olympics: Who Are the Real Winners and Losers?" *The Guardian,* July 19, 2016. Accessed July 20, 2016. https://www.theguardian.com/cities/2016/jul/19/rio-olympics-who-are-the-real-winners-and-losers

Wearing, Stephen. *Volunteer Tourism: Experiences That Make a Difference*. Wallingford, UK: CABI, 2001.

Wearing, Stephen, Deborah Stevensen, and Tamara Young. *Tourist Cultures: Identity, Place and the Traveller.* Los Angeles: Sage, 2010.

Webber, Derek. "Space Tourism: Its History, Future and Importance." *Acta Astronautica* 92 (2013): 138–43.

Weed, Julie. "Book Your 'Bud and Breakfast,' Marijuana Tourism Is Growing in Colorado and Washington." *Forbes*, March 17, 2015. Accessed June 16, 2016. http://www.forbes.com/sites/julieweed/2015/03/17/book-your-bud-and-breakfast-marijuana-tourism-is-growing-in-colorado-and-washington/#4af5ed2c73cf

Wichasin, Pimmada, and Nuntiya Doungphummes. "A Comparative Study of International Tourists' Safety Needs and Thai Tourist Polices' Perception towards International Tourists' Safety Needs." *International Journal of Social, Behavioral, Educational, Economic, Business and Industrial Engineering* 6, no. 7 (2012): 1938–44.

Williams, Stephen. *Tourism Geography*. London: Routledge, 1998.

Wilson, Erica, and Donna E. Little. "A 'Relative Escape'? The Impact of Constraints on Women Who Travel Solo." *Tourism Review International* 9 (2005): 155–75.

———. "The Solo Female Travel Experience: Exploring the 'Geography of Women's Fear'." *Current Issues in Tourism* 11, no. 2 (2008): 167–86.

Wong, Chak Keung Simon, and Fung Ching Gladys Liu. "A Study of Pre-Trip Use of Travel Guidebooks by Leisure Travelers." *Tourism Management* 32 (2011): 616–28.

Woods, Michael. *Rural Geography*. London: Sage, 2005.

World Commission on Environment and Development. *Our Common Future: Report of the World Commission on Environment and Development*. Oxford: Oxford University, 1987.

World Tourism Organization. *Collection of Tourism Expenditure Statistics, Technical Manual No. 2*. Madrid: World Tourism Organization, 1995.

World Travel Awards. "World Winners." Accessed June 16, 2016. https://www.worldtravelawards.com

World Travel & Tourism Council. "The Tohoku Pacific Earthquake and Tsunami." Accessed July 27, 2016. http://www.wttc.org/-/media/files/reports/special-and-periodic-reports/japan_report_march_update_v7.ashx

Xie, Philip Feifan. *Authenticating Ethnic Tourism*. Bristol: Channel View Publications, 2011.

Xu, Honggang, and Tian Ye. "Tourist Experience in Lijiang—The Capital of Yanyu." *Journal of China Tourism Research* 12, no. 1 (2016): 108–25.

Yardley, Jim. "For Chinese Police Officers, Light Duty on Tourist Patrol in Italy." *The New York Times*, May 12, 2016. Accessed August 1, 2016. http://www.nytimes.com/2016/05/13/world/europe/chinese-police-rome-italy.html

Yoo, Kyung-Hyan, and Ulrike Gretzel. "Use and Creation of Social Media by Travellers." In *Social Media in Travel, Tourism and Hospitality: Theory, Practice and Cases*, edited by. Marianna Sigala, Evangelos Christou, and Ulrike Gretzel, 189–205. Ashgate: Surrey, 2012.

Young Chung, Jin, Gerard T. Kyle, James F. Petrick, and James D. Absher. "Fairness of Prices, User Fee Policy and Willingness to Pay Among Visitors to a National Forest." *Tourism Management* 32 (2011): 1038–46.

Zillinger, Malin. "The Importance of Guidebooks for the Choice of Tourist Sites: A Study of German Tourists in Sweden." *Scandinavian Journal of Tourism and Hospitality* 6, no. 3 (2006): 229–47.

Index

Note: Page numbers of figures and tables are italicized.

About the Author

Velvet Nelson received her BS in business administration from West Liberty University, her MA in geography with a concentration in rural development from East Carolina University, and her PhD in geography from Kent State University. She joined the Department of Geography and Geology at Sam Houston State University in Huntsville, Texas, in 2006 and was promoted to associate professor in 2012. She is a human geographer with interests in cultural geography and human–environment interactions, but her primary research focus has been on tourism. She has conducted archival research on historical patterns of tourism in the Caribbean, as well as fieldwork on islands such as Dominica, Grenada, and St. Vincent to examine current issues. In 2010, she received a Fulbright Fellowship to conduct research and teach in the Faculty of Humanities and Social Sciences at the University of Primorska in Slovenia. In 2013 and 2014, she was invited back to the university as a visiting scholar in the Faculty of Tourism. Her most recent research examines the city of Houston, Texas, as an emerging food destination. She has published her research in peer-reviewed journal articles and presented it at regional, national, and international conferences both within geography and in the interdisciplinary field of tourism studies. In addition, she is a member of the *Tourism Geographies* editorial board. She strongly believes in direct experience through travel as a means of learning about new places. As such, she travels at every opportunity.